ARCHAEOLOGY OF EAST ASIA

AF707167

Agropastoralism and Languages Across Eurasia

Expansion, exchange, environment

EDITED BY
MARK HUDSON AND
MARTINE ROBBEETS

BAR INTERNATIONAL SERIES 3126

VOLUME 9

BAR PUBLISHING

Published in 2023 by
BAR Publishing, Oxford, UK

BAR International Series 3126

Archaeology of East Asia Volume 9
Agropastoralism and Languages Across Eurasia

ISBN 978 1 4073 6075 1 paperback
ISBN 978 1 4073 6076 8 e-format

DOI https://doi.org/10.30861/9781407360751

A catalogue record for this book is available from the British Library

© The editors and contributors severally 2023

COVER IMAGE *Horses graze at Erkhel Lake in northern Mongolia, overlooking a large Bronze Age archaeological complex. Photo by William Taylor.*

The Authors' moral rights under the 1988 UK Copyright, Designs and Patents Act are hereby expressly asserted.

All rights reserved. No part of this work may be copied, reproduced, stored, sold, distributed, scanned, saved in any form of digital format or transmitted in any form digitally, without the written permission of the Publisher.

Links to third party websites are provided by BAR Publishing in good faith and for information only. BAR Publishing disclaims any responsibility for the materials contained in any third-party website referenced in this work.

BAR
PUBLISHING

BAR titles are available from:

BAR Publishing
122 Banbury Rd, Oxford, OX2 7BP, UK
info@barpublishing.com
www.barpublishing.com

ARCHAEOLOGY OF EAST ASIA

A specialist sub-series in the BAR Series

Series Editor: Anke Hein (University of Oxford)

This series provides a platform for data-rich studies on a variety of topics and materials from all over East Asia as well as conference proceedings reflecting the newest research insights and trends. The series takes a regional approach to East Asia, encouraging projects that cross national borders, even into adjoining regions and/or cover areas usually overlooked in mainstream research. This includes all parts of China, Japan, Korea, Mongolia, the Russian Far East, the Tibetan Plateau as a whole and the northern reaches of Southeast Asia. Especially encouraged are submissions proposing and conducting new approaches and methods in all aspects of archaeology including scientific techniques, spatial analysis, various digital methods, but also theory and model-based or traditional chronology-focused studies.

If you would like to submit a proposal for the *Archaeology of East Asia* series, please contact editor@barpublishing.com

Editorial advisory board

Ursula Brosseder, University of Bonn

Katherine Brunson, Wesleyan University

Lothar von Falkenhausen, UCLA

Rowan Flad, Harvard University

Minku Kim, Chinese University of Hong Kong

Hongliang Lü, Sichuan University

Jade d'Alpoim Guedes, University of California Davis

Mark Hudson, Max Planck Institute for the Science of Human History, Jena

Xinwei Li, Academy of Social Sciences

Yangjin Pak, Chungnam National University

Gideon Shelach-Lavi, Hebrew University

Shinya Shoda, Nara National Research Institute for Cultural Properties

Sharon Wong, Chinese University of Hong Kong

Joshua Wright, University of Aberdeen

Titles in the Archaeology of East Asia Subseries

A Zooarchaeological Study of the Haimenkou Site, Yunnan Province, China
Juan Wang
BAR International Series **2902** | 2019 — Volume 1

The Origin of Cattle in China from the Neolithic to the Early Bronze Age
Chong Yu
BAR International Series **2959** | 2020 — Volume 2

Bronze Weapons of the Qin Terracotta Warriors
Standardisation, craft specialisation and labour organisation
Xiuzhen Li
BAR International Series **2992** | 2020 — Volume 3

Animal Classification in Central China
From the late Neolithic to the early Bronze Age
Ningning Dong
BAR International Series **3031** | 2021 — Volume 4

Painted Pottery Production and Social Complexity in Neolithic Northwest China
Ling-yu Hung†, edited by Anke Hein and Andrew Womack
BAR International Series **3045** | 2021 — Volume 5

Late Pleistocene and Early Holocene Microblade-based Industries in Northeastern Asia
A macroecological approach to foraging societies
Meng Zhang
BAR International Series **3056** | 2021 — Volume 6

Kilns in East and North Asia
The adoption of ceramic industries
Edited by Tomoko Nagatomo, Maria Shinoto and Daisuke Nakamura
BAR International Series **3082** | 2022 — Volume 7

Technological knowledge in the production of Neolithic Majiayao pottery in Gansu and Qinghai
Evgenia Dammer
BAR International Series **3122** | 2023 — Volume 8

Agropastoralism and Languages Across Eurasia
Expansion, exchange and environment
Mark Hudson and Martine Robbeets
BAR International Series **3126** | 2023 — Volume 9

Of Related Interest

Digging for Words
Archaeolinguistic case studies from the XV Nordic TAG Conference held at the University of Copenhagen, 16–18 April 2015
Edited by Rune Iversen and Guus Kroonen
BAR International Series **2888** | 2018

The 'Crescent-Shaped Cultural-Communication Belt': Tong Enzheng's Model in Retrospect
An examination of methodological, theoretical and material concerns of long-distance interactions in East Asia
Edited by Anke Hein
BAR International Series **2679** | 2014

The East European Plain on the Eve of Agriculture
Edited by Pavel M. Dolukhanov, Graeme R. Sarson and Anvar M. Shukurov
BAR International Series **1964** | 2009

The Transition from Foraging to Farming and the Origin of Agriculture in China
Tracey Lie Dan Lu
BAR International Series **774** | 1999

Archaeology of the Russian Far East: Essays in Stone Age Prehistory
Edited by Sarah M. Nelson, Anatoly P. Derevianko, Yaroslav V. Kuzmin and Richard L. Bland
BAR International Series **1540** | 2006

Transition from Foraging to Farming in Northeast China
(Peter) Wei Ming Jia
BAR International Series **1629** | 1999

Ethnoarchaeological Perspectives of Prehistoric Settlement Patterns of South-Central Ganga Valley
Shahida Ansari
BAR International Series **2242** | 2011

Stable Isotopic Analysis of Carbon and Nitrogen as an Indicator of Paleodietary Change among Pre-State Metal Age Societies in Northeast Thailand
Christopher A King
BAR International Series **1788** | 2008

Néolithisations
Proche et Moyen Orient, Méditerranée orientale, Nord de l'Afrique, Europe méridionale, Chine, Amérique du Sud
Olivier Aurenche et Jacques Cauvin
BAR International Series **516** | 1989

For more information, or to purchase these titles, please visit **www.barpublishing.com**

Contents

Contributors

Rasmus Bjørn, Archaeolinguistic Research Group, Max Planck Institute of Geoanthropology, Jena, Germany

David Bradley, Department of Linguistics, La Trobe University, Melbourne, Australia

Bingcong Deng, Archaeolinguistic Research Group, Max Planck Institute of Geoanthropology, Jena, Germany

Mark Hudson, Archaeolinguistic Research Group, Max Planck Institute of Geoanthropology, Jena, Germany

Takamune Kawashima, Hiroshima University Museum, Higashi-Hiroshima, Japan

Martijn Knapen, Archaeolinguistic Research Group, Max Planck Institute of Geoanthropology, Jena, Germany

Seiji Nakayama, Research Institute of Cultural Properties, Teikyo University, Yamanashi, Japan

Claudio Pelloli, Padua, Italy

Martine Robbeets, Archaeolinguistic Research Group, Max Planck Institute of Geoanthropology, Jena, Germany

Hui Wang, School of Sociology and Anthropology, Sun Yat-sen University, Guangzhou, China

Jie Zhang, School of Sociology and Anthropology, Sun Yat-sen University, Guangzhou, China

Jingming Zhang, School of Ethnology, North Minzu University, Yinchuan, China

1

Introduction

Mark Hudson and Martine Robbeets

The essays collected in this volume explore questions relating to human movements and exchange across the diverse environments of Eurasia. The authors examine the ways in which humans over the last ten thousand years have innovated or exchanged practices related to food production to reshape and adapt to their varied environments. The research presented is set in the human niche, the context in which humans and their communities make their lives and where they share socio-economic, cultural, linguistic, demographic and ecological histories. Therefore, many of the chapters adopt an interdisciplinary perspective. A number of the chapters combine archaeology and linguistics, an approach which we call 'archaeolinguistics'. Other chapters utilise insights from ethnography, art history, and the analysis of historical texts. The chronological time frame is from the Neolithic to the Middle Ages.[1]

Human sociocultural niche construction has led to profound and wide-ranging impacts on the biosphere (Ellis 2015). Agriculture has been an especially important factor in generating these changes (Boivin et al. 2016; Stephens et al. 2019). The expansion of the human niche has been associated with language dispersals, sometimes through the migration of human populations, sometimes by trade and exchange. 'Exchange' has been an important concept in anthropology, but the term can cover a broad range of phenomena. In early anthropology, exchange involving the diffusion of ideas or objects was often understood as an alternative to migration. Such thinking was especially influential in East Asian archaeology through the work of Heine-Geldern (1937), Karlgren (1942) and others (see Anke Hein (2014) for an excellent overview of this research). Today, we are in the midst of something of a re-evaluation of these concepts. On the one hand, new research in ancient DNA is transforming our understanding of past migrations. At the same time, innovative concepts such as 'ancient globalisation' are changing the way we approach cultural interaction (see e.g., Vandkilde 2016; Autiero and Cobb 2022).

Exchange can also have a contemporary geopolitical dimension. Several of the chapters here discuss Central Asia and contacts between west and east. In a fascinating analysis of the 1980 Sino-Japanese TV documentary *The Silk Road*, Yin (2022) shows how China attempted to use the series to represent the regions of Central Asia under its control as part of a 'multi-ethnic state' and to emphasise the uniting role of trade—both past and present. Japan, for its part, tried to move away from its twentieth-century history of colonialism in the region, presenting the perspective of a traveller along the Silk Road.

Part One of the present volume focuses on the Neolithic and on how agropastoralism led to new adaptive niches for human societies. We use 'agropastoralism' as a cover term for practices involving plant cultivation and the raising or management of animals in varied combinations. Not all of the case studies in the volume raised animals. In Neolithic Jōmon Japan, discussed in Chapters 2 and 3, the dog was the only domesticated animal, although there is also growing evidence for the management of wild boar, including transport to offshore islands (Kawamura et al. 2016). Some Jōmon sites on offshore islands have produced remains of other possibly transported species, including raccoon dog and pheasant. In such cases, however, it is hard to be sure whether this represents a natural distribution, live transport by humans, or the introduction of pieces of meat from dead animals and birds. Agropastoralism sometimes involved population expansions—for example those modelled by the farming/language dispersal hypothesis—but in other cases was expressed through exploitation of the environment in novel ways. Both of these changes could, of course, occur at the same time (see e.g., Ethier et al. 2017). The legume cultivation and nut tree management in Neolithic Japan discussed by Nakayama in Chapter 2 would have had profound impacts on landscape ecology around Jōmon settlements (see Nakayama's Fig. 2.8). It is unclear, however, to what extent this led to population dispersals. By contrast, the dispersals hypothesised by David Bradley in Chapter 7 appear to have occurred *before* farming, but the precise background remains to be elucidated.

The first chapter in Part One, by Seiji Nakayama, shows how Jōmon societies in parts of the Japanese Islands began to cultivate several plants, though as mentioned he leaves open the question of associated population dispersals. Nakayama has been a leading figure in the archaeobotany of Japan for many years and his chapter here summarises his most recent research results. The question of whether the often quite complex societies of the Jōmon period possessed farming has a long history of debate (for

[1] Most of the chapters included here derive from a project granted to co-editor Martine Robbeets, which received funding from the European Research Council (ERC) under the European Union's Horizon 2020 research and innovation programme (grant agreement No 646612). Chapters 2 and 3 were originally presented at a conference on the Anthropocene and World Heritage held in Shizuoka, Japan in February 2018.

overviews in English, see Pearson and Pearson 1978; Matsui and Kanehara 2006; Crawford 2008; Hudson 2020; Kawashima this volume). The new work by Nakayama and other scholars such as Obata (2016) makes it clear that plant cultivation was present, although the overall impact on Jōmon society requires further research. In Nakayama's synthesis model (Fig. 2.8), secondary vegetation around settlements includes a variety of economically useful plants, such as bracken, lilies, kudzu, soybean, adzuki bean, and nut producing trees. Forest clearance, burning and logging help to maintain and expand the area of secondary vegetation at the expense of the primary forest, leading in some cases to artificially created pure stands of chestnut trees.

In the following chapter, Takamune Kawashima discusses the role of stone tools in Neolithic Japan from the perspective of recent debates about the Anthropocene. This is a newly proposed unit of geological time, used to describe the period when human activity started to have a significant impact on our planet's climate and ecosystems (Ellis 2018). From Kawashima's perspective as an archaeologist, this relationship can be dated back over ten thousand years to the spread of cultivation and sedentism. Kawashima uses various archaeological and ethnographic studies of stone tools thought to be linked with forest clearance and plant cultivation. Axes, in particular, were key tools. Here, the focus is on chipped stone axes and adzes. Jōmon cultures also used polished stone axes, which are said to be stronger and easier to haft than chipped ones, though their functional role could be combined with a symbolic one given the attractive appearance of many polished axes (Bradley 1998: 44). Kawashima builds on the pioneering work of Japanese archaeologist Masaki Nishida to argue that the shift to greater sedentism, initiated by the Initial Jōmon phase, had such a significant impact on nature and society that it marks the beginning of the Anthropocene in Japan. Nishida (e.g., 1983) developed early analyses of the human niche in Japanese archaeology and it is exciting to see his work extended into Anthropocene studies in Kawashima's chapter. At the same time, there is also a need to consider the ways in which Jōmon societies attempted to resist the full-scale agricultural patterns already found on the East Asian mainland (cf. Hudson 2020; Hudson et al. 2021). The Japanese Islands were one of the last places in temperate Eurasia where full-scale agriculture was adopted, although this delay seems not to have had a significant impact on anthropogenic pathways in the archipelago (Hudson et al. 2022).

In Chapter 4, Martine Robbeets associates the three cases of linguistic interaction between Transeurasian and non-Transeurasian speech communities, discussed in Chapters 5, 6 and 8, with genetic admixture events in Neolithic and Bronze Age Northern Asia and assesses them in the light of socio-economic interaction. The results indicate that contact with speech communities with economically less productive strategies—such as when farmers meet hunter-gatherers—leads to the dispersal of the dominant language, while interaction between populations with comparably productive economies, such as between farmers and farmers or farmers and pastoralists, encourages linguistic borrowing. At the end of her chapter, Robbeets sets out further questions for new research on the inter-relationships between languages, economies and human populations in Northeast Asia.

In Chapter 5, Bingcong Deng examines lexical borrowings related to agriculture, which are estimated to have taken place in the Neolithic and the Bronze Age between Sino-Tibetan and Transeurasian, two ancient speech communities in what is today northeast China associated with farming. As the ancestral languages involved have never been recorded, Deng applies historical-comparative linguistics to reconstruct their agricultural vocabulary. The expansion of agriculture from north China to the west and southwest has been the subject of several recent archaeological studies (Dal Martello 2022; Liu et al. 2022; Tang et al. 2022). Deng's analysis adds a valuable linguistic perspective on this question.

A similar approach is applied by Martijn Knapen in Chapter 6 for the reconstruction of Tungusic and Amuric cultural vocabulary, used by the ancestors of a small group of Tungusic speakers living in Manchuria and Eastern Siberia, and Nivkh speakers, living today in the lower Amur region and on Sakhalin. Knapen attempts to trace back the earliest interactions between Amuric hunter-gatherers and Tungusic farmers and to infer where they could have occurred. Despite the sometimes highly technical arguments presented by Knapen, his chapter shows very clearly how linguists go about dating changes in language forms. The Amur-Sakhalin region discussed in this chapter is relatively poorly known from an archaeological perspective. Archaeologically, we know more about the edges of the area discussed by Knapen, especially the Primorye and Hokkaido, raising the question of possible connections with respect to farming and other cultural features. Knapen's chapter shows very clearly the importance of triangulating linguistic data with information from archaeology and ethnohistory.

Chapter 7, by David Bradley, examines the hypothesis of a linguistic relationship between Sino-Tibetan, Yeniseian and Na-Dene. As detailed by Bradley, this hypothesis has been discussed in the literature for some time (see also Ruhlen 1998). Yet new archaeological and especially genetic research enables the author to expand our understanding of the problem. The hypothetical split of a Yeniseian/Na-Dene branch from Sino-Tibetan (see Fig. 7.3 in Bradley's chapter) must have been quite ancient, and is suggested by Bradley to have occurred at least 6500 years ago. Such a date is not necessarily incompatible with agriculture. Millet cultivation in north China began by around 8000 BP and agricultural societies were established there by 6500 BP (Stevens and Fuller 2017). In the fourth millennium BC, millet farming spread to both the Korean peninsula and the Maritime (Primoyre) region of the Russian Far East (Li et al. 2020). However, Bradley argues that the languages concerned have no shared cognate vocabulary for any crops or domestic animals other than the dog. If this was a pre-farming dispersal, what caused such an apparently long-distance expansion? Further research is needed, but the proposed chronology suggests a possible link with the mid-Holocene climatic optimum (ca. 7000-5000 BP). The complex human prehistory of the Northern Eurasian Greenbelt and human movements therein is summarised in a recent paper by Uchiyama et al. (2020). How those movements connected to Holocene dispersals to the Americas is a subject for future work yet Bradley's chapter hints at fascinating new hypotheses.

In Part Two of the volume, the emphasis shifts to exchange between east and west across Eurasia in the Bronze Age and Middle Ages. How do we understand the geography of proto-globalisation in ancient East Asia? This question has been over-determined by Silk Road discourse, yet other perspectives exist and new research on the 'Global Middle Ages' is transforming the field (Heng 2021; Hermans 2020). Until the 1970s, the field of Chinese archaeology focused on the rise of civilisation in the Yellow River basin and its subsequent spread to other parts of China. This approach gradually began to break down through a new emphasis on local regions and their interactions (von Falkenhausen 1995). The 'Crescent-Shaped Cultural Communication Belt' proposed by Tong Enzheng has been one influential model in this respect (Hein 2014). More recently, the so-called 'Belt and Road Initiative' supported by the government of the People's Republic of China since 2013 has encouraged a new interest in the archaeology of long-range exchange by Chinese scholars (Storozum and Li 2020). This interest has even expanded to putative contacts between Alaska and Late Neolithic China (Qu 2014). Within Chinese archaeology, the whole topic of long-range interaction has become highly charged and fraught with political sensitivities in recent years.

In Chapter 8, Rasmus Bjørn begins Part Two with a detailed archaeolinguistic analysis of west Eurasian crops, livestock and technology in East Asia. To this end, Bjørn reconstructs interactions between speakers of Indo-European and Transeurasian languages in the Bronze Age when pastoralism was introduced from the western to the eastern steppes. While the steppe zone has long been considered as an important route of interaction between eastern and western Eurasia, recent research has also considered the role of the Northern Eurasian Greenbelt (Uchiyama et al. 2020) and the Inner Asian Mountain Corridor (Frachetti et al. 2017). In his chapter, Bjørn compares the role of the steppe and the Inner Asian Mountain Corridor in the spread of linguistic borrowings. Bjørn also notes the important synergy over the past decade or so between research in archaeobotany and historical linguistics.

The following three chapters then look at the archaeology of the northern and western borderlands of China. Chapters 9 and 10, by Jingming Zhang and Hui Wang, respectively, examine evidence for contact and exchange between China and western Eurasia across the steppes. Jingming Zhang does this from the perspective of gold and silverware. His analysis shows extensive technological and stylistic influences from the west over a long period. Gold was especially important in the nomadic cultures of the steppes since small yet highly precious items could be carried by an individual. The colour of gold represented durability and authority (Biran 2015: 4). Nevertheless, in the words of the title of Bunker's (1993) review article, gold in the ancient Chinese world has been seen as a 'cultural puzzle'. Jingming Zhang's chapter focuses on stylistic and technical considerations and contains little discussion of the social context of gold, a topic that has received some attention in Bronze Age studies in Europe (e.g., Herrero 2019), as well as recently in Han dynasty China (Wang 2021). Silver also has a long history from its first regular appearance in the archaeological record in West Asia in the fourth millennium BC, a topic recently treated by Susan Sherratt (2019).

Hui Wang discusses the archaeology of the foreign contacts of the medieval Liao dynasty, bringing in a fascinating range of materials, including watermelon cultivation and Buddhist beliefs. Wang makes extensive yet uncritical use of the term *hu*, found in early Chinese texts to refer to 'foreign barbarians' (see Di Cosmo 2002: 128-131 for a discussion of the origins of this word). One of Wang's research interests is the relationship between historical texts and archaeological problems, an issue with a huge literature in Europe with respect to terms like 'barbarian', 'Celts', and so forth. How views of 'western barbarians' in the Chinese dynastic histories can be combined with the archaeological record in a more nuanced fashion remains a topic for future research, yet Wang sets out aspects of the basic culture history in an approachable fashion.

In Chapter 11, Jie Zhang analyses Liao dynasty banquets and the drinking of liquor and tea as depicted in tomb murals. These murals, painted for the owners of the tombs, provide detailed scenes of upper class customs related to eating and drinking in the Liao. Jie Zhang's chapter also brings new information to the history of tea consumption, a topic covered exhaustively in a recent work by George van Driem (2019). The archaeology of alcohol consumption has become a popular topic in Chinese archaeology, though one which is sometimes accompanied by rather simplistic assumptions about ethnic customs. Thus, for example, Siyi Wang (2022: 461) informs us that 'The Jurchens were originally from very cold places where consuming alcohol helped them cope with the cold; therefore, the Jurchens had a long history of alcohol consumption'. Whether alcohol consumption was really higher in cold places in the past (or present) is an interesting empirical question, but here Jie Zhang shows the elaborate culture associated with drinking amongst Liao elites.

Chapters 9, 10 and 11 display the historiographic emphasis of Chinese archaeology (cf. von Falkenhausen 1995). While Chapter 11 considers the social context of patterns of diffusion *within* the Chinese cultural sphere, Chapters 9 and 10 provide little analysis with respect to the historical background of the broader, international exchanges. At the same time, these chapters are lavishly illustrated with important materials not easily available in Western-language sources. The art historical focus of these three chapters differs from—yet complements—the other chapters in the volume, while leaving room for further archaeological and historical analysis of the cultural factors behind the trans-Eurasian interactions that are discussed.

In the book's final chapter, Claudio Pelloli looks at east-west contacts from the perspective of the acceptance of rice in Italy and the Mediterranean. Rice spread widely from its original centre of domestication in East Asia, though the processes behind that dispersal were complex and historically contingent (Fuller et al. 2017; Spengler et al. 2021). A number of historical records enable Pelloli to discuss the western spread of rice in some detail, though many questions remain. The chapter provides an interesting counterpoint to those by Deng and Bjørn who discuss the spread of crops from a primarily linguistic perspective. Pelloli's chapter is also of interest in the light of recent debates over ancient 'food globalisation' and specifically whether that process was 'bottom up' or 'top down' (Jones et al. 2011; Liu and Jones 2014; Liu et al. 2019). The evidence from Italy supports conclusions that peasants were often conservative but could transform their crops and economy when needed. Parker (2013: 638, citing McArdle 1978) notes that in the Tuscan village of Altopascio, peasants experimented with rice after one famine in the seventeenth century but reverted to other cereals soon thereafter and introduced maize only after the catastrophic harvest of 1710. (That bad harvest was likely connected to the *Grand Hiver* or 'Great Frost' of the European winter of 1708–1709). In Spain, the cultivation of maize is also said to have spread only after a famine, in 1630–31 (Parker 2013: 638). Pelloli's argument, based on careful citation of the historical literature, is that late medieval population growth in Italy encouraged diversification of agricultural crops and led to wider adoption of rice, which could be grown on previously 'marginal' lands.

Brought together, the chapters collected here provide a diverse series of detailed case studies which will be of considerable interest to specialists in the relevant fields, yet which also highlight the rich history of cultural contacts across Eurasia and the importance of inter-disciplinary research.

Acknowledgement

We thank Matthias Donners for his editorial help and Michelle O'Reilly and Junzō Uchiyama for producing several of the maps. Three anonymous reviewers provided valuable feedback on the manuscript. We are also grateful to the whole BAR team for their professional assistance at every stage of the publication process.

References

Autiero, S. and M.A. Cobb (Eds.) (2022) *Globalization and Transculturality from Antiquity to the Pre-Modern World.* Routledge.

Biran, Michal (2015) Introduction: Nomadic culture. In R. Amitai and M. Biran (eds.), *Nomads as Agents of Cultural Change: The Mongols and their Eurasian Predecessors*, pp. 1–9. University of Hawai'i Press.

Boivin, Nicole, Melinda Zeder, Dorian Fuller, et al. (2016) Ecological consequences of human niche construction: examining long-term anthropogenic shaping of global species distributions. *Proceedings of the National Academy of Sciences USA* 113: 6388–6396.

Bradley, Richard (1998) *The Passage of Arms: An Archaeological Analysis of Prehistoric Hoards and Votive Deposits*, Second edition. Oxford: Oxbow.

Bunker, Emma C. (1993) Gold in the ancient Chinese world: a cultural puzzle. *Artibus Asiae* 53: 27–50.

Crawford, Gary W. (2008) The Jomon in early agriculture discourse: issues arising from Matsui, Kanehara and Pearson. *World Archaeology* 40: 445–465.

Dal Martello, Rita (2022) The origins of multi-cropping agriculture in southwestern China: archaeobotanical insights from third to first millennium BC Yunnan. *Asian Archaeology* 6: 65–85.

Ellis, Erle C. (2015) Ecology in an anthropogenic biosphere. *Ecological Monographs* 85: 287–331.

Ellis, Erle C. (2018) *Anthropocene: A Very Short Introduction.* Oxford University Press.

Ethier, J., E. Bánffy, J. Vuković, K. Leshtakov, K. Bacvarov, M. Roffet-Salque, R.P. Evershed and M. Ivanova (2017) Earliest expansion of animal husbandry beyond the Mediterranean zone in the sixth millennium BC. *Scientific Reports* 7: e7146.

Fuller, Dorian Q., C.C. Castillo and C. Murphy (2017) How rice failed to unify Asia: globalization and regionalism of early farming traditions in the monsoon world. In T. Hodos (ed.), *The Routledge Handbook of Archaeology and Globalization*, pp. 711–729. Routledge.

Hein, Anke (2004) Introduction: diffusionism, migration, and the archaeology of the Chinese border regions. In Anke Hein (ed.), *The 'Crescent-Shaped Cultural Communication Belt': Tong Enzheng's Model in Retrospect*, pp. 1–17. Oxford: BAR International Series 2679.

Heine-Geldern, Robert von (1937) *L'art prébouddhique de la Chine et de l'Asie du sud-est et son influence en Océanie.* Paris: Éditions d'art et d'histoire.

Heng, Geraldine (2021) *The Global Middle Ages: An Introduction.* Cambridge University Press.

Hermans, Erik (Ed.) (2020) *A Companion to the Global Early Middle Ages.* Leeds: Arc Humanities Press.

Herrero, Borja Legarra (2019) Gold, conspicuous consumption and prestige—a relationship in need of a review. The case of Early and Middle Bronze Age Crete. In X-L. Armada, M. Murillo-Barroso and M. Charlton (eds.), *Metals, Minds and Mobility: Integrating Scientific Data with Archaeological Theory*, pp. 107–119. Oxford: Oxbow.

Hudson, Mark (2020) Slouching toward the Neolithic: complexity, simplification and resilience in the Japanese archipelago. In Gwen Schug (ed.), *Routledge Handbook of the Bioarcheology of Climate and Environmental Change*, pp. 379–395. Routledge.

Hudson, Mark, Ilona R. Bausch, Martine Robbeets, Tao Li, J. Alyssa White and Linda Gilaizeau (2021) Bronze Age globalization and Eurasian impacts on later Jōmon social change. *Journal of World Prehistory* 34: 121–158.

Hudson, Mark, Junzo Uchiyama, Takamune Kawashima, Kati Lindström, Ian Reader, Tinka Delakorda Kawashima, Danièle Martin, Chris Gillam, Linda Gilaizeau, Ilona Bausch and Kara C. Hoover (2022) Global processes of anthropogenesis characterise the early Anthropocene in the Japanese Islands. *Humanities and Social Science Communications* 9: e84.

Jones, M.K., H.V. Hunt, E. Lightfoot, D. Lister, X. Liu and G. Motuzaite-Matuzevicuite (2011) Food globalization in prehistory. *World Archaeology* 43: 665–675.

Karlgren, Bernhard (1942) The date of the early Dong Son culture. *Bulletin of the Museum of Far Eastern Antiquities* 14: 1–28.

Kawamura, Ai, Chun-Hsiang Chang and Yoshinari Kawamura (2016) Middle Pleistocene to Holocene mammal faunas of the Ryukyu Islands and Taiwan: an updated review incorporating results of recent research. *Quaternary International* 397: 117–135.

Li, Tao, Chao Ning, Irina S. Zhushchikhovskaya, *et al.* (2020) Millet agriculture dispersed from Northeast China to the Russian Far East: integrating archaeology, genetics and linguistics. *Archaeological Research in Asia* 22: e100177.

Liu, Li, Jian Chen, Jiajing Wang, Yanan Zhao and Xingcan Chen (2022) Archaeological evidence for initial migration of Neolithic Proto Sino-Tibetan speakers from Yellow River valley to Tibetan Plateau. *Proceedings of the National Academy of Sciences* USA 119: e2212006119.

Liu, Xinyi and Martin K. Jones (2014) Food globalisation in prehistory: top down or bottom up? *Antiquity* 88: 956–963.

Liu, X., P.J. Jones, G.M. Matuzeviciute, H.V. Hunt, D.L. Lister, T. An, N. Przelomska, C.J. Kneale, Z. Zhao and M.K. Jones (2019) From ecological opportunism to multi-cropping: mapping food globalisation in prehistory. *Quaternary Science Reviews* 206: 21–28.

Matsui, Akira and Masaaki Kanehara (2006) The question of prehistoric plant husbandry during the Jomon period in Japan. *World Archaeology* 38: 259–273.

McArdle, F. (1978) *Altopascio: A Study in Tuscan Rural Society, 1587–1784*. Cambridge University Press.

Nishida, Masaki (1983) The emergence of food production in Neolithic Japan. *Journal of Anthropological Archaeology* 2: 305–322.

Obata, Hiroki (2016) *Tane o maku Jōmonjin*. Tokyo: Yoshikawa Kōbunkan.

Parker, Geoffrey (2013) *Global Crisis: War, Climate Change and Catastrophe in the Seventeenth Century*. Yale University Press.

Pearson, Richard and Kazue (1978) Some problems in the study of Jomon subsistence. *Antiquity* 52: 21–27.

Qu, Feng (2014) Eskimo art prototypes in the Chinese Neolithic: a comparison of Okvik/Old Bering Sea and Liangzhu ritual art. *Sibirica* 13: 45–78.

Robbeets, M., R. Bouckaert, M. Conte, et al. (2021) Triangulation supports agricultural spread of the Transeurasian languages. *Nature* 599: 616–621.

Ruhlen, Merritt (1998) The origin of the Na-Dene. *Proceedings of the National Academy of Sciences USA* 95: 13994–13996.

Sherratt, Susan (2019) Why was (and is) silver sexy? Silver during the 4th-3rd millennia in the Near East and Mesopotamia. In X-L. Armada, M. Murillo-Barroso and M. Charlton (eds.), *Metals, Minds and Mobility: Integrating Scientific Data with Archaeological Theory*, pp. 97–106. Oxford: Oxbow.

Spengler, Robert N., Sören Stark, Xinying Zhou, Daniel Fuks, Li Tang, Basira Mir-Makhamad, Rasmus Bjørn, Hongen Jiang, Luca M. Olivieri, Alisher Begmatov and Nicole Boivin (2021) A journey to the west: the ancient dispersal of rice out of East Asia. *Rice* 14: e83.

Stephens, Lucas, Dorian Fuller, Nicole Boivin, et al. (2019) Archaeological assessment reveals Earth's early transformation through land use. *Science* 365: 897–902.

Stevens, Chris and Dorian Fuller (2017) The spread of agriculture in eastern Asia: archaeological bases for hypothetical farmer/language dispersals. *Language Dynamics and Change* 7: 152–186.

Storozum, Michael J. and Yugi Li (2020) Chinese archaeology goes abroad. *Archaeologies: Journal of the World Archaeological Congress* 16: 282–309.

Tang, Li, H. Lu, X. Chen, H. Xu, N. Boivin, M. Storozum, F. Yang, S. Li, X. Liu and RN. Spengler (2022) Prehistoric agricultural decision making in the western Himalayas: ecological and social variables. *Antiquity* 96: 1214–1231.

Uchiyama, Junzo, J. Christopher Gillam, Alexander Savelyev and Chao Ning (2020) Population dynamics in Northern Eurasian forests: a long-term perspective from Northeast Asia. *Evolutionary Human Sciences* 2: e16.

Van Driem, George (2019) *The Tale of Tea: A Comprehensive History of Tea from Prehistoric Times to the Present Day*. Leiden: Brill.

Vandkilde, Helle (2016) Bronzisation: the Bronze Age as pre-modern globalization. *Praehistorische Zeitschrift* 91: 103–123.

von Falkenhausen, Lothar (1995) The regionalist paradigm in Chinese archaeology. In Philip L. Kohl and Clare Fawcett (eds.), *Nationalism, Politics, and the Practice of Archaeology*, pp. 198–217. Cambridge University Press.

Wang, Shengyu (2021) Exotica, fashion and immortality: the new use of gold in Han dynasty China (206 BCE – CE 220). *Cambridge Archaeological Journal* 32: 345–360.

Wang, Siyi (2022) An archaeological perspective of alcoholic beverages in the Song dynasty (960–1279). *Archaeologies: Journal of the World Archaeological Congress* 18: 436–467.

Yin, Zhixi (2022) Representing the Silk Road: literature and images between China and Japan during the Cold War. *Humanities* 12: e4.

Part I

Agropastoralism and Expanding Human Niches

2

The Utilisation of Domesticated Plants and Secondary Vegetation in the Japanese Islands in the Jōmon Period

Seiji Nakayama

Abstract: While the Jōmon culture is often considered to be one of hunter-gatherers, archaeobotanical research has shown that plant cultivation was an important element of the Jōmon economy. Rich forests emerged in the Japanese archipelago through climatic warming in the Holocene. These forests were managed by Jōmon groups who expanded their settlements. This chapter argues that the Jōmon plant cultivation system was fostered by the symbiosis of people and plants. However, Jōmon plant cultivation did not immediately continue to the emergence of mature agriculture; it coexisted variably with hunting, gathering and fishing according to the environment of each region and each age. That pattern can be considered as one cause of the stability and sustainability of Jōmon culture over 10,000 years.

Keywords: Jōmon, Neolithic, plant cultivation, domestication, landscape ecology, secondary vegetation

2.1. Introduction

The Jōmon was a Neolithic culture which lasted more than 10,000 years in the Japanese archipelago, and which was found in the four main islands of Kyushu, Shikoku, Honshu and Hokkaido, as well as on many small offshore islands. The beginning of Jōmon culture is marked by the emergence of pottery between 15,000 to 16,000 years ago. The Jōmon continued until the first millennium BC when an agricultural society based on paddy rice cultivation was established. The Jōmon period is generally divided into six phases: Incipient, Initial, Early, Middle, Late and Final (Fig. 2.1).[1]

The oldest style of Jōmon pottery has a characteristic pointed base. The name Jōmon comes from the Japanese word for cord-marked patterns, although the oldest style of Jōmon pottery does not display such cordmarking. The Incipient Jōmon phase corresponds to the final stage of the Last Glacial age and the Initial Jōmon and following phases correspond to the Post Glacial. About 11,000 years ago, the climate became rapidly warmer after the end of the Younger Dryas and rich forests appeared throughout the Japanese archipelago. It was previously believed that the economy of the Jōmon period was based primarily on food acquisition such as hunting, fishing, gathering, and that agriculture or food production did not exist. This economy was viewed as distinct compared to both West and East Asia in the Neolithic era when agriculture and animal domestication had developed. Through recent archaeobotanical research, however, it has been recognised that woody plants such as chestnut and lacquer trees and herbaceous plants like soybeans, adzuki beans, *Perilla* and barnyard millet were used and cultivated by Jōmon people.

2.2. Plant cultivation in the Jōmon period

2.2.1. Woody plants

2.2.1.1. Lacquer tree (Toxicodendron vernicifluum)

Objects painted with lacquer sap existed from the early stages of the Jōmon period. The oldest lacquer product in East Asia was unearthed at the Kakinoshima B site (Hokkaido) and dated to the Initial Jōmon (Minami Kayabechō 2002). Numerous lacquer products are found from many archaeological sites dating from the Early to the Final Jōmon (Kudō 2014b). Lacquer was used for various purposes such as waterproof paint, to repair vessels, and as an adhesive to connect arrowheads and other lithic artefacts. Lacquer products with red or black lacquer have significant decorative effects.

It used to be believed that lacquer trees belong to the genus *Rhus*, but they were identified as belonging to the genus *Toxicodendron* in recent molecular phylogenetic studies (e.g., Nie et al. 2009). There are several species in the genus *Toxicodendron*, including *T. vernicifluum*, *T. trichocarpum*, *T. succedaneum*, *T. sylvestre* and *T. radicans*. Trees of *Toxicodendron* and *Rhus* are divided into species by observation of tubes in annual rings (Noshiro and Suzuki 2004). Noshiro and Suzuki reanalysed the lacquer tree recovered from the Torihama site (Fukui) and identified it as *T. trichocarpum* (J. *yamaurushi*) (Noshiro et al. 2007). According to AMS dating, the lacquer tree from Torihama dates back to ca. 12,600 cal BP, a date consistent

[1] Editors' note: In the English literature, the terms 'Earliest' and 'Latest' are sometimes used for the Initial and Final Jōmon phases, respectively. The former terms were used by American archaeologist J. Edward Kidder (1922–2014) but Initial and Final have become more standard in recent years.

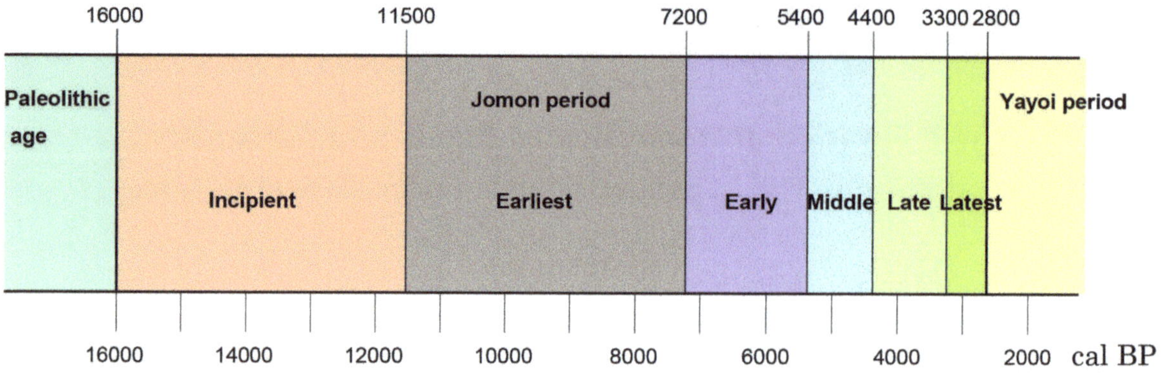

Figure 2.1. Chronology of the Jōmon period.

with the Incipient Jōmon period. Pollen analyses have revealed the existence of lacquer trees after the Initial Jōmon. These trees are also found in the sediments of the Sannai Maruyama site (Aomori) dating from 5600 to 4500 years ago (M. Yoshikawa 2011a). Lacquer fruit remains have been recovered at the Iwatokotani 4 site (Aomori) from the Early Jōmon (Yoshikawa and Itō 2004).

According to recent DNA analyses of lacquer chloroplasts, they can be divided into several types such as Hubei, Zhejiang and Japanese. It is possible that the Japanese type lacquer trees known from the Jōmon period were brought from Shandong Province in China for the purpose of making lacquer products, and that they were cultivated by Jōmon people (Suzuki 2014). Various kinds of lacquer products and traces of scratches for sap were discovered at the Shimo-yakebe site in the Late Jōmon period; moreover, it is recognised that the scratch intervals for the sap collection were changed depending on the wood thickness (Fig. 2.2, Chiba 2014). The foundation of lacquer crafts in Japan was largely completed in the Jōmon period when Jōmon people who understood this plant cultivated and managed lacquer trees around their settlements.

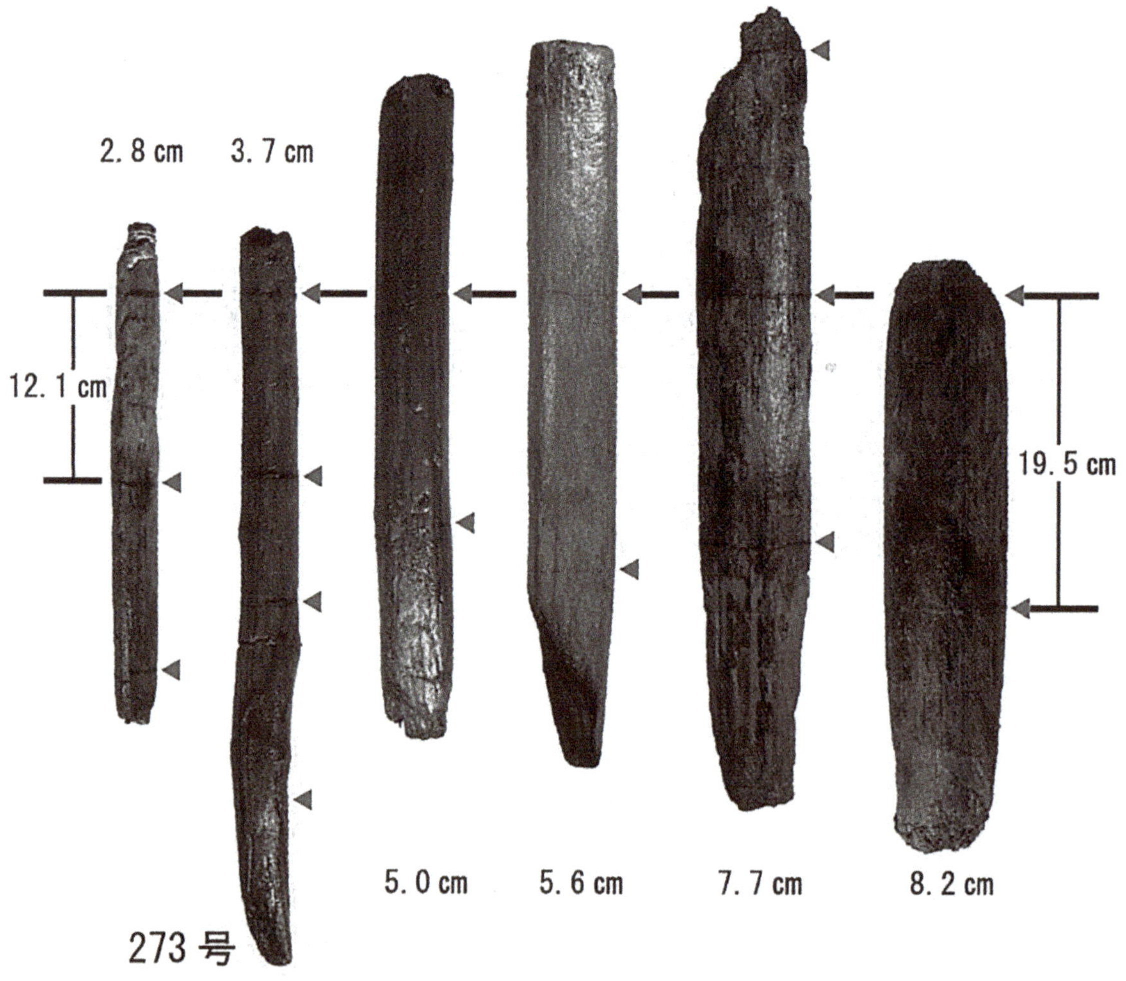

Figure 2.2. Lacquer trees with scratching traces for sap collection recovered from the Shimo-yakebe site. Source: Chiba (2014).

2.2.1.2. Chestnut (Castanea crenata)

Chestnut trees are woody plants which have long been suggested as candidates for cultivation by humans (Sakazume 1956). A comparison of wood ratios at the Late-Final Jōmon Akayama-jinja, Juno and Terano-higashi sites found that chestnut accounted for 50~80% of total wood (Noshiro and Sasaki 2007). Chestnut comprised 18-50% of all wood species recovered from waterlogged deposits at the Shimo-yakebe site. Chestnut trees were not cut down all together in the same year, but were selected based on appropriate size and were flexibly maintained and managed by Jōmon people (Noshiro 2014). Pollen analysis at Sannai Maruyama found that chestnuts were more common than other trees, comprising 40-90% of the whole pollen spectra during the phases of human settlement. After the Late Jōmon, however, as the settlement occupation came to an end, *Quercus serrata* and Japanese horse chestnut increased (Fig. 2.3, Yoshikawa et al. 2006; M. Yoshikawa 2011). From this, some pollen analysts suggest the existence of pure chestnut forests maintained around the settlement.

Chestnut fruits were very small in size during the Initial to Early Jōmon, but the fruit sizes gradually became larger, and large fruits about 4-5 cm in length similar to domesticated chestnuts appeared from the Middle Jōmon (J. Yoshikawa 2011). This is evidence that Jōmon people continuously managed chestnut trees and selected large-sized fruits. Chestnut fruits in the Jōmon period were kept and processed as dried chestnuts for use as important foods throughout the year (Nakui 2012).

2.2.2. Herbaceous Plants

Plant impression analysis is one of the methods for archaeobotanical research. A similar method has been known for about one hundred years, but new techniques using silicon resin for mould impressions and a scanning electric microscope to observe the mould replicas have dramatically improved the accuracy of identification (Ushino and Tagawa 1991). This type of replica-SEM analysis has the potential to obtain seed impressions from reliably dated pottery without risk of contamination. Plant

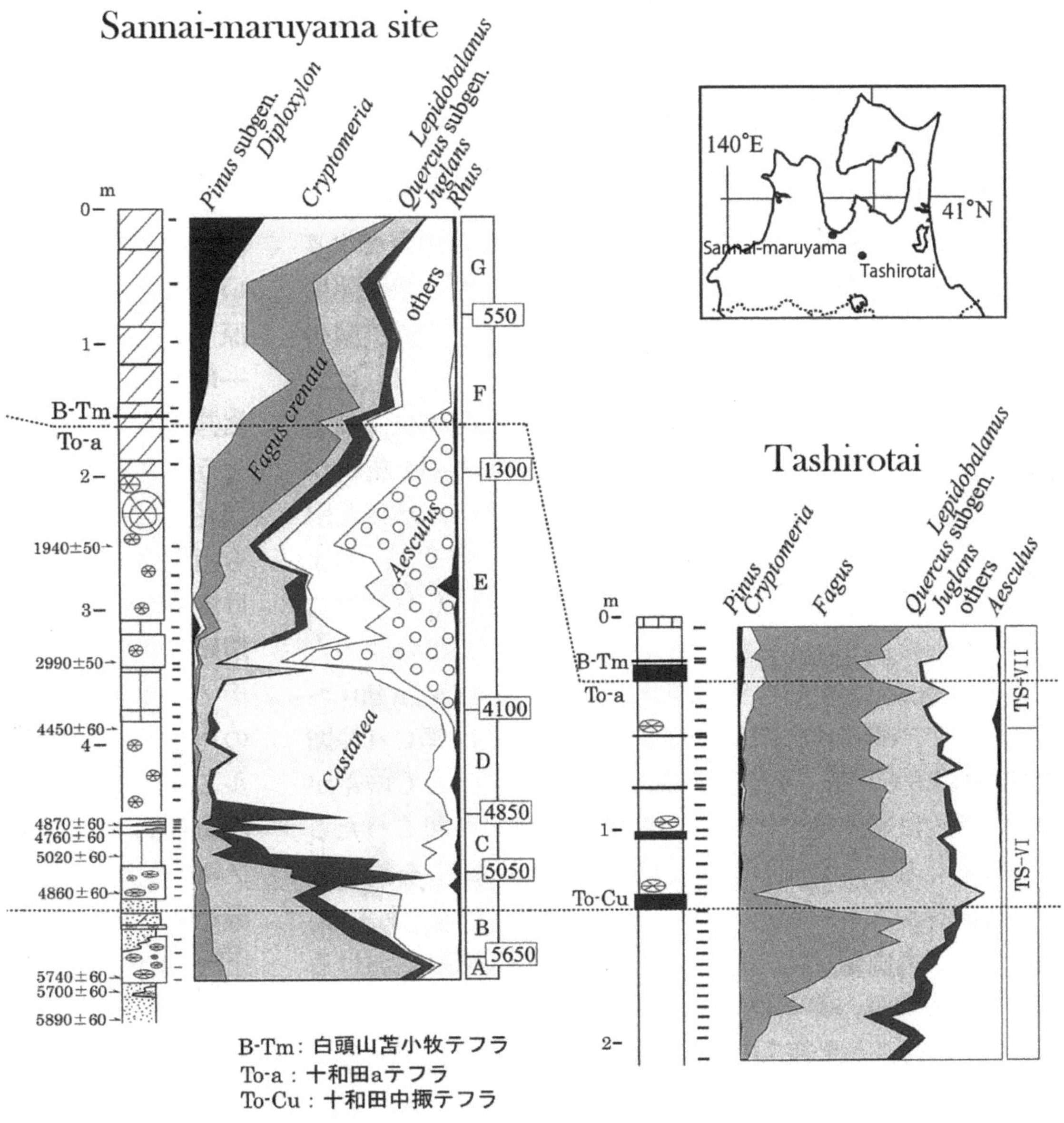

Figure 2.3. Change of woody pollen spectra around the Sannai Maruyama site and Tashirotai. Source: Yoshikawa et al. (2006).

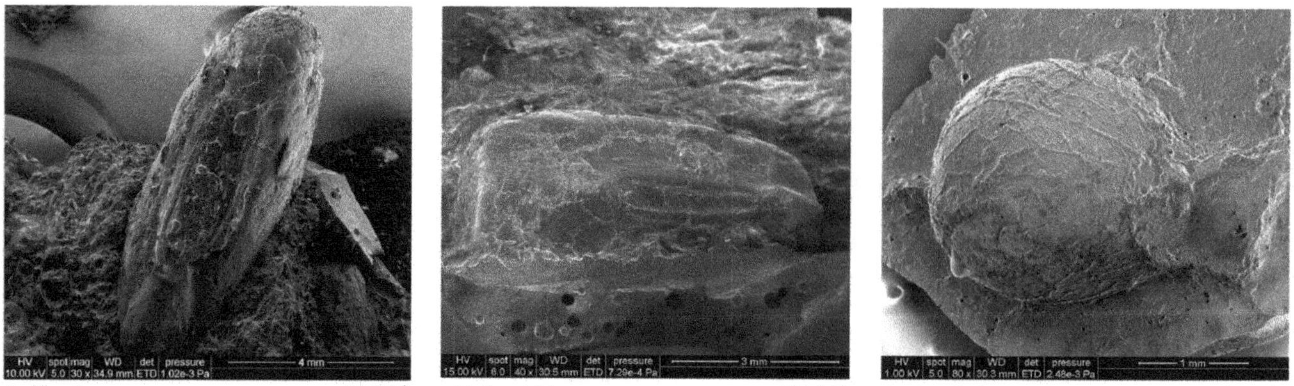

1. Soybean (×30) Sakenomiba site
2. Azuki bean (×40) Sakenomiba site
3. Beefsteak plant (×80) Daishi site

Figure 2.4. Plant impressions on Jōmon potsherds. Source: the author.

impression analyses have identified many kinds of plants such as rice, foxtail millet, broomcorn millet, beefsteak plant, Japanese pepper, Japanese red elder, genus *Perilla, Glycine max, Vigna ceratotropis, Toxicodendron, Rhus, Vitis* and *Setaria* from sites of the Jōmon period. Rice and foxtail and broomcorn millet came from the Korean peninsula at the end of the Final Jōmon and spread throughout the Japanese archipelago in the first millennium BC. At the same time, it has been shown by replica and SEM analysis that domesticated plants like soybean, adzuki bean and *Perilla* had already existed in the Middle Jōmon (Fig. 2.4, Nakayama 2014).

2.2.2.1. **Perilla** *sp.*

The mint herb *Perilla*, sometimes known in English as the beefsteak plant, is classified into two forms of the same species: *Perilla frutescens* var. *frutescens* (J. *egoma*) and *P. frutescens* var. *crispa* (J. *shiso*). These plants are able to naturally cross-fertilise (Nitta 2001), but they have distinct nutritional characteristics. In 100 grams, *shiso* contains 3.4g protein, 0.1g lipid, 8.9g carbohydrate and 4.1 kcal of energy, whereas *egoma* is much more nutritious with 17.7g protein, 43.4g lipid, 29.4g carbohydrate and 54.4 kcal of energy (MEXT 2015). The high lipid content of *egoma* is especially prominent.

Perilla remains from Jōmon archaeological sites are discovered not only as single fruits but also as chunks or carbonised 'cookies' (Nagasawa 1989). *Shiso* has a bactericidal action with a unique odour. *Egoma* was used for fuel and for admixing with lacquer, in addition to consumption as a food. The possible utilisation of *egoma* for its oil in the Jōmon period is still uncertain. *Perilla* seed impressions are found on potsherds from the beginning of the Early Jōmon phase and, together with legumes, have been frequently identified. These impressions can be divided into *shiso* and *egoma* through a comparison of the size of fruits (Fig. 2.5). *Shiso* has dormancy under natural conditions, but *egoma* is seeded by people and doesn't have dormancy (Nitta 2001). Since *egoma* requires human seeding and management, it is therefore possible that it was already cultivated by Jōmon people.

***2.2.2.2. Genus* Glycine**

Soybean is an annual herbaceous plant of the Fabaceae family, genus *Glycine*, subgenus *Soja*. *Soja* subgenus contains two species of soybean and wild soybean, and *Glycine* subgenus contains seven perennial wild species (Shimamoto 2003). It was previously believed that soybean reached the Japanese archipelago at the beginning of the Yayoi period, but evidence of Jōmon-period soybeans was revealed by replica-SEM analysis over a decade ago.

The oldest evidence of wild soybean in Japan is impressions on pottery at the Ojiyama site (Miyazaki), which dates back to 13,350-13,300 cal. BP (Obata and Manabe 2011; Kudō 2011). In Honshu, soybean impressions are found on potsherds from the Initial Jōmon period. From these finds there is no doubt that people in the Japanese archipelago began to use wild soybeans from the latest Pleistocene to the beginning of the Holocene. Numerous soybean remains have been discovered in central Honshu in the Middle Jōmon (Nakayama 2010b) and in Kyushu in the Late to Final Jōmon (Obata et al. 2007). Based on the volume of seed impressions, it has been revealed that the size of seeds changed over time (Fig. 2.6, Nakayama, 2009, 2015a). I have classified *Glycine* seed impressions into a wild soybean type of 60 mm^3 or less and a cultivated soybean type of 61 mm^3 or more. Until the Early Jōmon, the impressions are of a similar size to wild soybeans; but from the Middle to Final phases they increase in size and volume. I therefore suggest that large-type soybean seeds indicate one aspect of the domestication syndrome of soybeans in the Jōmon period. It is recognised that domesticated soybeans appeared in multiple regions of prehistoric East Asia and one such region seems to be the Japanese archipelago (Nakayama 2009, 2010).

2.2.2.3. Vigna *subgenus* Ceratotropis

Vigna subgenus *Ceratotropis*, also called the Asian *Vigna*, is divided into three sections and 21 species including six domesticated plants (Tomooka et al. 2002). Yamaguchi (2003) classifies subgenus *Azuki* (*Ceratotropis*) into adzuki bean and mung bean types based on the

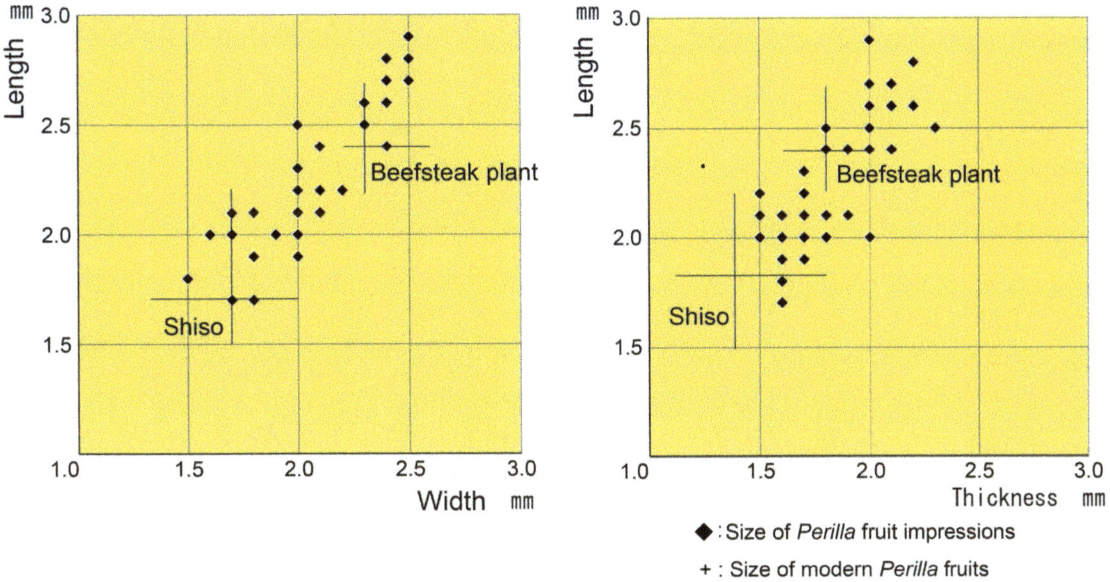

Figure 2.5. Comparison of *Perilla* fruit sizes in the Jōmon and modern eras. Source: the author.

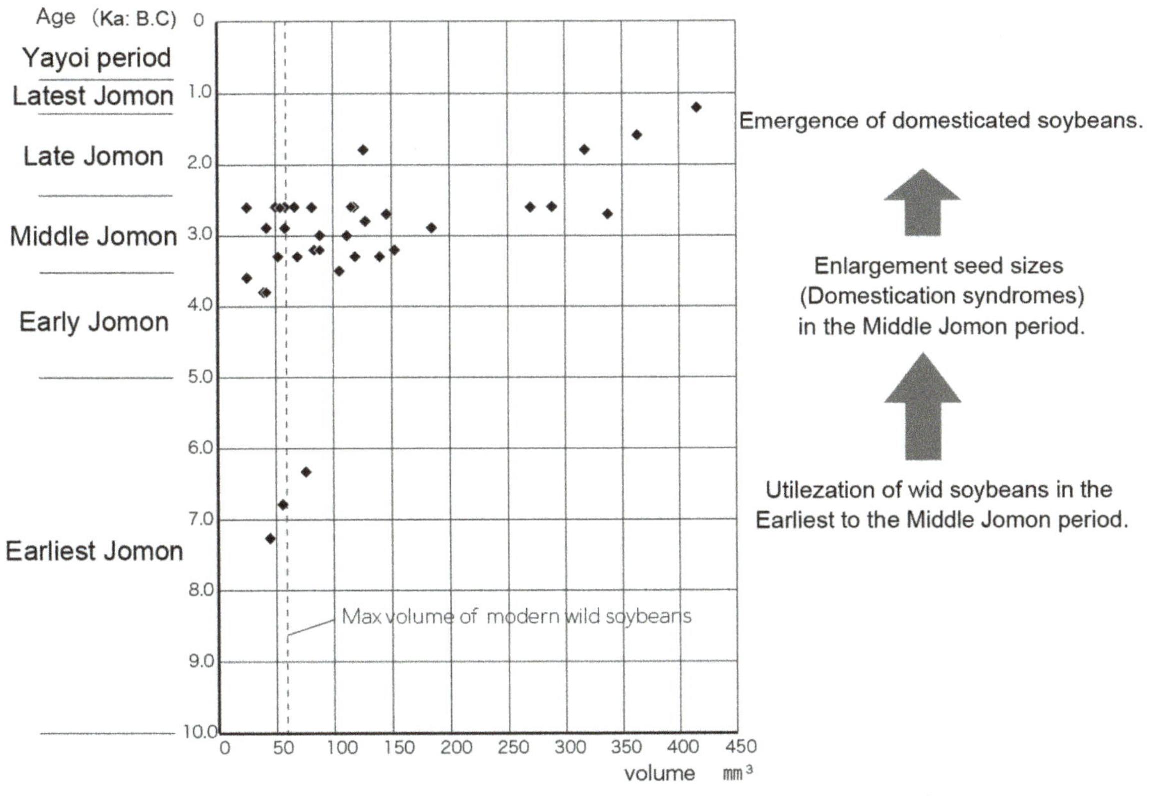

Figure 2.6. Changes in the size of *Glycine max* seeds in the Jōmon period.

nucleotide sequence of chloroplast DNA.[2] According to a morphological analysis of radicles and primary leaves conducted by Hokkaido University, a similar classification was made into adzuki (*Vigna angularis*) and mung bean (*Vigna radiata*) types. Yoshizaki (2001) applied this system of identification to legumes excavated from Jōmon period sites and found that almost all small beans of Jōmon age contained the adzuki type. Small beans found at the Nakaya, Otsuki, Shimoyakebe and Sakuramachi sites have also been identified as adzuki type or adzuki group (Matsutani 1997; Yoshizaki et al. 2003; Sasaki et al. 2007). Beans recovered from walnut mounds at Shimoyakebe have been AMS dated to ca. 5300–4800 cal BP, placing them in the Katsusaka phase of the Middle Jōmon (Kudō et al. 2007). The Awazu Lake Bed site (Shiga) has produced

[2] Editor's note: The Japanese word *azuki* is usually rendered as adzuki in English.

what is considered as the oldest wild adzuki bean in Japan, dating to the first half of the Initial Jōmon (Minamiki and Nakagawa 2000).

As a result of plant impression analysis in central Honshu, *Vigna* subgenus *Ceratotropis* beans have been identified at 17 sites from the Early to Late Jōmon (Nakayama 2010a, 2010b). I suggest that they can be strictly identified as adzuki species (*Vigna angularis*) through a comparison with the seed form and the hilar shape of modern Asian *Vigna* and seed impressions of adzuki type. The problem is whether they are wild species or cultivars. According to comparative research on seed sizes, the average volume of modern wild adzuki beans is 22.6 mm^3 and the maximum is about 45.0 mm^3. The volumes of impressions of adzuki beans from the Jōmon period have various sizes from 12.7 mm^3 to 83.1 mm^3 (Fig. 2.7). In particular, the materials dating from after the Middle Jōmon period are bigger than modern wild adzuki beans. It is therefore possible that the cultivar type of adzuki bean emerged from the Middle Jōmon in a similar way to soybeans (Nakayama 2015b).

2.2.2.4. Barnyard millet (Echinochloa utilis *Ohwi et Yabuno*)

The barnyard millet currently cultivated in Japan is thought to derive from *Echinochloa crus-galli* (J. *inubie*), thus differing from South Asian Indian barnyard millet (*E. frumentacea*) which is thought to have derived from *E. colona* (Yabuno 1962; Hoshikawa 1980). All the wild *Echinochloa* species in Japan are annuals and include *E. oryzicola* (J. *tainubie*) and *E. crus-galli* var. *formosensis* (J. *himetainubie*), which only grow in wet rice paddies, *E. crus-galli* var. *crus-galli* (J. *inubie*), which is found in wetlands including but not limited to rice paddies, and *E. crus-galli* var. *praticola* (J. *himeinubie*), which occurs on dry land including roadsides and fields (Yabuno and Yamaguchi 2001). Sakamoto (1988) argues that because there are no archaeological examples or ancient records of cultivated *Echinochloa* in China, it was probably cultivated independently in Japan.

The earliest example of *Echinochloa* from a Jōmon site is *E. crus-galli* from the Initial phase Nakano B site in Hokkaido. Between the Early and Final phases, at least 18 sites

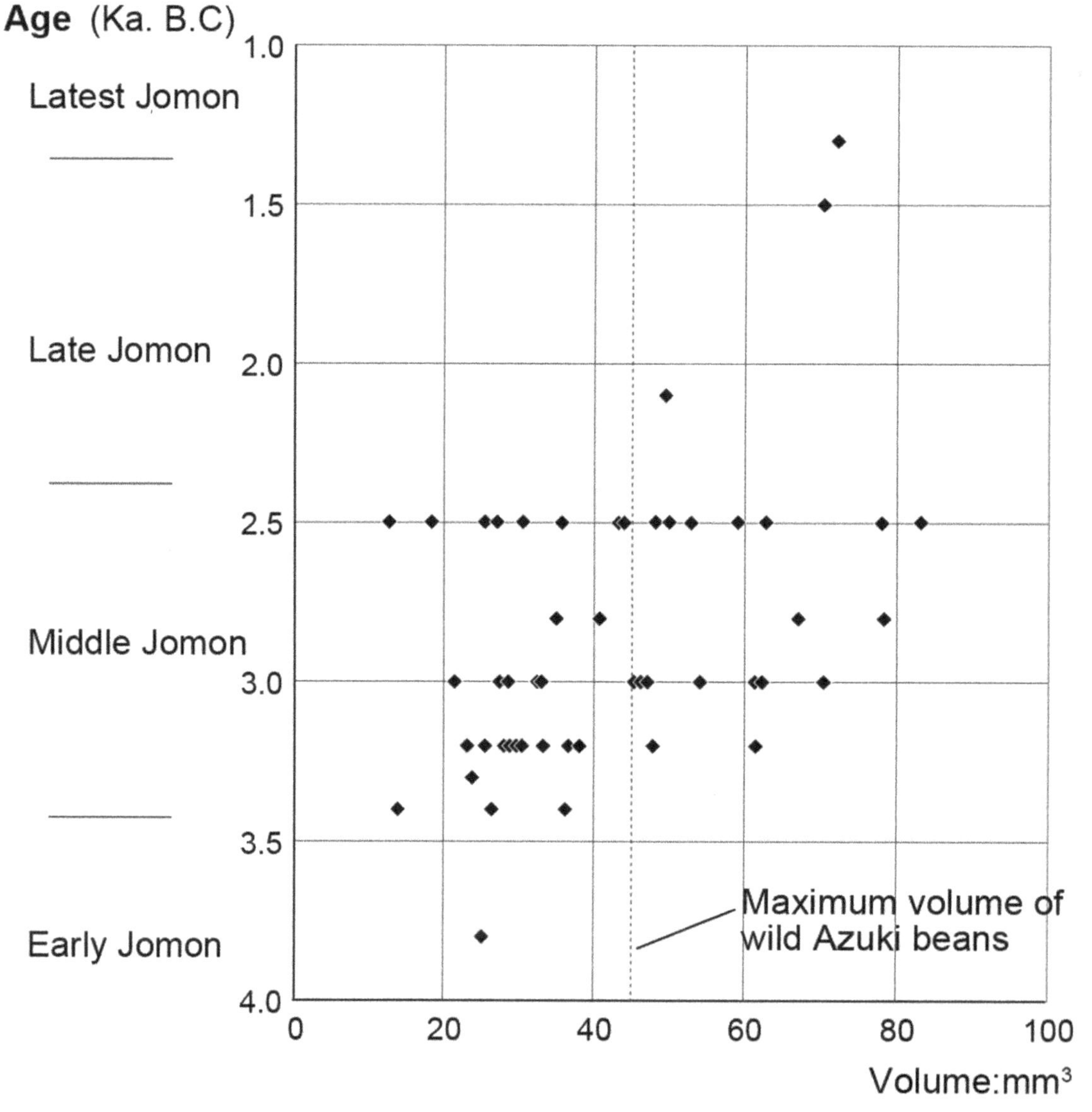

Figure 2.7. Changes in the size of adzuki bean seeds in the Jōmon period. Source: the author.

concentrated in Hokkaido and the Tohoku have produced *Echinochloa* (Nakayama 2010b: 87–89). Of this total of 19 sites, 11 can be classified as 'B rank', in other words they are materials where a microscope has been used to make a secure identification of the seed but which have only been dated by associated archaeological remains. Only one is 'A rank' with direct dates on two seeds from Middle Jōmon Tomi-no-sawa (Aomori); the dates are calibrated (OxCal 2020) to 3006–2703 BC and 2879–2851 BC. The other seven *Echinochloa* finds are C and D ranks with low confidence.

Masakazu Yoshizaki noticed that *Echinochloa* is common amongst carbonised seeds discovered from the floors of Jōmon dwellings. In addition to wild type *E. crus-galli*, Yoshizaki (1992) noticed a larger, rounder type which he called *Jōmon hie* or 'Jōmon *Echinochloa*'. From the fact that finds of cultivated *Echinochloa* in Hokkaido increase from the Epi-Jōmon phase onwards, it has been suggested that 'Jōmon *Echinochloa*' was a wild species under selection towards a cultivated species (Crawford 1983, 1992; Yoshizaki 1995). Building on the radiocarbon dates from the Tomi-no-sawa site, Sumida and colleagues (2008) classified *Echinochloa* into narrow and round types, suggesting that these represent weed and cultivated types, respectively. In any case, *Echinochloa* is very important as the only Gramineae in the Japanese archipelago to undergo local domestication. Furthermore, with the exception of the Middle Jōmon Midden 3 at the Awazu lake bottom site, almost all *Echinochloa* finds are concentrated in Hokkaido and the Tohoku, showing that Jōmon plant domestication could be extremely local in areal scope.

2.3. Plant cultivation in the Jōmon period

I have pointed out the existence of domesticated plants in the seed impressions detected on pottery from central Japan in the Middle Jōmon period. How were these herbaceous plants cultivated at that time? I would like to explain the process of the emergence of domesticated plants in the Jōmon by using the theory of field succession (Fukui 1983). Figure 2.8 shows a conceptual diagram of the system of utilisation of

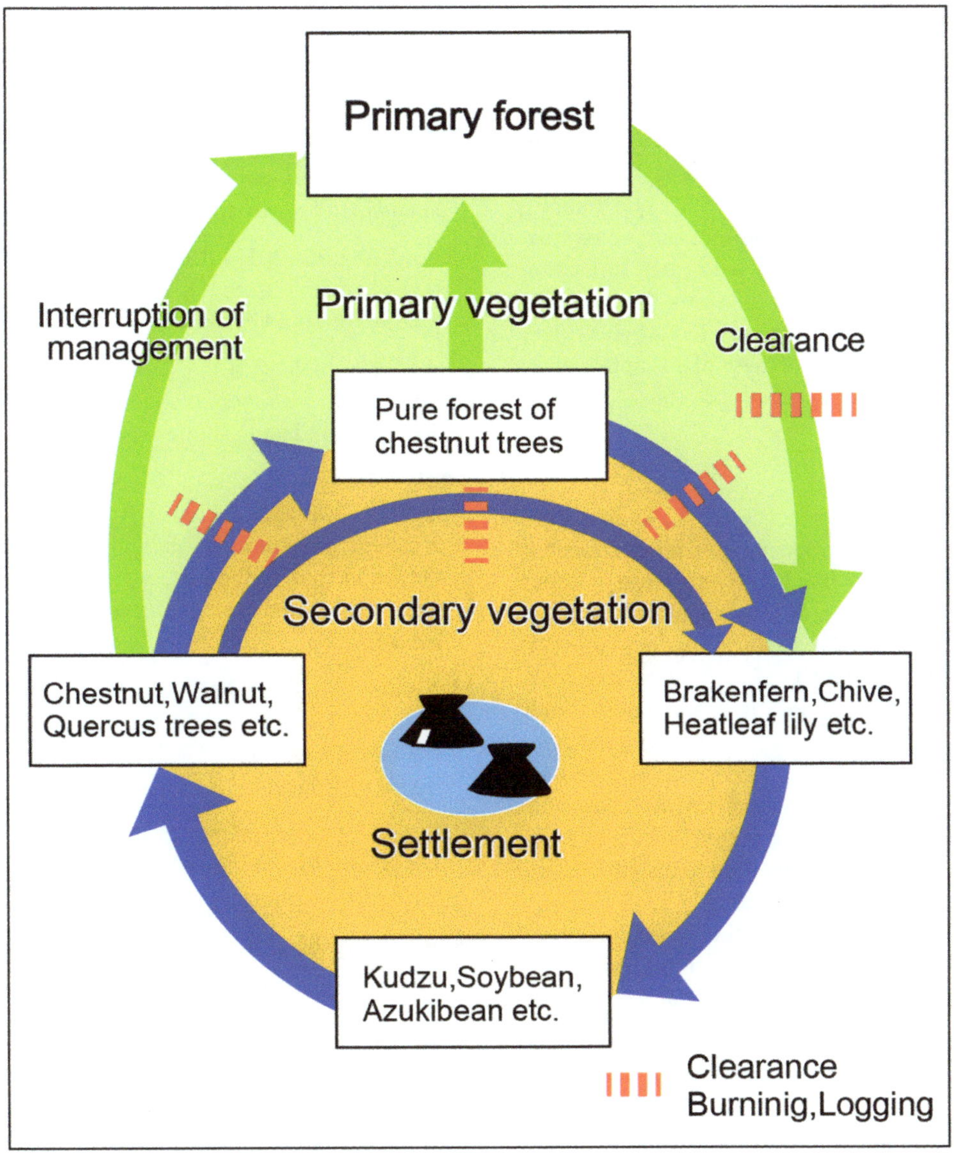

Figure 2.8. Utilisation model of cultivated plants and secondary vegetation in the Jōmon period. Source: the author.

plants from the perspective of archaeobotany (Nakayama 2015b). Areas of secondary vegetation appeared between the primary vegetation and human settlements when people cut and burnt the natural forest. Plants favouring open ground such as bracken fern, Asian royal fern and long-stamen chive flourished after the forest clearance. The vegetation changed to legumes and rhizome plants such as kudzu, Japanese yams, wild soybeans and adzuki beans on fertile land, as well as herbaceous plants frequently utilised by humans. Woody plants like chestnuts, walnuts and Japanese horse chestnuts were grown in these areas, producing several kinds of wood for use for tools and fuel by the Jōmon people. Chestnut and walnut trees in the secondary forest were consciously managed and almost pure chestnut forests were likely maintained. Through clearance by burning or felling trees, open ground plants continued to flourish and the cycle of secondary vegetation was maintained. When human settlements moved, however, the cycle was interrupted and the areas eventually returned to natural forest.

Tsuji (2009) argued that under natural ecosystems in the Jōmon period, laurel forests were spread in southern Japan and deciduous broad-leaved forests in northern Japan, but that anthropogenic ecosystems were associated with various human activities at this time. There were abundant plants for human utilisation in the anthropogenic ecosystem, and a rich forest was maintained by the human usage and selection of plants. At the same time, there were good feeding areas for many animals and hunting areas. Such management of the anthropogenic ecosystem probably dates back to the stage of wild plant utilisation by Jōmon people. Over thousands of years these human activities eventually produced cultivars like soybean and adzuki bean. The domestication of these plants appeared as a result of the management of the anthropogenic ecosystem.

2.4. Conclusion

It has been thought that the subsistence of the Jōmon culture was mainly one of food acquisition not food production, but according to archaeobotany plant cultivation was without doubt one important element of the Jōmon economy. Rich forests emerged in the Japanese archipelago through global warming in the Holocene, and were managed by Jōmon groups who expanded their settlements with the development of the forest. It has been argued here that the Jōmon plant cultivation system was fostered by the symbiosis of people and plants. However, the most significant characteristic of Jōmon culture was that plant cultivation did not immediately continue to the emergence of mature agriculture; it coexisted variably with hunting, gathering and fishing according to the environment of each region and each age. That pattern can be considered as one cause of the stability and sustainability of Jōmon culture over more than 10,000 years.

References

Abe, Jun and Yoshiya Shimamoto (2001) Daizu no shinka: tsurumame no hatashita yakuwari [Evolution of the soybean: the role of wild soybeans]. In H. Yamaguchi and Y. Shimamoto (eds), *Saibai shokubutsu no shizenshi* [The natural history of domesticated plants], pp.77–95. Sapporo: Hokkaido University Press.

Chiba, T. (2014) Jōmon people in the suitable place: Shimo-yakebe site. In *New Perspectives on the Plant Use of Jōmon People*, pp.118–137. Tokyo: Shinsensha (in Japanese).

Crawford, Gary W. (1983) *Paleoethnobotany of the Kameda Peninsula Jomon*. Ann Arbor: Museum of Anthropology, University of Michigan.

Crawford, Gary W. (1992) Prehistoric plant domestication in East Asia. In C.W. Cowan and P.J. Watson (eds), *The Origins of Agriculture: An International Perspective*, pp. 7–38. Washington DC: Smithsonian Institution Press.

Fukui, Katsuyoshi (1983) Yakihata nōkō no fuhensei to shinka [The universality and evolution of slash and burn agriculture]. In Yoshihiko Amino (ed.), *Nihon minzoku bunka taikei 5: sanmin to ama* [The Ethnographic Cultures of Japan 5: Mountain and Sea Peoples], pp.235–274. Tokyo: Shogakukan.

Fuller, Dorian Q. (2007) Contrasting patterns in crop domestication and domestication rates: recent archaeobotanical insights from the Old World. *Annals of Botany* 100(5): 903–924.

Hoshikawa, Kiyochika (1980) *Shin'hen shokuyō sakumotsu* [New edition food crops]. Yōkendō.

Kudō, Yūichirō (2011) Radiocarbon dating of carbonised plant remains from Ōjiyama site. In *Ōjiyama iseki*, pp.73–76. Tojō City Board of Education (in Japanese).

Kudō, Yūichirō (Ed.) (2014) *Koko made wakatta! Jōmonjin no shokubutsu riyō* [What We Know about Jōmon Plant Use]. Shinsensha.

Kudō, Yūichirō, Y. Sasaki, M. Sakamoto, K. Kobayashi and H. Matsuzaki (2007) A chronological study of archaeological remains showing plant utilisation during the second half of the Jōmon period at the Shimo-yakebe site, Tokyo, Japan. *Japanese Journal of Historical Botany* 15: 5–17 (in Japanese with English summary).

Matustani, Akiko (1997) Carbonised plants recovered from the Ōtsuki site. In Yamanashi Prefecture (eds), *Ōtsuki Site*, pp.115–117. Kōfu: Yamanashi Prefectural Board of Education (in Japanese).

MEXT (2015) *Standard Tables of Food Composition in Japan 2015* (Seventh Revised Ed.). Ministry of Education, Culture, Sports, Science and Technology.

Minami Kayabe-chō (2002) *Kakinoshima B iseki* [The Kakinoshima B site]. Hokkaido: Minami-kayabe Town.

Minamiki, M. and H. Nakagawa (2000) The plant macro remains. In *Awazu kōtei iseki*, pp.49–112. Shiga Prefectural Board of Education (in Japanese).

Nagasawa, H. (1989) Jōmon jidai ni okeru egoma no riyō ni tsuite [On the utilisation of *egoma* in the Jōmon period]. *Yamanashiken Kōkogaku Ronshū* II: 119–146.

Nakayama, Seiji (2009) An archaeobotanical study of utilization and cultivation of soybean (Genus Glycine) in the Jomon period. *Cultura Antiqua* 61(3): 40–59 (in Japanese with English summary).

Nakayama, Seiji (2010a) Identification of azuki beans (*Vigna angularis*) in the Jōmon period. *Tōkai Shigaku* 44: 83–103 (in Japanese).

Nakayama, Seiji (2010b) *Shokubutsu kōkogaku to Nihon no nōkō no kigen* [Archaeobotany and the Origin of Agriculture in Japan]. Dōseisha.

Nakayama, Seiji (Ed.) (2014) *The Origin of Grain Agriculture in the Korean Peninsula and the Japanese Archipelago.* Kōfu: Yamanashi Prefectural Museum (in Japanese).

Nakayama, Seiji (2015a) Domestication of the soybean (*Glycine max*) and morphological differentiation of seeds in the Jōmon period. *Japanese Journal of Historical Botany* 23(2): 33–42 (in Japanese with English summary).

Nakayama, Seiji (2015b) Utilization of domesticated plants and secondary vegetation in the central highlands of the Japanese archipelago in the Jōmon period. *The Quaternary Research* 54(5): 285–298 (in Japanese with English summary).

Nakui, F. (2012) *Inherited Techniques from the Jōmon Period.* Yoshikawa Kōbunkan (in Japanese).

Nie, Ze-Long, Hang Sun, Ying Meng and Jun Wen (2009) Phylogenetic analysis of *Toxicodendron* (Anacardiaceae) and its biogeographic implications on the evolution of north temperate and tropical intercontinental disjunctions. *Journal of Systematics and Evolution* 47: 416–430.

Nitta, Miyuki (2001) Shiso to egoma no bunka to tayōsei [The differentiation and diversity of *shiso* and *egoma*]. In H. Yamaguchi and Y. Shimamoto (eds.), *Saibai shokubutsu no shizenshi* [The Natural History of Domesticated Plants], pp. 165–175. Sapporo: Hokkaido University Press.

Noshiro, Shūichi and Suzuki, M. (2004) *Rhus verniciflua* Stokes grew in Japan since the Early Jōmon Period. *Japanese Journal of Historical Botany* 12(1): 3–11 (in Japanese with English summary).

Noshiro, Shūichi and Yuka Sasaki (2007) Use of timber resources at the Shimo-yakebe site, Tokyo, and its characteristics in the Kanto plain of the late to latest Jōmon period. *Japanese Journal of Historical Botany* 15: 19–34 (in Japanese with English summary).

Noshiro, Shūichi, Mitsuo Suzuki and Yuka Sasaki (2007) Importance of *Rhus verniciflua* Stokes (lacquer tree) in prehistoric periods in Japan, deduced from identification of its fossil woods. *Vegetation History and Archaeobotany* 16: 405–411.

Noshiro, Shūichi (2014) How did the Jōmon people use the forest? In *New Perspectives on the Plant Use of Jōmon People*, pp.50–67. Shinsensha (in Japanese).

Noshiro, Shūichi and Yuka Sasaki (2014) Management of forest resources during the Jōmon period in Japan deduced from excavated plant remains. *Bulletin of the National Museum of Japanese History* 187: 15–48 (in Japanese with English summary).

Obata, H., Y. Sasaki and Y. Senba (2007) Impressions on pottery revealed cultivation Glycine max subsp. max (soybean) in the late to latest Jōmon periods in the Kyushu island. *Japanese Journal of Historical Botany* 15: 97–114 (in Japanese with English summary).

Obata, H. and A. Manabe (2011) Replica analysis of impressions on the pottery of the Ōjiyama site. *Ōjiyama iseki*, p. 92. Tojō City Board of Education (in Japanese).

Sakamoto, Sadao (1988) *Zakkoku no kita michi: Yuurashia minzoku shokubutsushi kara* [The route of other grains: from the ethnobotany of Eurasia]. NHK.

Sakazume, Nakao (1956) Nihon genshi nōkō shiron [An essay on primitive agriculture in Japan]. *Kōkogaku Zasshi* 42(2): 1–12.

Sasaki, Y., Y. Kudo and A. Momohara (2007) Utilizaion of plant resources reconstructed from plant macrofossils during the latter half of Jōmon period at the Shimo-yakebe site,Tokyo. *Japanese Journal of Historical Botany* 15: 35–50 (in Japanese with English summary).

Shimamoto, Y. (2003) Daizu [Soybeans]. In *Shokuryō mamerui no kagaku: genjō to tenbō* [The Science of Edible Beans: Current Situation and Perspective], pp.2–14. Fuyodo.

Suzuki, M., S. Noshiro,T. Tanaka,K. Kobayashi, Y. Wang, J. Lui and Y. Zhengu (2014) Origin of Urushi (Toxicodendron vernicifluum) in the Neolithic Jōmon period of Japan. *Bulletin of the National Museum of Japanese History* 187: 49–72 (in Japanese with English summary).

Tomooka, Norihiko, Duncan A.Vaughan, Helen Moss and Nigel Maxted (2002) *The Asian Vigna: Genus Vigna Subgenus Ceratotropis Genetic Resouces.* Dordrecht: Kluwer Academic.

Tsuji, Seiichirō (2009) Jōmon jidai no shokuseishi [Vegetation history of the Jōmon period], in *Jōmon jidai no kōkogaku*, Vol. 3. Dōseisha.

Ushino, T. and H. Tagawa (1991) Repurika-hō ni yoru doki akkon no kansatsu [Observation of impressions on potsherds by replica analysis]. *Kōkogaku to Shizenkagaku* 24: 13–35.

Yabuno, Tomosaburō (1962) Cytotaxonomic studies on the two cultivated species and the wild relatives in the genus Echinocloa. *Cytologia* 27: 296–305.

Yabuno, Tomosaburō and Hirofumi Yamaguchi (eds) (2001) *Hie to iu shokubutsu* [The Plant Called Barnyard Millet]. Zenkoku Nōson Kyōiku Kyōkai.

Yamaguchi, H. (2003) Adzuki beans raised in the laurel forest culture and its ancestor species. In *Zakkoku no shizenshi* [Natural History of Miscellaneous Cereals], pp.128–142. Sapporo: Hokkaido University Press (in Japanese).

Yoshikawa, Junko (2011) Change in the size of *Castanea crenata* fruits during the Jōmon period. *Japanese Journal of Historical Botany* 18(2): 57–64 (in Japanese with English summary).

Yoshikawa, Junko and Y. Itō (2004) Plant macrofossils detected from Iwato-kotani (4) site, Aomori prefecture. *Iwato-kotani (4) site*. Aomori Prefectural Board of Education (in Japanese).

Yoshikawa, M., S. Suzuki, S. Tsuji, K. Goto and T. Murata (2006) Vegetation history and human activities at the Sannai-maruyama site, Aomori prefecture. *Japanese Journal of Historical Botany*, special issue No. 2: 49–82 (in Japanese with English summary).

Yoshikawa, M. (2011) Dispersal of *Castanea crenata* pollen and distribution of *C. crenata* forest around the Sannai-maruyama site during the Jōmon period. *Japanese Journal of Historical Botany* 18(2): 65–76 (in Japanese with English summary).

Yoshizaki, Masakazu (1992) Fugoppe kaizuka kara shutsudo shita shokubutsu itai to hie zoku shushi ni tsuite no shomondai [Plant remains excavated from the Fugoppe shell midden and problems connected to the *Echinochloa* seeds]. In Hokkaido Buried Cultural Properties Centre (eds.), *Fugoppe kaizuka* [Fugoppe shell midden], pp. 535–547. Sapporo: Hokkaidō Maizō Bunkazai Sentā.

Yoshizaki, Masakazu (1995) Nihon ni okeru saibai shokubutsu no shutsugen [The appearance of cultivated plants in Japan]. *Kikan Kōkogaku* 50: 18–24.

Yoshizaki, Masakazu (2003) Senshijidai no zakkoku [Miscellaneous cereals in the prehistoric period]. In *Zakkoku no shizenshi* [The Natural History of Miscellaneous Cereals], pp.52–72. Sapporo: Hokkaido University Press.

Yoshizaki, Masakazu and T. Tsubakisaka (2001) Senshi jidai no mamerui nit suite [Legumes in the prehistoric era]. *Mamerui Jihō* 24: 1–9.

3

When Did We Choose the Way to the Anthropocene? An Examination of the Jōmon Period

Takamune Kawashima

Abstract: The Anthropocene has become an important issue, even in Japanese archaeology. There are multiple definitions regarding the beginning of the Anthropocene, which is often set in the last few centuries with the industrial revolution. However, considering the history of industry, the first important trigger of the evolution of human society could have traced back to the emergence of sedentism. In previous studies, some archaeologists have asserted that agriculture existed in the Jōmon period and seed impression analysis now shows that some plants were cultivated at that time. Although agriculture may have had only limited significance in the overall subsistence of the Jōmon, radical social changes are noteworthy with sedentary lifeways. This chapter examines the possibility of the existence of agriculture by analysing chipped stone adzes and their impact on sedentary lifeways in the Jōmon against the background of debates on the early Anthropocene.

Keywords: Anthropocene, Jōmon Period, agriculture, sedentary lifeway, stone adze, prehistory, Japan

3.1. The Anthropocene from an Archaeological Perspective

It seems that the Anthropocene has become an important issue even in Japanese archaeology. Takigawa (2016) summarised previous studies regarding the definition and beginning of the Anthropocene in *Kōkogaku Kenkyū*, one of the most influential archaeological journals in Japan. As the word 'Anthropocene' spreads in various media in Japan, archaeologists will be aware of its significance. However, probably because most archaeologists are engaged in work on sites and remains of earlier times, their direct interest does not cover the Anthropocene. In this paper, I will argue that their work is nevertheless deeply connected with the definition of the Anthropocene.

In recent studies the beginning of Anthropocene is set within the last few centuries, for example at the Orbis spike dip of CO_2 in 1610 or the bomb spike peak in 1964 (Lewis and Maslin 2016). The focus has been on globally synchronous markers, which can be observed as global-scale atmospheric gases. Crutzen and Stoermer (2000), who first advocated the word Anthropocene, suppose that the increase of CO_2 gas emissions caused significant impacts on human society, meaning that atmospheric gas can be a *Merkmal*. On the other hand, some researchers are focusing on the cause of changing atmospheric gases. Ruddiman (2003) described the relationship between the emission of CO_2 gas and past deforestation. From archaeological and anthropological perspectives, this relationship can be dated back thousands of years ago, for example to approximately 10,000 BP (Erlandson and Braje 2013).

The innovation of atomic weapons and experiments can also be an important marker. However, in all the cases mentioned above, the beginning of the Anthropocene is still recent, and the meaning of the period often remains ambiguous. Thus, we need more discussion over defining the Anthropocene as an epoch. In this paper, I approach this task by examining social changes in Jōmon society in the Japanese Islands, focusing on human influence on nature.

3.2. Jōmon Agriculture

Since the end of the nineteenth century, Jōmon society has been regarded as one of hunter-gatherer-fishers, with archaeologists stressing the contrast with the agricultural society of the following Yayoi period (Table 3.1). However, recent research has shown the existence of cultivated plants in the Jōmon (Kudō 2014; Nakayama this volume). The emergence of agriculture or horticulture implies, to some extent, the modification of natural environments around settlements. While in the Incipient Jōmon period some kinds of cultivated plants such as *Perilla frutescens* and gourd were discovered, the calorie-based ratio of cultivated plants could have increased in later periods, most probably in the Middle Jōmon.

Table 3.1. Jōmon chronology (after Kobayashi 2008).

	cal BP
Incipient	16,000-11,500
Initial	11,500-7000
Early	7000-5470
Middle	5470-4420
Late	4420-3220
Final	3220-2850/2350

The 'Jōmon agriculture theory' was proposed by scholars such as Fujimori (1970) and Kagawa (1972). Recent studies of plant seed impressions on pottery have examined the emergence and spread of cultivation in the Jōmon period (Kudō 2014; Hamada and Nakazawa 2014; Obata 2015). While some cereal seed impressions were previously thought to belong to the Middle and Late Jōmon, a recent re-evaluation found no evidence for the existence of cereals before the Final Jōmon (Obata 2018). In previous studies on agriculture in central Japan, plants such as beans were suggested to have been domesticated from the Middle Jōmon (see Nakayama 2019, this volume). Soybean *(Glycine max subsp. max)* and adzuki bean *(Vigna angularis var. angularis)* including their wild species, *Glycine max subsp. soja* and *Vigna angularis var. nipponensis*, were used from the Initial and possibly cultivated from the Middle Jōmon (Nakayama 2015, this volume). While the scale of this agriculture is still unknown, archaeobotanical research suggests that Middle Jōmon societies in some areas cleared land for cultivation. For clearing forests, stone axes and adzes could have been used, accelerating the expansion of cleared land, especially in the latter half of the Jōmon period. To evaluate the extent of land use, this chapter focuses on the chipped stone adzes of the Late and Final Jōmon phases in western Japan. These adzes have been considered as cultivation tools in previous research.

3.3. Axes and Adzes for Jōmon Agriculture

Yamazaki (2005) suggested that slash-and-burn agriculture appeared in the Late Jōmon period in western Japan because of the increase of polished and chipped stone adzes, in addition to some finds of possible huts located apart from settlements of this period. Yamazaki asserts that plants could have been cultivated by slash-and-burn, and possibly in permanent cultivated fields following an increase in chipped stone adzes.

For slash-and-burn agriculture, axes were used in many cases in ethnographical records. Polished axes seen in ethnographical records were used for various purposes, not only for cutting trees (Sahara 1994). Axes were involved in cultivation in the sense of slash-and-burn agriculture, while they were not used as cultivating tools such as hoes. On the other hand, chipped stone adzes have been generally regarded as tools for cultivating (Kagawa 1972; Sahara 1994). Both polished stone axes and chipped stone adzes increased in Japan after the Late Jōmon. Although the Jōmon culture of western Japan is thought to have had fewer artefacts overall as compared to the eastern part of the archipelago, some objects such as chipped stone adzes dramatically increased in the Late Jōmon.

The chipped stone adze cultivation hypothesis relies on the abundance of chipped stone adzes, especially in Middle Jōmon central Honshu and Late and Final Jōmon Kyushu. While the hypothesis was widely supported by numerous researchers and mentioned in many articles and books, it lacks direct archaeological evidence proving that adzes were used for cultivation. In ethnological records, it is rare to find examples of chipped stone adzes used for cultivation (Itakura 2009). Moreover, Itakura states that according to observations of the type of blades of chipped stone adzes from southern Kyushu, there could have been multiple functions for chipped stone adzes even in the same region.

In Yamaguchi Prefecture, located at the western edge of Honshu Island, there seems to be less variability in the form of Jōmon chipped stone adzes. At the Kokushū site (Fig. 3.1), 184 chipped stone adzes belonging to the early Final Jōmon were found (Iwasaki et al. 1992). The site is located on slightly high ground on the edge of a polje, through which several small streams run. As some traces of former streams after the Jōmon were found at the site, the land was not stable and was affected by flooding. Jōmon settlements were not found at Kokushū. Figure 3.2 shows some locations where adzes were concentrated; these could have been working areas with adzes (Kawashima 2017). Of 184 adzes, 47 examples retained their whole shape. Including ten almost whole adzes, the percentage of whole and almost whole adzes is 31 per cent. Broken adzes are categorised into three groups: 58 shoulders, 14 middle parts, and 55 blade parts. Although adzes at Kokushū were not associated with structures like pit dwellings, they include a certain amount of whole adzes. Ethnological records would be useful to consider the context of the uncovered adzes.

3.4. Axes in Ethnological Records

As mentioned above, there is little information about chipped stone adzes in ethnological records. However, such records have been collected in various areas; although the differences between examples are not small, we can see the tendency of discarded locations of the axes, information which can be useful to understand the distribution of adzes left in archaeological sites.

The distribution of used and abandoned stone adzes was examined based on ethnological data from New Guinea. As the Duna tribal group in highland New Guinea obtained stone axes from other groups by exchange (Sahara 1994; White and Modjeska 1978), their adzes were used longer than other groups. The fact that they transmitted their axes from generation to generation also extended the period of use. According to the data by White and Modjeska, abandoned axes found from the ground were small and of low quality. Sixty-two percent of axes found from the ground (and probably from abandoned huts) were smaller than 10 cm. This means these low quality or old axes were not recovered once they were lost during use. Although the data reported by White and Modjeska were polished stone axes used for cultivation, the way of abandonment of the axes suggests similarities with Kokushū.

Figure 3.1. Adze distribution (black circles) at Kokushū (34.272566, 131.262217) (modified from GSI Tiles). Source: the author.

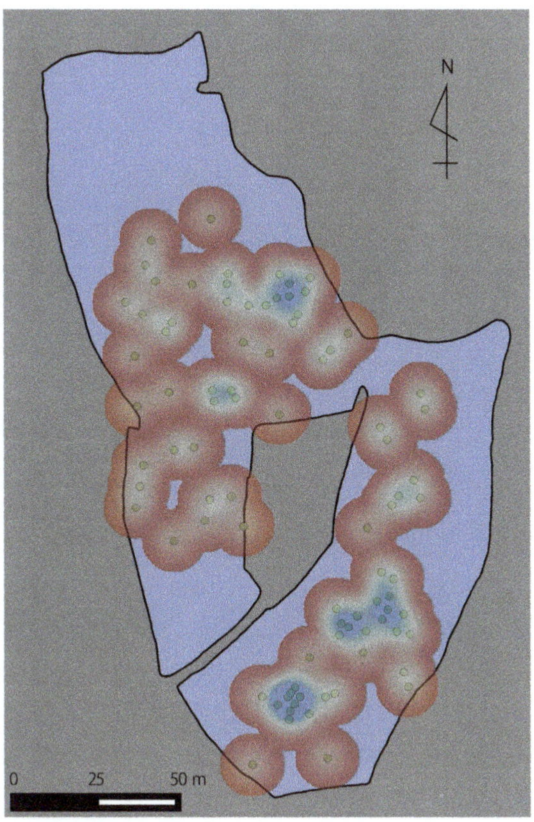

Figure 3.2. Heatmap of adze distribution (green circles) at Kokushū. Source: modified from Iwasaki et al. (1992: Fig. 27).

In the case of Kokushū, adzes show uniformity in size and material. As the material was obtained locally, the period of adze use may have been shorter than that of the Duna axes. Taking account of abandoned Duna axes in cultivating land, most adzes found at the Kokushū site can be seen as used-up or worthless ones. The location of dense adze distribution (Fig. 3.2) implies that the land was used as a working space, probably for cultivation.

3.5. Jōmon Adzes in Yamaguchi Prefecture

While Jōmon site density is low in Yamaguchi prefecture, there are some similar Late and Final Jōmon sites regarding the relative abundance of chipped stone adzes. Twenty-six adzes, as well as 44 pits and numerous postholes, were found at Nakakomida (34.170296, 131.468332) in the Yamaguchi Basin, a site which belongs to the Final Jōmon (Horinaga et al. 2004). Although it is reported that some adzes were found from pits, such adzes are seen at the surface of the pits, as seen in the profiles of Pit SK43 and SK44. It is difficult to determine whether they were placed in the pits and the top of pits was destroyed in a later period, or if they belonged to the upper layers.

At Shimomigita (34.077215, 131.562282) on the alluvial plain of the Saba River, 111 chipped stone adzes were found from a former stream with middle Final Jōmon pottery (Sawashita 2000). Although details of the context of the adzes are unknown and the site locations are different from Kokushū, the number of adzes suggests that adzes were densely distributed at particular sites on alluvial plains of any size.

The Iwata site (33.893126, 132.083171) is located on a small coastal alluvial plain (Kawagoe 1981; Nakagoshi 2000; Shiomi 1960). From the third and fourth excavations, more than 50 chipped stone adzes were found with pottery belonging to the end of the Late Jōmon to the early Final Jōmon (Shiomi 1960: 138). Twenty-eight adzes were also found from locations A and E in the sixth and seventh excavations, suggesting that they were used after the end of the Late Jōmon period. The distance between locations C and D is approximately 100 m, and at locations A and E 30 wet-type storage pits for acorns were distributed between C and D (Kawagoe 1981). As asserted in the Okayama Plain (Yamamoto 2005), storage of acorns was thought to have been a part of subsistence which supplemented unstable agriculture after the Late Jōmon. At Iwata, as the size of the cultivated land was limited, acorn storage could have occupied a large part of the subsistence economy.

Adzes might be related to digging storage pits, but the relationship is not straightforward considering the situation at the Tanoura site (33.792099, 132.034403) on the western coast of Nagashima Island. At Tanoura, while 14 acorn storage pits of the Late Jōmon period were found, only 3 adzes and 8 possible unfinished adzes were found. As the storage pits at Iwata belonged to a later period than those at Tanoura, few adzes could have been needed at Tanoura.

3.6. Changing Land Use in the Jōmon

Although the Jōmon agriculture hypothesis has focused on chipped stone adzes as a cultivating tool, prehistoric chipped stone adzes could have had various uses, as seen in the ethnographic record. In North America, the adzes of the Dalton culture (ca. 10,500 - 9300 cal BC) were regarded as heavy woodworking tools (Morrow 2011). Yerkes and Gaertner (1997) support this interpretation by microwear analysis, and conclude that they were used for charred wood. Microware analysis on a Palaeolithic chipped stone adze in China found that the adze was used for woodworking (Zhang et al. 2010). On the other hand, in some agricultural societies, chipped stone adzes are thought to have been used as hoes. In the Late Archaic period in Mesoamerica (ca. 3000 BC), adzes are regarded as a cultivating tool as well as a tool for cutting trees for forest clearance (Sharer and Traxler 2006: 159). In the Mississippi culture of North America (AD 800–1600), chert hoes were used for cultivation (Pauketat 2004; Hammerstedt and Hughes 2015). Hammerstedt and Hughes conducted an experimental study. They used a replica Mill Creek hoe, and proved that the hoe could be used for cultivating upland prairie. They mention that many excavated hoes have a polished surface, which was also observed on the replica hoes. In the Middle and Late Neolithic period of Korea, microwear analysis demonstrates that chipped stone adzes were used for cultivation (Yamada and Kim 2011).

Thus, while the use of chipped stone adzes is undetermined, from the examples of some sites it can be said that the adzes were distributed mostly at sites located on alluvial plains near the bottom of hill slopes. Considering the site locations, the most probable use could be for cultivating plants. While further analysis such as microwear will be needed to determine the use of the chipped stone adzes as done by Yamada and Kim (2011), the changing stone tool assemblage with an increase in adzes suggests that Late Jōmon subsistence shifted to become more dependent on cultivation. Taking into consideration the increase of axes as well as adzes, in the Late Jōmon of western Japan land clearance for cultivation could have been accelerated (Kawashima 2015a).

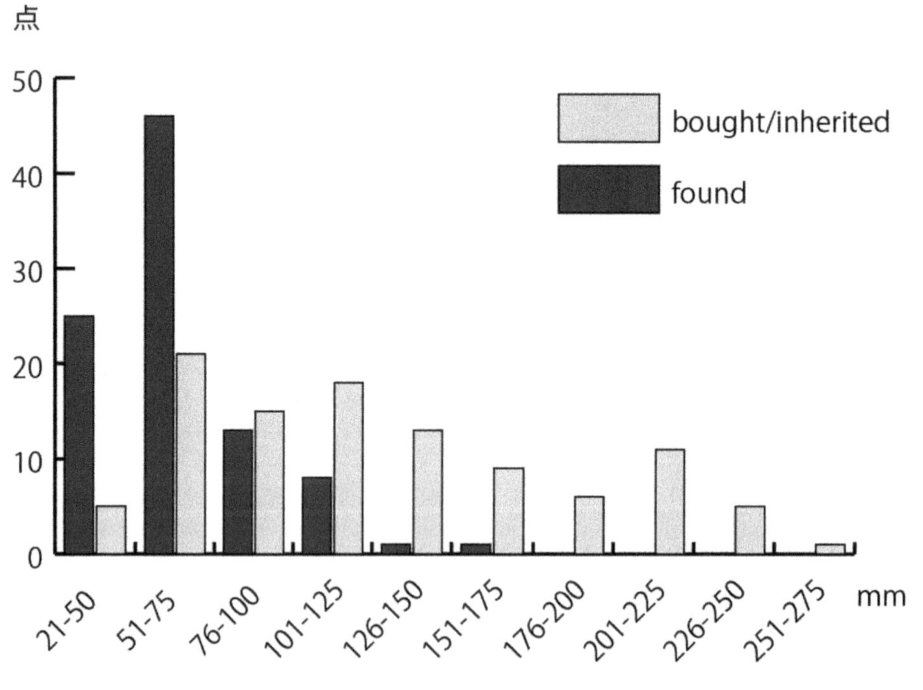

Figure 3.3. Duna axes found in different contexts. Source: based on data in White and Modjeska (1978: Table 5).

I have argued that the Late and Final Jōmon society of eastern Japan became more sedentary and developed social complexity (Kawashima 2015b). The sedentary lifeway could have caused an expansion of the exploitation of natural resources and land use. In some areas, such a lifeway was likely connected to agriculture. As Nishida (1986, 2007) noted, sedentary lifeways played a key role in human evolution preceding the emergence of agriculture, a transformation which he called the 'sedentary revolution'. While some sedentism was found in the Palaeolithic period, the Jōmon period can be characterised by pottery, arrowheads and pit houses. With the increase of settlements in the Initial phase, Jōmon societies in the Holocene became more sedentary than those in the Pleistocene. Some scholars regard the turning point within the Jōmon period as the beginning of the Initial Jōmon (Mizunoe 2012; Taniguchi 2011), which is approximately the beginning of the Holocene.

As mentioned above, the increase of chipped stone adzes in the Late Jōmon period could have been connected to land clearance for cultivation, probably by the slash and burn method. Considering the results of seed impression analyses, in western Japan this trend was accelerated in the Late Jōmon. In Late Jōmon society in eastern Japan, several specialised activities occurred, such as salt production and the fabrication of ritual objects and ornaments (Kawashima 2010, 2012). Such specialisation could be seen in some areas of western Japan, supported by the increase of axes, adzes and pit houses. In western Japan, harvesting acorns became an important subsistence activity in the Late Jōmon, as wet-type storage pits rapidly increased (Kawashima 2016). Although some regional differences could exist and a linear evolutionary model is unlikely, social change into a sedentary society gradually proceeded in the Japanese archipelago.

3.7. Conclusion

Where can we put the starting point for this social change to sedentism? As Nishida (1986, 2007) asserts, the social change when a sedentary lifeway was introduced could be the most important event regarding the impact on nature and society during the Jōmon and later periods including the present time. This chapter considered the Jōmon, but the same idea can be applied to human history more broadly. The data of changing atmospheric gases or the invention of atomic weapons can be explicit markers of the beginning of the Anthropocene. However, it is also important to consider the long process in which various kind of human societies became somehow unified. Although the Anthropocene has become an essential topic for present-day society, several hundred years are not long enough, if we consider human history. The discussions regarding the Anthropocene should lead to a rethinking of past societies—to which we can never go back.

References

Crutzen, P.J. and E.F. Stoermer (2000) The 'Anthropocene'. *IGBP Newsletter* 41: 17–18.

Erlandson, J. M. and T.J. Braje (2013) Archeology and the Anthropocene. *Anthropocene* 4: 1–7.

Fujimori, E. (1970) *Jōmon nōkō* [Jōmon Farming]. Tokyo: Gakuseisha.

Hamada, T. and M. Nakazawa (2014) Nishi Nihon tottaimon doki bunpuken ni okeru saibai shokubutsu no shutsugen. In S. Nakayama (ed), *Nikkan ni okeru kokumotsu nōkō no kigen* [The Origins of Cereal Agriculture in Japan and Korea], pp. 318-323. Kōfu: Yamanashi Kenritsu Hakubutsukan Chōsa Kenkyū Kiyō 9.

Hammerstedt, S. W. and E.R. Hughes (2015) Mill Creek chert hoes and prairie soils: implications for Cahokian production and expansion. *Midcontinental Journal of Archaeology* 40(2): 149–165.

Horinaga, T., T. Kawahara and A. Fujiwara (2004) *Nakakomida iseki II* [The Nakakomida Site II]. Yamaguchi: Yamaguchi City Board of Education.

Itakura, Y. (2009) Kyūshū nanbu Jōmon jidai kō-banki daseisekifu rui no kishu bunrui: keitai to shiyōkon no kanten kara [Classification of Late-Final Jōmon chipped stone axes in southern Kyushu]. In Shintō Kōichi daihyō kanreki kinen ronbunshū kankōkai (ed), *Minami no Jōmon chiikibunka ronko jō* [The Regional Cultures of the Southern Jōmon], pp. 195–204. Aira: Shintō Kōichi daihyō kanreki kinen ronbunshū kankōkai.

Iwasaki, H., H. Shiraoka and M. Muraoka (1992) *Kokushū* [Kokushū]. Yamaguchi: Yamaguchiken kyōiku zaidan and Yamaguchi Prefecture Board of Education.

Kagawa, M. (1972) *Nōkō no kigen: Nihon bunka no genryū o saguru* [The Origins of Agriculture: In Search of the Roots of Japanese Culture]. Tokyo: Kōdansha.

Kawagoe, T. (1981) Yamaguchiken Hiraochō Iwata iseki [The Hiraochō Iwata site, Yamaguchi]. *Nihon Kōkogaku Nenpō* 21/22/23: 47–52.

Kawashima, Takamune (2010) Jōmon jidai dokiseien ni okeru rōdō keitai. *Tsukuba Archaeological Studies* 21:1–34.

Kawashima, Takamune (2012) Emerging craft production and local identity: a case of the Late Jōmon Period. *Documenta Praehistorica* 39: 263–268.

Kawashima, Takamune (2015a) An adze collected at Shitomi, Tsushima Island. *Yamaguchi Daigaku Bungaku Kaishi* 65: 63–72 (in Japanese).

Kawashima, Takamune (2015b) *Seisan to kyōen kara mita jōmon jidai no shakaiteki fukuzatsuka*. Tokyo: Rokuichi shobō.

Kawashima, Takamune (2016) Food processing and consumption in the Jomon. *Quaternary International* 404: 16–24.

Kawashima, Takamune (2017) Yamaguchiken ni okeru iseki bumpu kara mita Jōmon jidai kenkyū no genjō to kadai. In Chūshikoku Jōmon kenkyūkai (ed), *Dai 28 kai Chūshikoku Jōmon kenkyūkai happyō yōshishū*, pp. 1–13. Takamatsu: Chūshikoku Jōmon kenkyūkai.

Kobayashi, K. 2008 Jōmon doki no nendai: higashi Nihon [The date of Jōmon pottery: east Japan]. In Tatsuo Kobayashi (ed), *Sōran Jōmon doki* [Overview of Jōmon Pottery], pp. 896-903. Tokyo: UM Promotion.

Kudō, Y. (2014) *Kokomade wakatta Jōmonjin no shokubutsuriyō* [What We Now Know About Jōmon Plant Use]. Tokyo: Shinsensha.

Lewis, S. L. and M.A. Maslin (2016) Defining the Anthropocene. Nature 519: 171–180.

Mizunoe, K. 2012 *Kyūshū Jōmon bunka no kenkyū: Kyūshū kara mita Jōmon bunka no wakugumi* [A Study of Jōmon Culture in Kyushu: The Framework of Jōmon Culture Seen from Kyushu]. Tokyo: Yūzankaku.

Morrow, J. E. (2011) The Sloan Dalton Site (3GE94) assemblage revisited: Chipped-stone raw material procurement and use in the Cache Basin. *The Missouri Archaeologist* 71: 5–40.

Nakagoshi, T. (2000) Iwata iseki [The Iwata site]. In Yamaguchi Prefecture (ed.), *Yamaguchikenshi shiryōhen kōko I* [History of Yamaguchi Prefecture, Materials, Archaeology I], pp. 102-116. Yamaguchi: Yamaguchi Prefecture.

Nakayama, Seiji (2015) Utilization of domesticated plants and secondary vegetation in the central highlands of the Japanese archipelago in the Jomon period. *The Quaternary Research* 54(5): 285–298 (in Japanese).

Nishida, Masaki (1986) *Teijū kakumei: yūdō to teijū no jinruishi*. [The Sedentary Revolution: A Human History of Mobility and Sedentism]. Tokyo: Shin'yōsha.

Nishida, Masaki (2007) *Jinruishi no naka no teijū kakumei* [The Sedentary Revolution in Human History]. Tokyo: Kōdansha.

Obata, Hiroki (2015) *Tane o maku Jōmonjin: saishin kagaku ga kutsugaesu nōkō no kigen* [The Jōmon People Sowed Seeds]. Tokyo: Yoshikawa kōbunkan.

Obata, Hiroki (2018) Cereal composition in the Initial Yayoi period in the Kumamoto plain seen from impressions on pottery. *Bungakubu Ronsō* 109: 1–13 (in Japanese).

Pauketat, P. R. (2004) *Ancient Cahokia and the Mississippians*. Cambridge University Press.

Ruddiman, William F. (2003) The anthropogenic greenhouse era began thousands of years ago. *Climatic Change* 61: 261–293.

Sahara, Makoto (1994) *Ono no bunkashi* [A Cultural History of the Axe]. Tokyo University Press.

Sawashita, T. (2000) Shimomigita iseki [The Shimomigita site]. In Yamaguchi Prefecture (ed.), *Yamaguchikenshi shiryōhen kōko I* [History of Yamaguchi Prefecture, Materials, Archaeology I], pp. 131–132. Yamaguchi: Yamaguchi Prefecture.

Sharer, R. J. and L.P. Traxler (2006) *The Ancient Maya* (6th ed). Stanford University Press.

Shiomi, H. (1960) Yamaguchi-ken Iwata iseki shutsudo Jōmon jidai ibutsu no kenkyū [Research on Jōmon artefacts excavated from the Iwata site, Yamaguchi prefecture]. *Hiroshima University Studies, Literature Department* 18: 90–148.

Takigawa, Wataru (2016) The future of the Anthropocene. *Kōkogaku Kenkyū* 63(2): 23–27 (in Japanese).

Taniguchi, Y. (2011) *Jōmon bunka kigenron no saikōchiku*. [Re-constructing a Theory of Jōmon Origins]. Dōseisha.

White, J. Peter and N. Modjeska (1978) Acquirers, users, finders, losers: the use axe blades make of the Duna. *Mankind* 11(3): 276–287.

Yamada, M. and Kim, S. (2011) Shoki nōkō kaishiki no dasei sekifu ni kansuru nikkan kyōdō kenkyū [Joint Japan-Korea research on chipped stone axes and the beginning of early agriculture]. Electronic document, http://www.jfe-21st-cf.or.jp/jpn/hokoku_pdf_2010/asia05.pdf. Last accessed 26.7.22.

Yamamoto, Etsuyo (2005) Chūshikoku ni okeru Jōmon kō-banki no nōkō: Okayamaken ni okeru sekki no yōsō kara [Late-Final Jōmon agriculture in Chugoku and Shikoku]. In Kansai Jōmonbunka kenkyūkai, Chūshikoku Jōmon bunka kenkyūkai and Kyūshū Jōmon bunka kenkyūkai (eds), *Nishinihon Jōmon bunka no tokuchō* [The Characteristics of Jōmon Culture in Western Japan], pp. 47-57. Kobe: Nishinihon Jōmonbunka kenkyūkai.

Yamazaki, S. 2005 Nishinihon Jōmon nōkōron: shushi akkon to jōmon minkō no gaiyō [The theory of Jōmon agriculture in west Japan]. In Kansai Jōmon kenkyūkai, Chūshikoku jōmon kenkyūkai, Kyūshū jōmon kenkyūkai (eds.), *Nishinihon jōmon bunka no tokuchō* [The Characteristics of Jōmon Culture in Western Japan], pp. 59–68. Kobe: Nishinihon Jōmonbunka kenkyūkai.

Yerkes, R. and L.M. Gaertner (1997) Microwear analysis of Dalton artifacts. In D.F. Morse (ed), *Sloan: A Paleoindian Dalton Cemetery in Arkansas*, pp. 58–71. Washington DC: Smithsonian Institution Press.

Zhang, X., C. Shen, X. Gao, F. Chen and C. Wang (2010) Use-wear evidence confirms the earliest hafted chipped-stone adzes of Upper Palaeolithic in northern China. *Chinese Science Bulletin* 55(3): 268–275.

4

Prehistoric Interaction Between Transeurasian and Non-Transeurasian Speakers

Martine Robbeets

Abstract: The Transeurasian family consists of Turkic, Mongolic, Tungusic, Koreanic and Japonic daughter branches. Recent research suggests that the earliest dispersals of this family in the Neolithic can be explained by the Farming/Language Dispersal Hypothesis. However, this hypothesis by itself cannot explain later Transeurasian dispersals from the Bronze Age onwards. Here I compare three different cases of interaction between Transeurasian and non-Transeurasian speakers in the Late Neolithic and Bronze Age, whereby Transeurasian farmers meet Amuric hunter-gatherer-fishermen, interact with Sino-Tibetan farmers or exchange with Indo-European pastoralists from the steppe. The interactions are assessed in terms of genetic admixture, socio-economic interaction and linguistic outcome. The results indicate that contact with speech communities with economically less productive strategies, such as between farmers and hunter-gatherers, leads to language shift and the dispersal of the dominant language, while interaction between populations with comparably productive economies, such as between farmers and farmers or pastoralists stimulates linguistic borrowing.

Keywords: Transeurasian, Amuric, Sino-Tibetan, Indo-European, farming/language dispersal hypothesis, farmers, pastoralists, hunter-gatherers, archaeolinguistics, triangulation

4.1. Introduction

Northern Eurasia extends from the Pacific coast in the east to the Atlantic coast in the west and from the Arctic coast in the north to the Mediterranean, the Black Sea, the Caspian Sea, Turkmenistan, Uzbekistan, Tajikistan, across Northern China to the Japanese Islands in the south. It is spanned by two major language families with an extensive east-west gradient: Indo-European and Transeurasian. The Indo-European family comprises most of the languages of Europe together with those of the Iranian Plateau and northern India. The Transeurasian family stretches across Northern Eurasia from the Pacific in the east to the Baltic, the Black Sea and the Mediterranean in the west and includes five well-established linguistic families: Japonic, Koreanic, Tungusic, Mongolic and Turkic.[1]

In addition, Northern Eurasia is home to a variety of neighbouring families, some large ones such as Sino-Tibetan and Uralic, some smaller ones such as North-West Caucasian, Nakh-Daghestanian, Kartvelian and Chukotko-Kamchatkan and some minor isolates with only a single surviving member such as Yeniseian, Yukaghiric, Ainuic and Amuric. The map in Figure 4.1 shows the distribution of the Transeurasian languages along with that of neighbouring families in Northern Eurasia.

One of the explanations of the wide east-west distribution of the Transeurasian family *vis-à-vis* most other language families in Northern Eurasia may be the adoption of agriculture by its early speakers. Robbeets et al. (2021) suggest that the common ancestors of the speakers of Transeurasian were millet farmers who lived about 9000 BP near the West Liao River, in present-day northeastern China. These farmers moved northeastward to the Amur basin and the Russian Far East around 5000 BP, as well as southeastward via the Liaodong peninsula to the Korean Peninsula around 6000 BP and then via Kyushu into the Japanese Islands around 3000 BP, an expansion consistent with the so-called Farming/Language Dispersal Hypothesis (Bellwood and Renfrew 2002; Bellwood 2005). This hypothesis posits that many of the world's major language families have their homelands in nuclear centres where plant or animal domestication emerged and owe their primary dispersals to the adoption of agriculture by their early speakers.

[1] The term 'Transeurasian' is used to replace the classical label 'Altaic' in reference to these languages, while 'Altaic' is reserved for the unity of Turkic, Mongolic and Tungusic languages only. In the past, some linguists raised substantial criticism against the genealogical relatedness of the Transeurasian languages (Clauson 1956; Doerfer 1963–1975; Austerlitz 1982; Kiyose 2002; Shōgaito 2002; Vovin 2005; Georg 1999, 2007; Campbell and Poser 2008; Fuente 2017). Even if some of the objections are reasonable and a part of the evidence provided by past scholarship (Ramstedt 1952–1957; Murayama 1957, 1958; Poppe 1960, 1965, 1975; Miller 1971; Baskakov 1981; Tekin 1994; S. Starostin et al. 2003) is indeed questionable, there is nonetheless a core of reliable evidence for the classification of Transeurasian as a valid genealogical grouping. As a result, my previous argumentation that the Transeurasian languages are genealogically related (Robbeets 2005, 2015, 2020a/b/c/d) has recently gained greater acceptance in the field (Gözaydin 2006; Rozycki 2006; Büyükmavi 2007; Kara 2007; Décsy 2007; Johanson 2010; Koch 2014; Ross 2014; Dybo 2016; Eliasson 2016; G. Starostin 2016; Blažek 2019; Chaubey and van Driem 2019; Bellwood 2021; Bradley this volume).

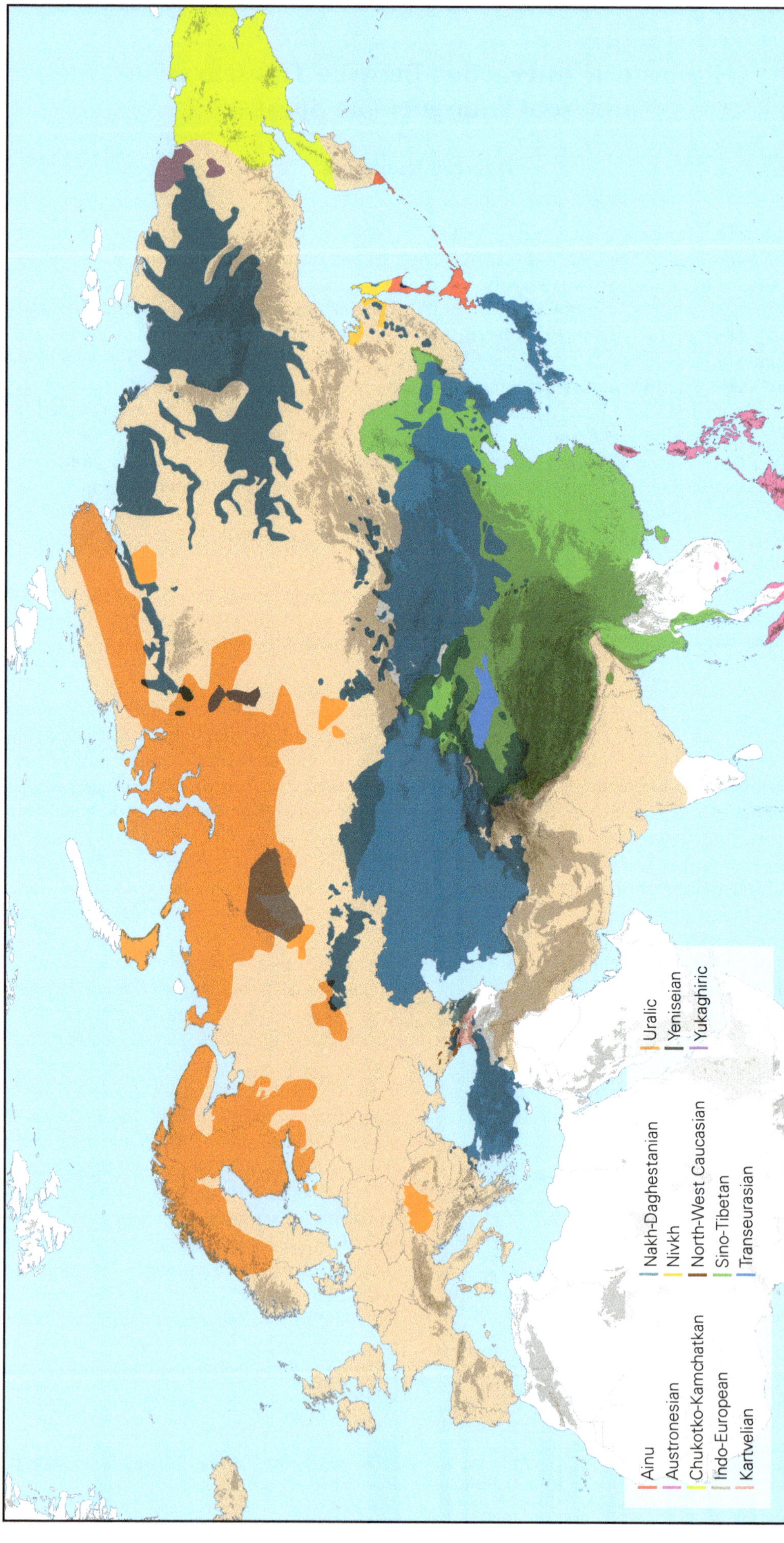

Figure 4.1. Language families in Northern Eurasia neighbouring the Transeurasian family. Map by M. O'Reilly, MPI-GEA, based on information by the author.

Even if the Farming/Language Dispersal Hypothesis can explain most of the earliest Transeurasian dispersals in the Neolithic, it cannot by itself explain later dispersals from the Bronze Age onwards. Admitted that it had a certain agricultural gloss (Savelyev 2017), the spread of the Turkic languages was driven by increased mobility on the steppes through the introduction of horse-driven pastoralism from around 3200 BP onwards (Taylor et al. 2020). The Mongolic languages owe their present-day distribution to military raids under Jinghis Khan in the thirteenth century AD. The dispersal of Northern Tungusic languages such as Even, Evenki and Negidal in the seventeenth century is associated with the spread of reindeer breeding, similar to the dispersal of Yakut, a North Siberian Turkic language in medieval times (Hudson 2020; Robbeets and Oskolskaya 2022).

However, the complexity of these post-Neolithic movements does not need to contradict the Farming/Language Dispersal Hypothesis. First, the hypothesis posits that 'many' of the world's major language families owe their dispersals to agriculture, not 'all of them'. Some widely-spread language families such as Pama-Nyungan in Australia or Eskimo-Aleut in North America never developed farming in the first place (Robbeets 2017b). Second, the hypothesis refers to 'primary' dispersals, for Transeurasian meaning deep nodes in the family such as 'Altaic', 'Japano-Koreanic' and their early break-ups in the Neolithic. Finally, the hypothesis presumes the 'adoption' of agriculture by populations with relatively less productive subsistence strategies. By contrast, when groups of comparable socio-economic status meet, the need to radically shift language and culture is less urgent. In such encounters, we expect to find a different linguistic outcome than the one predicted by the Farming/Language Dispersal Hypothesis.

In this chapter, I will compare the linguistic outcomes of three different cases of genetic admixture between Transeurasian and non-Transeurasian speakers in the Late Neolithic and Bronze Age and assess them in the light of socio-economic interaction. In the first case, discussed by Martijn Knapen in Chapter 6 in this volume, Transeurasian farmers meet Amuric hunter-gatherer-fishermen. In the second case, explained in detail by Bingcong Deng in Chapter 5, Transeurasian farmers interact with Sino-Tibetan farmers. Finally, in the third case, presented by Rasmus Bjørn in Chapter 8, they exchange with Indo-European pastoralists from the steppe. I will argue that where farming is introduced to populations with relatively less productive subsistence strategies, we expect language to shift and spread with farming. By contrast, when groups of comparable socio-economic status meet, whereby certain crops are added to a pre-existing agricultural package or farming is complemented by pastoralism, the encounters are expected to result in language maintenance with linguistic borrowing. For each case study, I will first provide a linguistic perspective and then align it with a summary of relevant findings from archaeology and genetics. 'Triangulating' the evidence in this way, I hope to reach a more balanced and richer understanding of the prehistoric interactions involved than each discipline can provide us with individually.

4.2. Farmers meet Hunter-Gatherer-Fishermen

4.2.1. Linguistics

The Proto-Tungusic language is thought to have separated from Proto-Mongolic around 4491 BP (2599-6373 95% HPD), while it broke up in its primary branches around 1950 BP (1499-2412 95% HPD) (Robbeets et al. 2021; Knapen this volume). During this time span of roughly two and a half millennia, the ancestral Tungusic language was situated in the mid-Amur region and in the border area between present-day Northern Heilongjiang and the Russian Primorye (Robbeets et al. 2021). Even if the internal diversification of the surviving varieties of Nivkh provides only an estimate of around 500 BP for the breakup of Proto-Amuric (Janhunen 2010: 290; Knapen 2021: 9) and we lack surviving sister languages that can us bring back deeper in time, the language family must have been contemporaneous with Proto-Tungusic, given the borrowings at the proto-level between both language families. Janhunen (1996: 231, 2016: 5) situates the homeland of Amuric in the Sungari-Amur basin, but a homeland closer to the sea is more likely given the reconstructed maritime vocabulary and its borrowing from Proto-Amuric into Proto-Tungusic (Knapen this volume).

Out of 50 potential loanwords between Amuric and Tungusic, Knapen (2021: 48-49) accepts 12 as loanwords from Proto-Tungusic (PTg), while 9 are accepted as loanwords from Proto-Amuric (PAm). The former are frequently related to animal husbandry, tools and kinship relations (e.g., PTg **mụrịn* 'horse' >> PAm **murŋ* 'horse', PTg **kọnịn* 'sheep' >> PAm *χoɲ* 'sheep'; PTg **pụụ-pOOn* 'saw-NML.INST' >> Am **pʰuvŋ* 'saw', PTg **ʤapa-pOOn* 'grab- NML.INST' >> PAm **capŋ* 'pincers'; PTg **ama-kAA* 'grandfather, father-in-law' >> PAm **aχma(-)l-* 'to be a father-in-law', pTG **akaa* 'elder brother' >> PAm **aka-* 'to be older'), while the latter are associated with maritime vocabulary and shamanism (e.g., PAm **la-mo* 'sea' >> PTg **laamo* 'sea', PAm **keŋ* 'whale of intermediate size' >> PTg **kịaŋa* 'whale', PAm **qʰalm* 'whale of small size' >> PTg **kalịmu* 'whale'; PAm **tal(-)q* 'stroganina (dish consisting of raw, long-sliced frozen fish)' >> PTg **tala(kka)* 'stroganina', PAm **cʰamŋ* 'shaman, eagle' >> **samaan* 'shaman', PAm **qoŋqŋ* 'bell, jaw harp' >> PTg **kọọŋ(g)aa(n)* 'bell').

Korovina (2011) finds that fish species that inhabit the Pacific Ocean are not well distributed across the Tungusic languages and are often borrowed from Nivkh into one or more individual daughter languages (e.g. Oroqen *lokko* 'flounder' from Nivkh lok 'flounder', Oroqen *la:kka* 'herring' and Orok la:qqa 'herring' from Nivkh *laku* 'herring', etc.). This suggests that some maritime vocabulary was not acquired until long after the separation of Proto-Tungusic around 1950 BP.

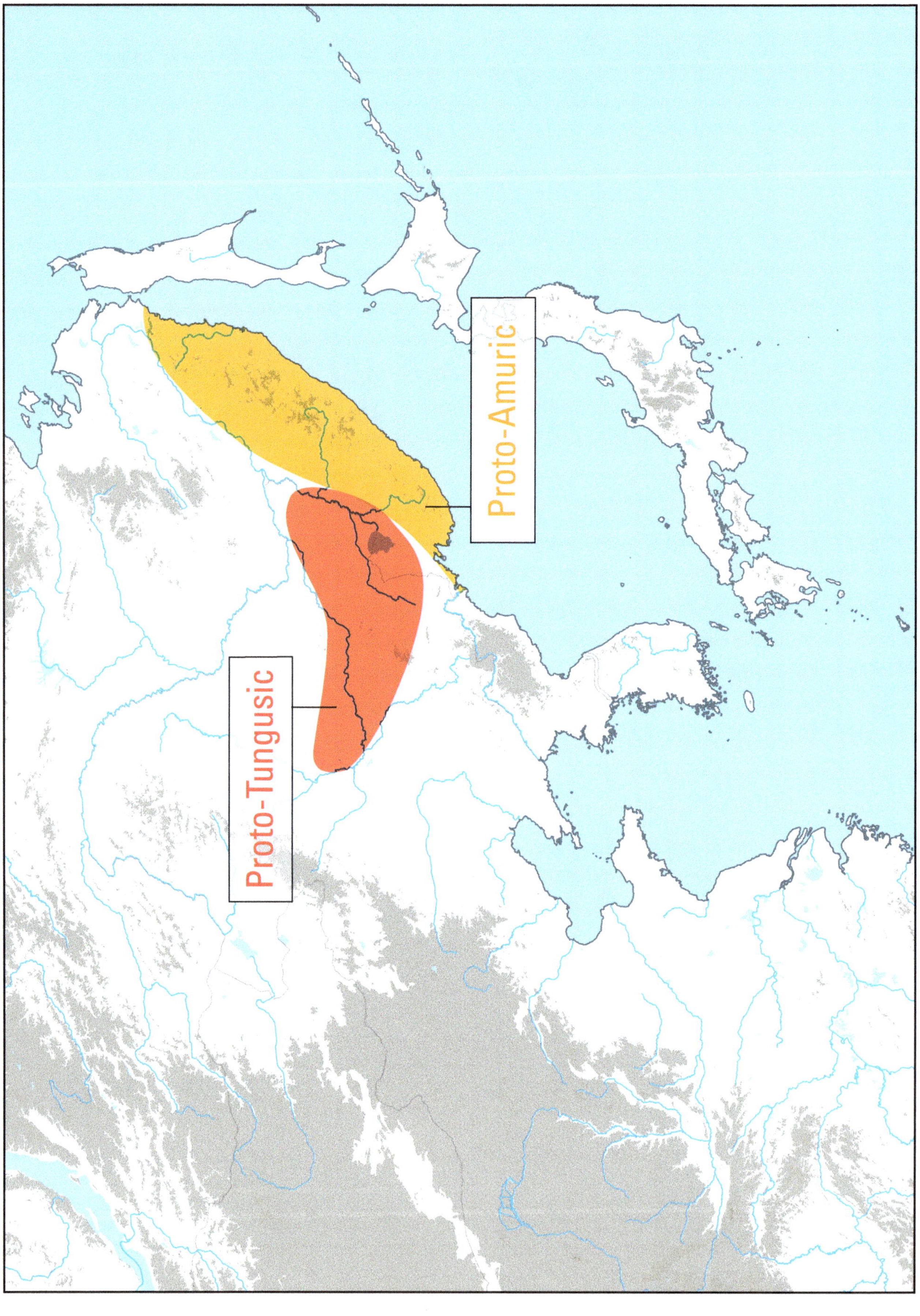

Figure 4.2. The estimated location of the Proto-Tungusic and Proto-Amuric homelands before 1950 BP. Map by M. O'Reilly, MPI-GEA, based on information by the author.

In contrast to the relatively low number of prehistorical loans that can be identified from Proto-Amuric into Proto-Tungusic, we find a good number of shared structural features that suggest substratum interference from Amuric into Tungusic. The evidence comes from structural features in Tungusic that are atypical for the Transeurasian languages and likely to have developed through imperfect learning from proto-Amuric (Robbeets 2020d; Li et al. 2020). Among others, these include the development of an initial velar nasal in Tungusic through secondary assimilation of pTg **g-* (Poppe 1964: 4); the development of a palatal nasal in Tungusic through the assimilation of a following *i*-diphtong (Benzing 1955: 974, 986); the grammaticalisation of an inclusive-exclusive distinction on pronouns (Benzing 1955: 1017–1018); the development of head marking in possessive noun phrases; the grammaticalization of indexing on converbs; the use of different numeral collective suffixes for counting distinct semantic classes following numerals from 'two' to 'ten', and the grammaticalization of causative suffixes into pluractionality markers on the verb. Table 4.1 illustrates these features, providing an example from Nivkh along with an example of the presumed carry-over feature in a contemporary Tungusic language. The proto-typical status for the feature in Transeurasian in general is given in the third column.

Thomason and Kaufman (1988: 50) summarised the linguistic results of language contact as following. In the case of borrowing, languages will first borrow vocabulary and in a later stage of more intensive contact, they will also add phonological and syntactic features. In the case of substrate interference, the outcome is the opposite: first phonological and syntactical structural features and vocabulary only in a later phase. Examining the borrowing from Proto-Amuric into Proto-Tungusic, we find only a restricted number of ancient loanwords, while Tungusic developed a variety of structural features under influence of Amuric. The direction of the diffusion is supported by the fact that these features are more characteristic of Amuric than of Transeurasian because they are direct or indirect implicational tendencies of the head-marking or polysynthetic type to which Amuric belongs. It is also clear that these features have changed early in Tungusic because they diverge from the expected structural pattern in Transeurasian and preserve traces of phonological change or grammaticalisation in Proto-Tungusic. Therefore, these linguistic observations suggest a situation of language shift whereby some ancestral speakers of Proto-Amuric abandoned their native language and adopted the proto-Tungusic target language.

4.2.2. Triangulation

Given the estimation of the timing, the location of the homelands and the reconstruction of the cultural and natural environment of the ancestral speakers, it is likely to associate the Proto-Tungusic language with incoming farmers who brought millet from Northeast China ca. 5000 BP, while Proto-Amuric can be connected with the language of pre-existing coastal populations in the Primorye, such as the Devil's Gate (7750-7450 BP) and Boisman (7500-6000 BP) people (Li et al. 2020; Robbeets and Oskolskaya 2022). A part of the speakers of Proto-Amuric shifted their language to that of the Proto-Tungusic farmers, leaving substratum interference in the form of structural features through imperfect learning. Another part of the speakers of Proto-Amuric did not shift to Proto-Tungusic and after 5000 BP continued to live from fishing, hunting and gathering in coastal areas neighbouring the farmers in the Mid-Amur and the area around Lake Khanka. The resulting interactions led to ongoing borrowings from the Late Neolithic into the Bronze and Iron Ages in two directions: maritime and shamanistic vocabulary from Amuric into Tungusic and animal husbandry, tools and kinship vocabulary from Tungusic into Amuric.

At the end of the third century AD, there was a sharp cooling of the climate, which led to a worsening of the conditions for agriculture (Yanuševič et al. 1990). This provided the impetus for a gradual migration of millet farmers to coastal regions across most of the Primorye. Based on archaeobotanical data (Sergusheva and Vostretsov 2009: 214–215) it appears that the coastal groups ceased to cultivate millets, barley and wheat and returned to a subsistence strategy of hunting, fishing and gathering plant resources. This change could explain why southern Tungusic populations on the Lower Amur such as the Nanai, Oroch and Udehe people were traditionally predominantly fishers and gatherers, rather than farmers. It also accounts for the fact that some maritime vocabulary was not acquired until well after 1950 BP when the individually separated speech communities came in contact with indigenous Amuric speakers on the Pacific coast.

Recent genome-wide DNA analyses (Pugach et al. 2016; Siska et al. 2017; Wang and Robbeets 2020; Wang et al. 2021; Robbeets et al. 2021) show that early Holocene hunter-gatherers from the southern Primorye, such as Devil's Gate (7750-7450 BP) and Boisman (7500-6000 BP), have high proportions of Amur ancestry. These ancient genomes cluster with contemporary southern Tungusic speakers, such as Ulcha, Negidal and Nanai in the lower Amur Region, as well as with Nivkh speakers in nearby Sakhalin island. Contemporary northern Tungusic populations, such as Evenki and Even, cluster with surrounding Siberian populations because they received gene flow from West Eurasian related populations. Since the spread of agriculture to the Russian Far East went over the region northeast of the West Liao River (Li et al. 2020) and ancient DNA from the Middle and Late Neolithic in that region suggests high proportions of Amur ancestry, we can infer that the people bringing agriculture to the Russian Far East displayed high proportions of Amur ancestry. As the Tungusic- and Amuric-speaking populations share the same Amur ancestry, there are no traces of genetic admixture indicating population migration at the time of the agricultural dispersals. Given the vast geographical reach of the Amur genetic profile, including the West Liao River region as well as the Russian Far East, a genetic admixture between incoming Proto-Tungusic farmers and local Proto-Amuric fishers would have led to an admixture of two similar Amur genomes, like mixing two white paints together. Bringing the linguistic, archaeological and genetic evidence together thus makes a case for imperceptible genetic admixture.

Table 4.1. Illustration of potential substratum interference from Amuric into Tungusic.

	Amuric feature	Tungusic feature	Prototypical Transeurasian feature
1	Initial velar nasal *ŋ- Nivkh *ŋaarla* 'very fat'	Secondary initial velar nasal pTg **g*- > non-Manchu *ŋ*-/ followed by *n*, *r* , *l*, *m*, *y*, pTg **ge:le*- in Evk. *ŋe:le*-, Ma. *gele*- 'to fear'	No initial velar
2	Palatal nasal **ɲ* Nivk *aɲ* 'year'	Secondary palatal nasal pTg **n* > *ɲ* / followed by -i-diphtong pTg **nia:ru* in Evk. *ɲa:rut* 'lake', Ma. *ɲari*, *Ud. ɲau*, Na. *nʲarõ* 'swamp'	No palatal nasal
3	Nivkh *ɲyŋ* 'we exclusive' vs. *mer* ~ *mir* 'we inclusive'	Grammaticalization of inclusive/exclusive distinction from the collective or augmented plural pTg **bö-(x)e* [1PL-PL] > Evk. *bu*, Ma. *be* 'we exclusive' pTg **bö-ti* [1PL-COLL] > Evk. *mut* ~ *mit* 'we inclusive' pTg **bö-sa* [1PL-COLL] > Ma. *muse* 'we inclusive'	No inclusive/ exclusive distinction
4	Possessive relations are head-marked Nivkh *vit-ɣaŋ* [2SG-book] 'your book'	Possessive relations are head-marked except for Manchu Even *svinija ulrə-n* [swine meat-3SG.POSS] 'swine's meat, pork'	Possessive relations are dependent-marked Khal. *min-ii eej* [I-GEN mother] 'my mother'
5	Imperative marked with suffix Nivkh *Ra-ja*! [drink-IMP.2SG] 'Drink!' (Gruzdeva 2012)	Imperative marked with suffix except for Manchu and Sibe Ud. *Diga-ja* [eat- IMP.2SG] 'Eat!'	Bare imperative Khal. *ir* 'come!'
6	Indexing on converbs Nivkh *k'ə-ɣe-t vi-d.*[axe-take-CONV.1SG.NFUT go-IND.SG] 'Taking the axe, I went.' (Gruzdeva 2012)	Indexing on converbs grammaticalized from nominal plural and possession Evk. *D'u-la-ver emem-mi-l d'ep-čo-tin.* [house-ALL-REFL.PL come-CONV-PL eat-PAST-3PL] 'Having come home, they ate' (Nedjalkov 1995: 445)	No indexing on converbs Tk. Ali *gel-ince şaşır-dı* [Ali come- CONV surprise-PAST.3SG] 'When Ali came, he was surprised' (Johanson 1995: 314)
7	Different numeral forms for counting 26 semantic classes Nivkh *men* '2 men' *mor* '2 animals', *meqr* '2 objects of irregular shape', *min* '2 boats', *mik* '2 small round objects', etc. (Gruzdeva 1998:23–25)	Different numeral collective suffixes for counting distinct semantic classes following numerals from 'two' to 'ten' Evk. *-kt*(e) and *-ni* for counting people (e.g. *d'u-kte* 'two (people together' Evk. *-gdA/ -ngnA* for counting objects, *Evk. -llA* for counting the number of days (e.g. *nada-lla* 'seven days, a week'), Evk. *-nu / -pu* for counting the number of tents (e.g. *ilan-nu* 'three tents') and Evk. *-musa* denoting the number of places or directions (e.g., *dygin-muse* 'four places/directions' (Nedjalkov 1997: 283)	No different numerals for counting distinct semantic classes
8	Pluractional marking on the verb by various suffixes and reduplication Nivkh *vi*- 'go' → *vi-tha*- 'repeatedly go', *pas*- 'throw (one object)' → *pa-ja*- 'scatter (objects in different places)', *zu*- 'wash' → *zu-zu*- 'wash one after another', *kor̆*- 'boil' → *kor̆-kor*- 'boil for a long time' (Gruzdeva 1997)	Grammaticalization of causative suffix into pluractional pTg *-*t*- ~ *-ti*- causative > pluractional in Evk. *bu:*- 'to give (tr.)' → *bu:t*- 'to distribute, give out, hand out (tr.)', *Evk. luptu*- 'to pull out (e.g. a tooth) (tr.)' → *luptut*- 'pull out, pluck (e.g. a fowl) (tr.)', Evk. *wa:*- 'kill (tr.)' → *wa:t*- 'massacre, slay (tr.)'; Ud. *tingme*- 'to fall (intr.)' → *tingmesi*- 'to fall (of several people) (intr.)', Ud. *bukta*- 'to break (tr.)' → *buktasi*- 'to break (several objects) (tr.)', Ud. *xuine*- 'to dive (intr.)' → *xuinesi*- 'to dive several times (intr.)' (Robbeets 2015).	No pluractional marking on the verb

4.3. Farmers meet Farmers

4.3.1. Linguistics

By contrast to the Transeurasian affiliation debate, the existence of the Sino-Tibetan or Trans-Himalayan family is relatively uncontroversial. Nevertheless, the last decades have seen an ongoing debate about the location of the homeland and the exact time-depth of this family, ranging from the Himalayan Plateau at 4000 BC (Matisoff 2016) to southwest Sichuan at 7000 BC (van Driem 2005) and northeast India at 7000 BC (Blench and Post 2013), with recent opinions converging on the Yellow River Basin, be it at different time-depths, i.e. 6000-4000 BP (LaPolla and Thurgood 2016), 5900 BP (M. Zhang et al. 2019), 7200 BP (Sagart et al. 2019) and 8000 BP (H. Zhang et al. 2020).

Bingcong Deng (this volume) follows H. Zhang et al. (2020) in dating and locating Sino-Tibetan, as illustrated in Figure 4.3. This implies that it was contemporaneous with and adjacent to the Altaic speech community, which lived in the West Liao River region between 9200 BP and 6800 BP and the Japano-Koreanic speech community, which had moved to the area of the Bohai Sea and the Liaodong Peninsula, where it existed between 9200 BP and 5500 BP (Robbeets et al. 2021). If Bradley (this volume) is correct in hypothesizing a linkage between Sino-Tibetan, Yeniseian and Na-Dene that existed before 8000 BP to the west of Transeurasian, contact would presumably have arisen with Transeurasian only after the Sino-Tibetan groups moved further east down the Yellow River and adopted agriculture.

Although ancient Sino-Tibetan loanwords for millets diffused into ancestral families south of the Yellow River that were more familiar with rice farming, Deng finds no good candidates for loans relating to millets between Sino-Tibetan and Transeurasian in the north. Nevertheless, she suggests that at least one agricultural tool for scooping soil (PSTB **ts(h)rop* 'spade' >> pTEA **sarpa* 'spade') and a secondary product (PSTB **ʔam* 'to eat/drink (liquid food such as gruel); liquid food' >> Proto-Transeurasian **amʊ* 'cooked cereal, millet gruel') may have been borrowed from Sino-Tibetan into Transeurasian at an early stage. She further finds evidence for the borrowing of a certain field type (PS **noŋ* 'agriculture' >> PJK **non* 'field for agriculture'), perhaps related to innovative irrigation techniques, which was borrowed from Sinitic into Japano-Koreanic after the break-up of Sino-Tibetan around 8000 BP. However, there are no plausible candidates for agricultural borrowing between Sino-Tibetan and Altaic. This may indicate that in the Early to Middle Neolithic there was some exchange relating to food production and cultivation technology between the farmers in the Yellow River and Liao River basins but there was no need for borrowing millet terms, perhaps because both farmer groups had developed millet cultivation independently from each other. It also suggests that in the Late Neolithic the interactions in the Liaodong-Shandong area were more intensive than those between the Yellow River and Liao River basins.

In addition, some studies (Hashimoto 1986; Comrie 2008; Robbeets 2017a) have argued that the Transeurasian languages left a serious mark on the linguistic structure of Sinitic from the Old Chinese period (ca. 1000 BC-1 AD) onwards but so far, we find no clear indications of structural influence between both language families that may predate this period (see also Bradley this volume). Therefore, these linguistic observations suggest a situation of language maintenance with linguistic borrowing, whereby the Sino-Tibetan and Transeurasian farmers each maintained their native language and just exchanged some specialised vocabulary.

4.3.2. Triangulation

From the viewpoint of archaeology, the traditional view of the Yellow River Basin as the 'cradle of Chinese civilization', a core area from where agriculture spread to the rest of China, has been challenged over the last decades (Shelach and Teng 2013; Miller et al. 2016; Stevens and Fuller 2017; Leipe et al. 2019). It is more likely that millet-based cultivation was developed before 8000 BP independently in the Yellow River and Liao River reaches (Shelach-Lavi et al. 2019; Stevens et al. 2020), giving rise to two distinct domestication centres, i.e., the so-called foxtail-prevalent Yangshao (7000–4800 BP) and broomcorn-prevalent Xinglongwa (8200–7400 BP) cultures in the respective regions. This scenario is consistent with the lack of ancient loanwords for millet terminology between Sino-Tibetan and Transeurasian but it does not rule out the possibility of early interactions between both regions, resulting in the borrowing of the specialised technology mentioned above.

Neolithic Yellow River and West Liao River cultures shared certain aspects in common, including tools associated with the management of cereals such as hoes, spades and shovels (Liu and Chen 2012: 127–152; Shelach-Lavi et al. 2019; Stevens et al. 2020) and there are some minor correlations in pottery shape and painting (Zhu 2007; Guo 2017). However, large-scale Neolithic migration of Yellow River populations into the Liao River basin or vice-versa is not supported by the archaeological evidence so far. Contacts within the Shandong-Liaodong interaction sphere, however, are thought to have been somewhat more intensive, taking place as early as 7000 BP, as indicated by various types of pottery including footed and tripod vessels and painted canoe-shaped jars (Xu 1995: 78–79, 85–86). In terms of influence, the Shandong Peninsula had more impact upon Liaodong in the Neolithic than the other way around, a point corroborated by the direction of the linguistic borrowings above. It is not until the Bronze Age that the material evidence for regional interaction becomes significant including, for example, the practice of oracle bone divination, different types of bronze artefacts, types of ceramic cowrie shells, and jades (Shelach 2009; Flad 2008). This is consistent with the observation that the majority of the ancient borrowings between Sino-Tibetan and Transeurasian languages took place at a time in the Bronze Age when the five Transeurasian daughter branches had already separated (Deng this volume).

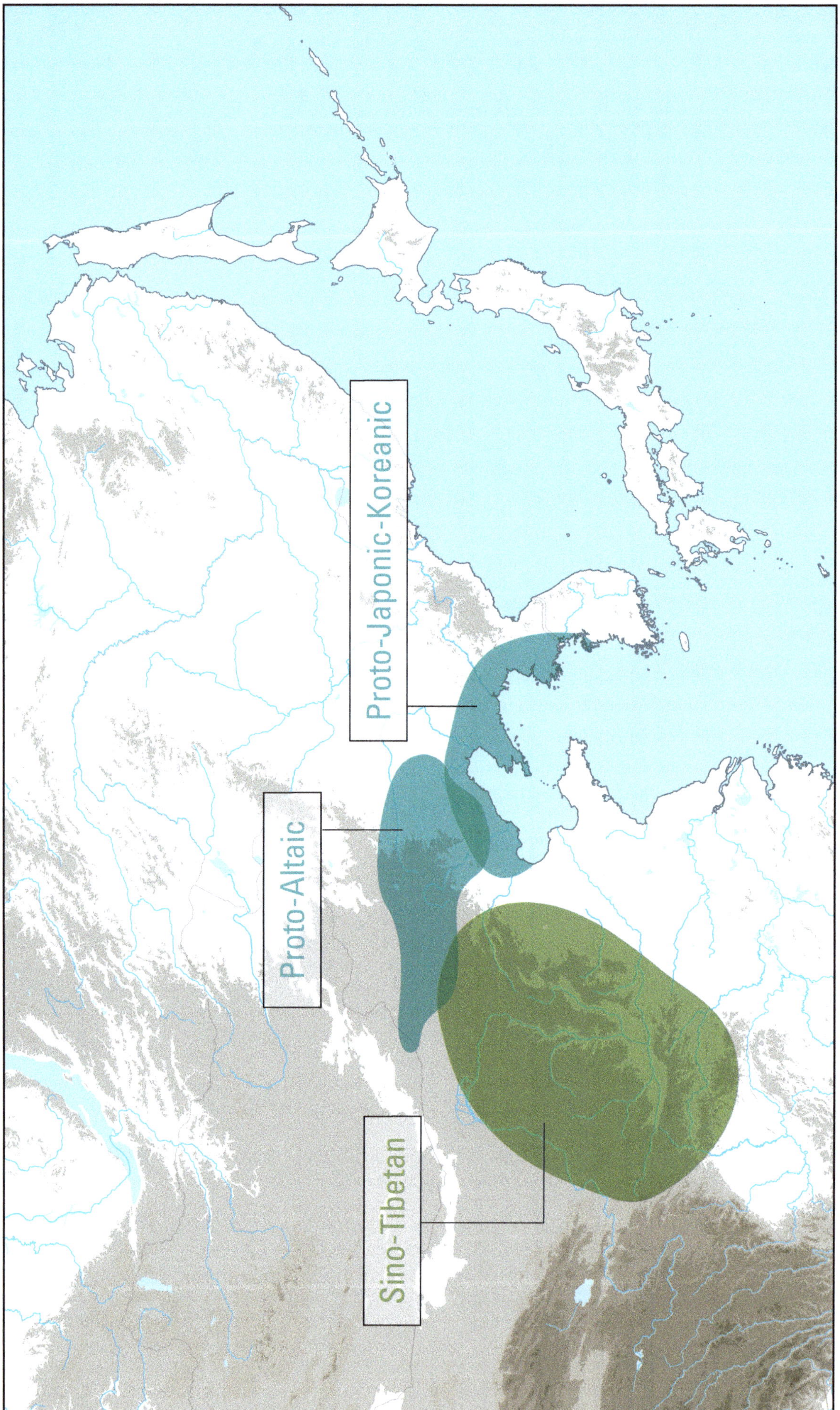

Figure 4.3. The estimated location of the Proto-Sino-Tibetan, Proto-Altaic and Proto-Japano-Koreanic homelands before 8000 BP. Map by M. O'Reilly, MPI-GEA, based on information by the author.

From a genetic point of view, the Yangshao culture in the Yellow River basin can be associated with Yellow River ancestry, reflected in the speakers of contemporary Sino-Tibetan languages (Ning et al. 2020; Wang et al. 2021; Robbeets et al. 2021; Yang et al. 2020), whereas the Xinglongwa through Hongshan (4500–2900 BC), cultures in the Liao River basin can be associated with Amur-related ancestries, reflected in the contemporary speakers of Transeurasian languages (Robbeets et al. 2021). These studies suggest a wide-scale genetic influence of Yellow River populations in West Liao River populations, gradually increasing from the Late Neolithic onwards. This is striking because the archaeological and linguistic evidence for early interaction between both regions is much less impressive until the Early Bronze Age.

4.4. Farmers meet Pastoralists

4.4.1. Linguistics

In recent studies (Robbeets et al. 2020; Robbeets et al. 2021), we aimed at identifying the ancestral homelands of Proto-Turkic, Proto-Mongolic and Proto-Tungusic. In the Bronze Age, Proto-Tungusic was situated in the mid-Amur region and in the border area between present-day Northern Heilongjiang and the Russian Primorye. Around that time, the common ancestor of the contemporary Mongolic languages and the now extinct para-Mongolic languages spoken by the Khitan of the Liao Empire (907–1125), the Tabghach of the Northern Wei (386–534), the Xianbei or 'Serbi' (208 BC–235 AD) and the Donghu (the first millennium BC), stretched from the West Liao River to the Nen river region. It was probably not until the Late Bronze Age that the Mongolic languages started to move westwards into present-day Mongolia. The Proto-Turkic speech community extended from the Ordos Plateau in Shanxi and Inner Mongolia to the Sayan-Altai Mountains, where it probably came into contact with speakers of Old Iranic, descendants of the Andronovo culture (4000-3400 BP), who had spread from the Aral Sea into the steppe zone and the Central Asian mountains. Due to the pressure of the Iranic and Turkic languages in the Sayan-Altai region, the language(s) of the Afanasievo culture (5300-4500 BP) had suffered almost complete extinction, but a likely descendant, Proto-Tocharian, migrated south into the Tarim Basin. Figure 4.4 illustrates the approximate homelands of these speech communities around 3000 BP.

The interactions between speakers of Indo-European and Transeurasian languages in the Bronze Age set the scene for pastoralism to be introduced from the western to the eastern steppes and are mirrored in the borrowing of pastoral vocabulary. Bjørn (2022, this volume) discusses borrowing chains for livestock (Common Iranic **bāra-ka-* 'beast of burden' >> Old Chinese **mˁraʔ* 'horse' >> PMo**morï-n* 'horse' >> PTg **murin* 'horse', Proto-Indo-Iranic **gā́w-* and Proto-Tocharian **keu* 'cow' >> Proto-Mo **geün* 'bull' >> Ma. geo 'mare; cow'), bee exploitation (Proto-Tocharian **ḿätä* > Tocharian B mīt 'honey' >> Old Chinese **mit* 'honey' >> Old Japanese mitu 'honey'; Proto-Iranic **madu-* >> Avestan maδu 'honey'>> PTk **bal* 'honey' >> PMo **bal* 'honey') and metallurgy (Proto-Tocharian **ẃäsa* 'gold' >> PTk **yez* 'copper (alloy); bronze') that were transmitted from Iranic and Tocharian into Turkic, Mongolic and Tungusic, occasionally through the intermediary of Old Chinese. Taken together with the repurposing in Mongolic and Turkic of inherited agricultural vocabulary relating to fermentation as dairying terminology and of inherited words for 'pig' as pastoral animal names such as 'cattle,' 'camel' or 'horse' (Savelyev 2017; Robbeets et al. 2021), these linguistic observations suggest a situation of language maintenance with linguistic borrowing, whereby Turkic and Mongolic farmers adopted pastoralism from Indo-European speakers to the west.

4.4.2. Triangulation

The earliest evidence for herding of sheep and cattle and dairying being introduced from the western to the eastern steppes can be traced to ca. 5000 BP, but the origins of horse-based pastoral economy on the eastern steppe goes only as far back as the Late Bronze Age (ca. 3500-2700 BP) (Taylor et al. 2020). While horses first played a comparatively limited role in Early and Middle Bronze Age pastoral economies across both Mongolia and Central Asia, around 3200 BP horses started to take an important role due to the innovation of mounted horseback riding. Horses were reported from sites in the West Liao, Ordos and Heilongjiang regions in the Late Bronze Age but are not known in the Primorye until the Iron Age and reach Japan only from the end of the fourth century AD (Robbeets et al. 2021). This relatively late adoption of horses in Northeast Asia is consistent with the dating of the above borrowings to the Late Bronze Age and beyond.

The ancestral gene pools of present-day Mongolic and Turkic populations are likely to have been formed through gene flow within the last 3000 years, as a mixture of an underlying Amur-related gene pool, common to all speakers of Transeurasian languages (Robbeets et al. 2021) with that of local populations in Mongolia and the Altai-Sayan region, respectively (Jeong et al. 2018; Ning et al. 2020). The relatively late westward dispersal of the Mongolic languages into present-day Mongolia in the Late Bronze Age appears consistent with an isotope analysis which finds an increase in millet consumption in Mongolia after ca. 2800 BP (Wilkin et al. 2021).

The identification of Yamnaya ancestry in the Afanasievo culture in the Altai-Sayan region (Honeychurch et al. 2021), suggests that a language ancestral to Tocharian was spoken already there around 4500 BP, while Tocharian speakers entered the Tarim Basin only after 3700 BP (Zhang et al. 2021).

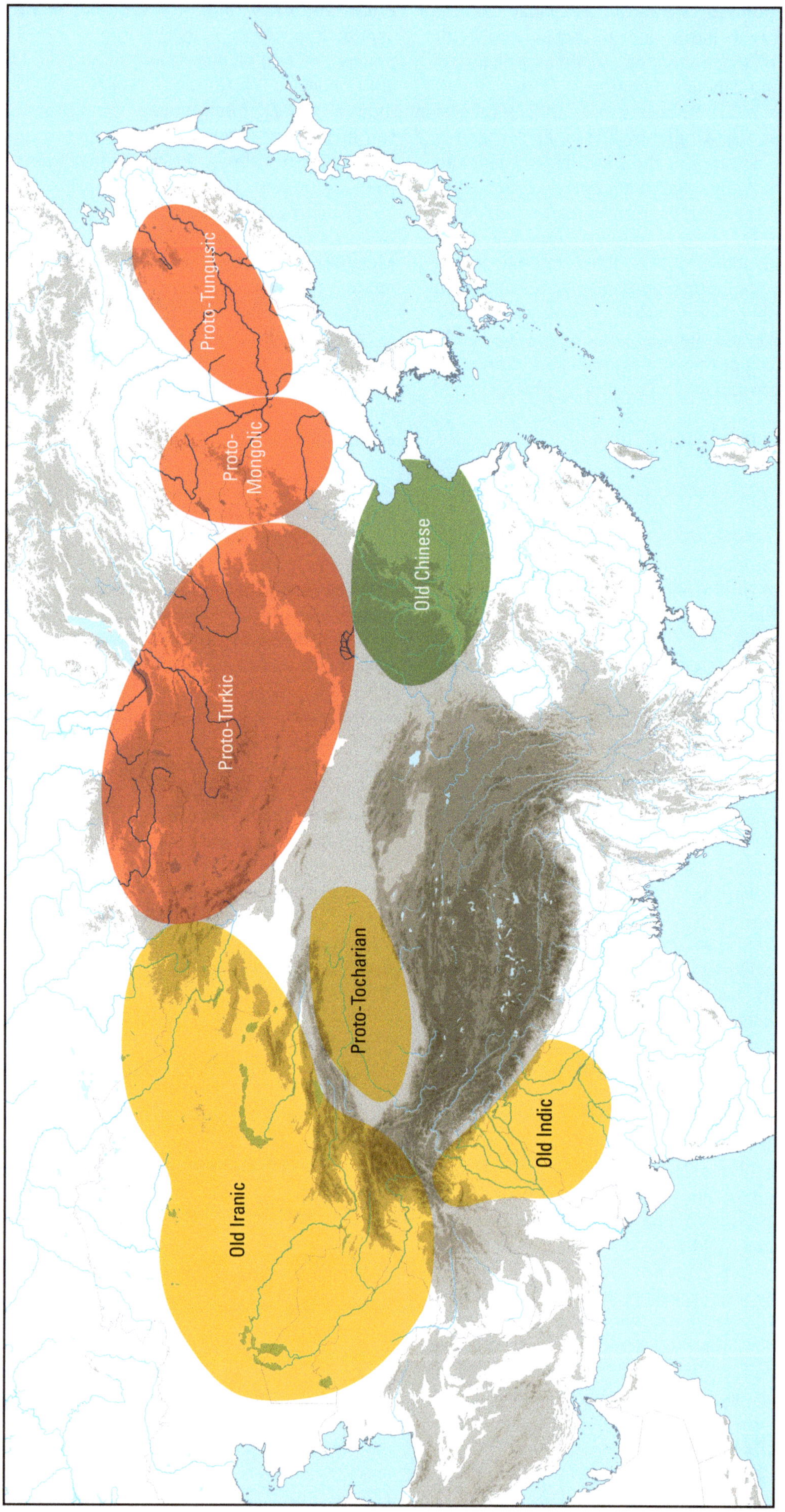

Figure 4.4. The estimated location of the Proto-Iranic, Proto-Tocharian, Proto-Turkic, Proto-Mongolic, Proto-Tungusic and Old Chinese around 3000 BP. Map by M. O'Reilly, MPI-GEA, based on information by the author.

The Xiongnu, a multi-ethnic steppe people, dominating the area from the Ordos Plateau to the Altai Mountains to south of Lake Baikal between the third century BC and the second century AD, included Turkic speakers in addition to Eastern Iranian, Yenisean, Tocharian and Mongolic elements. Xiongnu subsistence was dependent on nomadic pastoralism, probably introduced from their Eastern Iranian neighbours, with a strong agricultural component relating to millet farming. Genetically, the Xiongnu samples are diverse, showing various degrees of admixture between Amur ancestry, Slab Grave people from present-day Eastern Mongolia and Saka-related populations from southeastern Kazakhstan and the Tian Shan mountains (Damgaard et al. 2018; Savelyev and Jeong 2020).

Taken together with the relatively late introduction of horse-based pastoralism on the Eastern Steppe, the genetic evidence for east-west genetic admixture starting around 3000 years ago is consistent with the setting and the timing of the pastoral borrowings proposed by Bjørn in this volume.

4.5. Discussion

In this chapter, I brought together three cases of genetic admixture between Transeurasian-speaking and non-Transeurasian speaking populations in the Neolithic and Bronze Age. In two cases, notably the admixture between Yellow River and Amur lineages in Northeast China in the Neolithic and the admixture between Amur and western Eurasian ancestries on the Eastern Steppe in the Bronze Age, the genetic admixture is clearly measurable. The genetic admixture between Amur ancestries from the West Liao area and Amur ancestries from the Russian Far East in the Late Neolithic and Bronze Age, however, is imperceptible but implied by the archaeological and linguistic facts.

Based on the contributions by Knapen, Deng and Bjørn in this volume, I compared the alleged linguistic outcome of these three cases of genetic admixture. To this end, I associated the Amur-Amur admixture with Tungusic-Amuric language contact, Yellow River-Amur admixture with language contact between Sino-Tibetan, Altaic and Japano-Koreanic, and Amur-western Eurasian admixture with contacts between Old Iranic, Tocharian, Turkic and Mongolic languages. There is a difference between the linguistic outcome of these admixture events because in the case of the Tungusic-Amuric contact associated with the Amur-Amur admixture, the evidence for structural diffusion is at least as strong as the evidence for lexical borrowing, whereas in the other cases the evidence for lexical borrowing is more prominent than that for structural diffusion. Based on empirical observations of contact situations (Thomason and Kaufman 1988; Thomason 2001: 66ff; Thomason 2009), we expect that in situations of language maintenance with borrowing, vocabulary will be borrowed before phonological and syntactic features, while in situations of language shift with substratum interference, phonological and syntactical structural features will be carried over before vocabulary.

This model implies that at least some ancestral speakers of Proto-Amuric abandoned their native language and adopted the proto-Tungusic target language. Speakers of Amuric, first becoming bilingual in Tungusic, would have much lexical borrowing but only moderate morphosyntactic and phonological borrowing from Tungusic in their native language, while in their Tungusic, they would have weak lexical borrowing and moderate to strong phonological and morphosyntactic interference due to imperfect learning. This scenario is reminiscent of the spread of contemporary dominant languages among speakers of minority languages, such as for instance Chinese among Salar Turkic speakers in Qinghai (Dwyer 2007).

By contrast, in the other contact situations described above, the native Sino-Tibetan, Transeurasian and Indo-European languages were maintained and merely served as a model and/or recipient for borrowing. Lexical borrowing can be found in the total absence of bilingualism of any kind. One example is the cultural borrowing of the Min Chinese word 茶 *tê* for 'tea' into Malay *teh* and Dutch *thee* into English *tea*. Thus, when the Transeurasian speakers borrowed the word for 'spade' from Sino-Tibetan, or the Turkic speakers borrowed 'honey' from Iranic, they were not necessarily bilingual and maintained their native language.

For an answer to the question as to what determines the different linguistic outcomes of the admixture events, we can turn to the cultural context in which the interactions took place. Indeed, the ratio of economic dominance appears to influence the linguistic outcome of the genetic admixture.

Associating the Proto-Tungusic language with incoming farmers who brought millet from Northeast China in the late Neolithic and Proto-Amuric with pre-existing fisher-hunter-gatherers in the Primorye, there is an imbalance of economic productivity of the subsistence strategies. The socio-economic benefits will motivate the fisher-hunter-gatherers to adopt agriculture, a process facilitated by becoming bilingual in local Amuric and the target Tungusic language.

By contrast, I associated the Proto-Sino-Tibetan and Proto-Transeurasian speakers with foxtail-prevalent and broomcorn-prevalent millet farmers, having economically more or less equally viable strategies. Here some agricultural techniques or products may be transferred from one society into the other, but this process did not lead to language shift as both speech communities maintained their native languages. A similar scenario is true for the addition of pastoralism from the west to the originally agricultural lifestyles of the ancestors of the Turkic and Mongolic speakers when they entered the eastern steppes. The addition of new strategies to the pre-existing package led to lexical—and perhaps also some structural—borrowing but there was no language shift involved. The direction of the borrowing was from the technologically more advanced or more prestigious speech community into the community with the lower status.

Similar cases can be made for other admixture events involving other Transeurasian and non-Transeurasian speech communities, such as the admixture between Neolithic Jōmon populations and Bronze Age Yayoi migrants on the Japanese Islands (Hanihara 1991; Jinam et al. 2012; Hudson et al. 2020; Robbeets et al. 2021; Cooke et al. 2021). The Jōmon can be associated with a variety of languages spoken before the adoption of millet and rice agriculture in the Japanese Islands, some of them related to the ancestral Ainuic language, while the incoming Yayoi farmers are generally linked with Japonic language. Whereas some features that are prototypical of Ainu but atypical of Transeurasian languages, such as the occurrence of prefixing morphology in spite of a verb-final structure, productive noun incorporation, distinction between intentional and non-intentional action in certain verb affixes and absence of core-case marking, seem to have influenced the Japonic language, there is only very restricted evidence for prehistoric loanwords from Ainuic into Japonic (Vovin 2013). Reminiscent of the Amuric-Tungusic contacts discussed above, this implies language shift whereby some of the ancestral speakers of Ainu abandoned their native language and adopted the Japonic language of the Yayoi, leaving some structural learning imperfections in their newly acquired language. Here too, the economic dominance of the Yayoi over the Jōmon may explain why the outcome of the interaction was language shift rather than mere lexical borrowing.

4.6. Conclusion

In this chapter, I associated three cases of linguistic interaction between Transeurasian and non-Transeurasian speech communities with genetic admixture events in Neolithic and Bronze Age Northern Asia. Based on the socio-economic setting in which the prehistoric interactions are thought to have taken place, I hypothesised that the ratio of economic dominance determined the linguistic outcome of the genetic admixture in these cases. Contact with speech communities with economically less viable strategies led to dispersal of the dominant language through shift with substratum interference, while interaction between populations with comparably viable strategies resulted in language maintenance with borrowing from the more to the lesser advanced community.

Future study will need to test this hypothesis more carefully. While recent research has focused on Neolithic and Bronze Age interactions between farmers and hunter-gatherers from a genetic (Rivollat et al. 2020), archaeological (Bondetti et al. 2021) or linguistic (Blasi et al. 2019; Güldemann et al. 2020) perspective, it would be enlightening to triangulate these different perspectives into a single, integrative approach and to compare farmer-forager interactions with those between farmers and pastoralists and farmers with different specialisations. The prehistoric interaction between speakers of Transeurasian languages and their non-Transeurasian neighbours may continue to yield useful case studies because of the variety of regional histories and the complexity of subsistence strategies involved.

References

Austerlitz, Robert (1982) On comparing language families. *Études finno-ougriennes* 15: 45–54.

Baskakov, Nikolaj A. (1981) *Altaiskaja sem'ja jazykov i ee izučenie* [The Altaic language family and its origins.] Moscow: Nauka.

Bellwood, Peter (2005) *First Farmers*: The Origins of Agricultural Societies. Malden: Blackwell.

Bellwood, Peter (2021) Tracking the origin of Transeurasian languages. *Nature* 599: 557–558.

Bellwood, Peter and Colin Renfrew (eds) (2002) *Examining the Farming/Language Dispersal Hypothesis*. Cambridge: McDonald Institute for Archaeological Research.

Benzing, Johannes (1955) Die tungusischen Sprachen: Versuch einer vergleichenden Grammatik. *Abhandlungen der geistes- und sozialwissenschaftlichen Klasse* 11: 949–1099. Wiesbaden: Verlag der Akademie der Wissenschaften und der Literatur in Mainz.

Blasi, Damián, Steven Moran, Scott R. Moisik, Paul Widmer, Dan Dediu and Balthasar Bickel (2019) Human sound systems are shaped by post-Neolithic changes in bite configuration. Science 363: eaav3218.

Blažek, Vaclav (2019) *Altaic Languages*. Brno: Masaryk University Press.

Blench, Roger and Mark W. Post (2013) Rethinking Sino-Tibetan phylogeny from the perspective of North East Indian languages. In T. Owen-Smith and N.W. Hill (eds.), *Trans-Himalayan Linguistics*, pp. 71–104. Berlin: Mouton de Gruyter.

Bjørn, Rasmus (2022) Indo-European loanwords in Bronze Age Central and East Asia: Six new perspectives on prehistoric exchange in the Eastern Steppe Zone. *Evolutionary Human Sciences* 4: e23.

Bondetti, Manon, Lara González Carretero, Ekaterina Dolbunova, Krista McGrath, Sam Presslee, Alexandre Lucquin, Viktor Tsybriy, Andrey Mazurkevich, Andrey Tsybriy, Peter Jordan, Carl Heron, John Meadows and Oliver E. Craig (2021) Neolithic farmers or Neolithic foragers? Organic residue analysis of early pottery from Rakushechny Yar on the Lower Don (Russia). *Archaeological and Anthropological Sciences* 13: 141

Büyükmavi, Maud (2007) Rezension von Robbeets, Martine 2005. Is Japanese related to Korean, Tungusic, Mongolic and Turkic? *Oriens Extremus* 46: 306–310.

Campbell, Lyle and William J. Poser (2008) *Language Classification: History and Method.* Cambridge University Press.

Chaubey, Gyaneshwer and George van Driem (2020) Munda languages are father tongues, but Japanese and Korean are not. *Evolutionary Human Sciences* 2: e19.

Clauson, Gerard (1956) The case against the Altaic theory. *Central Asiatic Journal* 2: 181–187.

Comrie, Bernard (2008) The areal typology of Chinese: Between North and Southeast Asia. In Djamouri, R., B. Meisterernst & R. Sybesma (eds.), *Chinese Linguistics in Leipzig* (Collection des Cahiers de Linguistique Asie Orientale, 12.), 1-21. Paris: École des Hautes Études en Sciences Sociales, Centre de Recherches Linguistiques sur l'Asie Orientale.

Cooke, Niall P., Valeria Mattiangeli, Lara M. Cassidy, Kenji Okazaki, Caroline A. Stokes, Shin Onbe, Satoshi Hatakeyama, Kenichi Machida, Kenji Kasai, Naoto Tomioka, Akihiko Matsumoto, Masafumi Ito, Yoshitaka Kojima, Daniel G. Bradley, Takashi Gakuhari and Shigeki Nakagome (2021) Ancient genomics reveals tripartite origins of Japanese populations. *Science Advances* 7: eabh2419.

Damgaard, Peter, Rui Martiniano, Jack Kamm *et al.* (2018) The first horse herders and the impact of Early Bronze Age steppe expansions into Asia. *Science* 360: eaar7711.

Décsy, Gyorgy (2007) Review of Martine Robbeets (2005) Is Japanese related to Korean, Tungusic, Mongolic and Turkic? *Eurasian Studies Yearbook* 79: 157.

Doerfer, Gerhard (1963–1975) *Türkische und Mongolische Elemente im Neupersischen, unter besonderer Berücksichtigung älterer neupersischer Geschichtsquellen, vor allem der Mongolen- und Timuridenzeit*, vols. 1-4. [Turkic and Mongolic Elements in New Persian with Special Attention to Old New Persian Historical Sources, Mainly of the Period of the Mongols and the Timurids]. Wiesbaden: Franz Steiner.

Dwyer, Arienne (2007) Salar: *A Study in Inner Asian Language Contact Processes: Part I: Phonology* (Turcologica, Band 37). Wiesbaden: Harrassowitz.

Dybo, Anna (2016) New trends in European studies on the Altaic problem. *Journal of Language Relationship* 14: 41–106.

Eliasson, Stig (2016) Review of Gardani, F., Arkadiev, P. And Amiridze, N. (eds) 2015. Borrowed morphology. *Studies in Language* 40(3): 722–731.

Flad, Rowan K. (2008) Divination and power: a multiregional view of the development of oracle bone divination in early China. *Current Anthropology* 49 (3): 403–437.

Fuente, José Andrés Alonso de la (2016) Review of Robbeets, Martine 2015. Diachrony of verb morphology: Japanese and the Transeurasian languages. *Diachronica* 33(4): 530–536.

Georg, Stefan (1999) Haupt und Glieder der Altaischen Hypothese: die Körperteilbezeichnungen im Türkischen, Mongolischen und Tungusischen. *Ural-Altaische Jahrbücher* (Neue Folge) 16: 143–182.

Georg, Stefan (2007) Review of Robbeets, Martine 2005. Is Japanese related to Korean, Tungusic, Mongolic and Turkic? *Korean Studies*: 247–278.

Gözaydin, Nuretim (2006) Dergi ve kitap dünyasïndan. *Türk Dili* 653: 467–471.

Gruzdeva, Ekaterina (1997) Plurality of situations in Nivkh. In Viktor S. Xrakovskij (ed.) *Typology of Iterative Constructions*, pp. 164-188. München: LINCOM.

Gruzdeva, Ekaterina (1998) Nivkh. *Language of the world/ Materials* 111. München: LINCOM.

Gruzdeva, Ekaterina (2012) On syntactic definition of finiteness: Evidence from Nivkh. *Per Urales ad Orientem. Iter polyphonicum multilingue.* Festschrift for *Juha Janhunen. Mémoires de la Société Finno-Ougrienne* 264: 65–75.

Güldemann, Tom, Patrick McConvell and Richard Rhodes (Eds.) (2019) *The Language of Hunter-Gatherers.* Cambridge University Press.

Guo, Dashun (2017) Yangshao wenhu yu hongshan wenhua guanxi zai guancha [Reexamination of the relationship between the Yangshao and the Hongshan cultures]. *Zhengzheng Daxue Xuebao* [Journal of Zhengzhou University] 50(4): 103–107.

Hanihara, Kazuro (1991) Dual structure model for the population history of the Japanese. *Japan Review* 2: 1–33.

Hashimoto, Mantaro (1986) The Altaicization of northern Chinese. In J. McCoy and T. Light (eds.), *Contributions to Sino-Tibetan Studies*, pp. 76-97. Leiden: Brill.

Honeychurch, William *et al.* (2021) The earliest herders of East Asia: examining Afanasievo entry to Central Mongolia. *Archaeological Research in Asia* 26: 1–19.

Hudson, Mark J. (2020) Language dispersals and the 'Secondary Peoples' Revolution': a historical anthropology of the Transeurasian unity. In Martine Robbeets and Alexander Savelyev (eds.), *The Oxford*

Guide to the Transeurasian Languages, pp. 806–813. Oxford University Press.

Hudson, Mark J., Shigeki Nakagome and John Whitman (2020) The evolving Japanese: The dual structure hypothesis at 30. *Evolutionary Human Sciences* 2: e6.

Janhunen, Juha (1996) *Manchuria: An Ethnic History*. (Mémoires de la Société Finno-Ougrienne 222.) Helsinki: Suomalais-Ugrilainen Seura.

Jinam, Timothy, Nao Nishida, Momoki Hirai, Shoji Kawamura, Hiroki Oota, Kazuo Umetsu, Ryosuke Kimura, Jun Ohashi, Atsushi Tajima, Toshimichi Yamamoto, Hideyuki Tanabe, Shuhei Mano, Yumiko Suto, Tadashi Kaname, Kenji Naritomi, Kumiko Yanagi, Norio Niikawa, Keiichi Omoto, Katsushi Tokunaga and Naruya Saitou (2012) The history of human populations in the Japanese Archipelago inferred from genome-wide SNP data with a special reference to the Ainu and the Ryukyuan populations. *Journal of Human Genetics* 57: 787–795.

Johanson, Lars (1995) On Turkic converb clauses. In M. Haspelmath and E. König (eds.), *Converbs in Cross-Linguistic Perspective. Structure and Meaning of Adverbial Verb Forms - Adverbial Participles, Gerunds* (Empirical Approaches to Language Typology 13), pp. 313–347. Berlin: Mouton de Gruyter.

Johanson, Lars (2010) The high and low spirits of Transeurasian language studies. In L. Johanson and M. Robbeets (eds), *Transeurasian Verbal Morphology in a Comparative Perspective: Genealogy, Contact, Chance* (Turcologica 78), pp. 7–20. Wiesbaden: Harrassowitz.

Kara, Gyorgy (2007) Review of Robbeets, Martine 2005. Is Japanese related to Korean, Tungusic, Mongolic and Turkic? *Anthropological Linguistics* 49: 95–98.

Kiyose, Gisaburō Norikura (2002) The relationship among the languages of the Korean Three Kingdoms, Proto-Japanese and Tungusic, based on historical sources. *Transactions of the 47th International Conference on Eastern Studies* 121–122.

Knapen, Martijn (2021) *The Oldest Layer of Amuric-Tungusic Lexical Contacts*. MA dissertation: University of Helsinki.

Koch, Harold (2014) Morphololological Borrowing and Genetic Relationship. A review article of Johanson and Robbeets (eds). 2012: 'Copies Versus Cognates in Bound Morphology' *Journal of Language Contact* 7: 408–424.

Korovina, Evgeniya (2011) *Leksika prirodnogo i kul'turnogo okruženiya v Tunguso-Man'čžurskich jazykach (v istoriko-tipologičeskom osveščenii)* [Tungus-Manchu vocabulary related to natural environment and cultural activities (in historical and typological perspective)]. Moscow: Russian State University of Humanities, BA dissertation.

LaPolla, Randy J. and Graham Thurgood (2016) *Sino-Tibetan Languages*. Routledge.

Leipe, Christian, Tengwen Long, Elena A. Sergusheva *et al.* (2019) Discontinuous spread of millet agriculture in eastern Asia and prehistoric population dynamics. *Science Advances 5*: eaax6225.

Li, Tao, Chao Ning, Irina S, Zhushchikhovskaya, *et al.* (2020) Millet agriculture dispersed from Northeast China to the Russian Far East: integrating archaeology, genetics and linguistics. *Archaeological Research in Asia* 22: e100177.

Liu, Li and Xingcang Chen (2012) *The Archaeology of China: From the Late Paleolithic to the Early Bronze Age*. Cambridge University Press.

Matisoff, James A. (2016) *The Sino-Tibetan Etymological Dictionary and Thesaurus*. Accessed via http://stedt.berkeley.edu/search/> on 2020-02-04.

Miller, Naomi F., Robert N. Spengler and Michael Frachetti (2016) Millet cultivation across Eurasia: Origins, spread, and the influence of seasonal climate. *Holocene* 26: 1566–1575.

Miller, Roy Andrew (1971) *Japanese and the Other Altaic Languages*. University of Chicago Press.

Murayama, Shichiro (1957) Vergleichende Betrachtung der Kasus-Suffixe im Altjapanischen, in Omeljan Pritsak (ed.), *Studia Altaica. Festschrift für Nicholaus Poppe zum 60. Geburstag am 8 August 1956*, pp. 126–131. Wiesbaden: Otto Harrassowitz.

Murayama, Shichiro (1958) Einige Formen der Stammverkürzung in den altaischen Sprachen. *Oriens* 11: 225–230.

Nedjalkov, Igor (1995) Converbs in Evenki. In M. Haspelmath and E. König (eds.), *Converbs in Cross-Linguistic Perspective. Structure and Meaning of Adverbial Verb Forms - Adverbial Participles, Gerunds* (Empirical Approaches to Language Typology 13), pp. 97–136. Berlin: Mouton de Gruyter.

Nedjalkov, Igor V. (1997) *Evenki. Descriptive Grammar*. Routledge.

Ning, Chao, Tao Li, K. Wang, T. Li, F. Zhang, X. Wu, S. Gao, J. Krause, M. Robbeets, C. Jeong, Y. Cui *et al.* (2020) Genomic history of northern China for the last 7,500 years. *Nature Communications* 11: e2700.

Poppe, Nicholas (1960) *Vergleichende Grammatik der altäischen Sprachen* [Comparative Grammar of the Altaic

Languages]. Teil I: Vergleichende Lautlehre. Wiesbaden: Otto Harrassowitz.

Poppe, Nicholas (1964) Der altaische Sprachtyp. In Bertold Spuler *et al.* (eds.), *Mongolistik.* (Handbuch der Orientalistik 5.2.), pp. 1-16. Leiden: Brill.

Poppe, Nicholas (1965) *Introduction to Altaic Linguistics.* Wiesbaden: Harrassowitz.

Poppe, Nicholas (1975) Altaic linguistics: An overview. *Gengo no kagaku [Sciences of language]* 6: 130–186.

Pugach, Irina, Rostislav Matveev, Viktor Spitsyn, Sergey Makarov, Innokentiy Novgorodov, Vladimir Osakovsky, Mark Stoneking and Brigitte Pakendorf (2016) The complex admixture history and recent southern origins of Siberian populations. *Molecular Biology and Evolution* 33(7): 1777–1795.

Ramstedt, Gustaf J. (1952–1957) *Einführung in die altaische Sprachwissenschaft I: Lautlehre; II: Formenlehre* [Introduction in Altaic Linguistics. I: Phonology, II: Morphology]. Helsinki: Suomalai-Ugrilainen Seura.

Rivollat, Maïté, *et al.* (2020) Ancient genome-wide DNA from France highlights the complexity of interactions between Mesolithic hunter-gatherers and Neolithic farmers. *Science Advances* 6: e5344.

Robbeets, Martine (2005) *Is Japanese related to Korean, Tungusic, Mongolic and Turkic?* Wiesbaden: Harrassowitz.

Robbeets, Martine (2015) *Diachrony of verb morphology. Japanese and the Transeurasian languages.* (Trends in Linguistics 291). Berlin: De Gruyter Mouton.

Robbeets, M. (2017a) Areality: Japanese, Korean and the Transeurasian languages. In R. Hickey (ed.), *The Cambridge Handbook of Areal Linguistics* (Cambridge Handbooks in Language and Linguistics), pp. 586–626. Cambridge University Press.

Robbeets, Martine (2017b) Language farming dispersal: food for thought. In M. Robbeets and A. Savelyev (eds.), *Language Dispersal Beyond Farming*, pp. 1–23. Amsterdam: Benjamins.

Robbeets, Martine (2020a) The classification of the Transeurasian languages. In M. Robbeets and A. Savelyev (eds.), *The Oxford Guide to the Transeurasian Languages*, pp. 31–39. Oxford University Press.

Robbeets, Martine (2020b) Verbal morphology in Transeurasian. In M. Robbeets and A. Savelyev (eds.), *The Oxford Guide to the Transeurasian Languages*, pp. 511–521. Oxford University Press.

Robbeets, Martine (2020c) Basic vocabulary in Transeurasian. In M. Robbeets and A. Savelyev (eds.), *The Oxford Guide to the Transeurasian Languages*, pp. 645–649. Oxford University Press.

Robbeets, Martine (2020d) The typological heritage of Transeurasian. In M. Robbeets and A. Savelyev (eds.), *The Oxford Guide to the Transeurasian Languages.*, pp. 127–144. Oxford University Press.

Robbeets, Martine and Sofia Oskolskaya (2022). Proto-Tungusic in time and space. In Andreas Hölzl & Thomas E. Payne (eds.), Tungusic languages: Past and present, pp. 263–294. Berlin: Language Science Press.

Robbeets, Martine, Juha Janhunen, Alexander Savelyev and Evgeniya Korovina (2020) The homelands of the individual Transeurasian proto-languages: where, when and what? In M. Robbeets and A. Savelyev (eds.), *The Oxford Guide to the Transeurasian Languages.*, pp. 754–771. Oxford University Press.

Robbeets, Martine, Remco Bouckaert, Matthew Conte, Alexander Savelyev , Tao Li , ... Mark Hudson and Chao Ning (2021) Triangulation supports agricultural spread of the Transeurasian languages. *Nature* 599: 616–621.

Ross, Malcolm (2014) Affixes in language change. Copying, inheritance and accommodation. Review Essay of Johanson and Robbeets (eds). 2012: Copies Versus Cognates in Bound Morphology. *Language Dynamics and Change* 4: 271–284.

Rozycki, William (2006) Review of Robbeets, Martine 2005. Is Japanese related to Korean, Tungusic, Mongolic and Turkic? *Mongolian Studies* 28: 114–115.

Sagart, Laurent, Guillaume Jacques, Yunfan Lai, Robin J. Ryder, *et al.* (2019) Dated language phylogenies shed light on the ancestry of Sino-Tibetan. *Proceedings of the National Academy of Sciences USA* 116: 10317–10322.

Savelyev, Alexander (2017) Farming-related terms in Proto-Turkic and Proto-Altaic. In M. Robbeets and A. Savelyev (eds.), *Language Dispersal Beyond Farming*, pp. 123–154. Amsterdam: John Benjamins.

Savelyev, Alexander and Choongwon Jeong (2020) Early nomads of the Eastern Steppe and their tentative connections in the West. *Evolutionary Human Sciences* 2: e20.

Sergusheva, Elena and Yury Vostretsov (2009). The advance of agriculture in the coastal zone of East Asia. In A. Fairbairn and E. Weiss (eds), *From Foragers to Farmers: Papers in Honour of Gordon C. Hillman*, pp. 205–219. Oxford: Oxbow.

Shelach, Gideon (2009) *Prehistoric Societies on the Northern Frontiers of China.* Cambridge University Press.

Shelach, Gideon and Mingyu Teng (2013) Earlier Neolithic economic and social systems of the Liao River Region,

Northeast China. In A.P. Underhill (ed.), *A Companion to Chinese Archaeology*, pp. 37–54. Oxford: Blackwell.

Shelach-Lavi, Gideon, Mingyu Teng, Yonaton Goldsmith *et al.* (2019) Sedentism and plant cultivation in northeast China emerged during affluent conditions. *PLoS ONE* 14: e0218751.

Shōgaito, Masahiro and Yoshizō Itabashi (2002) Arutai shogengo (gan Chōsengo) to Nihongo ni taisuru komento. *Gengo kenkyū*, 121: 112–121.

Siska, Veronika, *et al.* (2017) Genome-wide Data from Two Early Neolithic East Asian Individuals Dating to 7700 years ago. *Science Advances* 3: e1601877.

Starostin, George (2016) Altaic languages. *Oxford Research Encyclopedia of Linguistics*, pp. 1–16. Oxford University Press.

Starostin, Sergei, Anna Dybo and Oleg Mudrak (2003) *Etymological Dictionary of the Altaic Languages*, I–III. Leiden: Brill.

Stevens, Chris J. and Dorian Fuller (2017) The spread of agriculture in eastern Asia: archaeological bases for hypothetical farmer/language dispersals. *Language Dynamics and Change* 7: 152–186.

Stevens, Chris J., Gideon Shelach-Lavi, Hai Zhang, Mingyu Teng and Dorian Fuller (2020) A model for the domestication of *Panicum miliaceum* (common, proso or broomcorn millet) in China. *Vegetation History and Archaeobotany* 30: 21–33.

Taylor, William T. T., Julia Clark, Jamsranjav Bayarsaikhan, Tumurbaatar Tuvshinjargal, Jessica Thompson Jobe, *et al.* (2020) Early pastoral economies and herding transitions in eastern Eurasia. *Scientific Reports* 10: 1001.

Tekin, Talat (1994) Altaic languages. In R. E. Asher (ed.), *The Encyclopedia of Language and Linguistics. Vol. 1.*, pp. 82–85. Oxford: Pergamon Press.

Thomason, Sarah G. and Terrence Kaufman (1988) *Language Contact, Creolization, and Genetic Linguistics.* University of California Press.

van Driem, George (2005) Sino-Austronesian vs. Sino-Caucasian, Sino-Bodic vs. Sino-Tibetan, and Tibeto-Burman as default theory. In Y.P Yadava, G. Bhattarai, R. R. Lohani, B. Prasain and K. Parajuli (eds.), *Contemporary Issues in Nepalese Linguistics,* pp. 285–338. Kathmandu: Linguistic Society of Nepal.

Vovin, Alexander (2005) The end of the Altaic controversy (In Memory of Gerhard Doerfer). *Central Asiatic Journal* 49: 71–132.

Vovin, Alexander (2013) *Man'yōshū. Book 20. A New English Translation Containing the Original Text, Kana Transliteration, Romanization, Glossing, and Commentary.* Folkestone: Global Oriental.

Wang, Chuan-Chao and Martine Robbeets (2020). The homeland of Proto-Tungusic inferred from contemporary words and ancient genomes. *Evolutionary Human Sciences* 2: e8.

Wang, Chuan-Chao, Hui-Yuan Yeh, Alexander N. Popov, Hu-Qin Zhang, Hirofumi Matsumura, Kendra Sirak, Olivia Cheronet, ..., Martine Robbeets, Stephan Schiffels, Douglas J. Kennett, Li Jin, Hui Li, Johannes Krause, Ron Pinhasi, David Reich (2021) Genomic insights into the formation of human populations in East Asia. *Nature* 591: 413–419.

Wilkin, Shevan *et al.* (2021) Economic diversification supported the growth of Mongolia's nomadic empires. *Scientific Reports* 10: e3916.

Xu, Yu-lin (1995) The Houwa site and related issues. In Sarah Nelson (ed.), *The Archaeology of Northeast China. Beyond the Great Wall,* pp. 65–88. Routledge.

Yang, Melinda A., Xuechun Fan, Bo Sun, *et al.* (2020) Ancient DNA indicates human population shifts and admixture in northern and southern China. *Science* 369: 6501.

Yanuševič, Zoya, Yuri Vostretsov, and Sofiya V. Makarova. (1990) *The Palaeo-ethnobotanical finds in Primorye.* Vladivostok: Nauka. [in Russian]

Zhang, Fan, Chao Ning, Ashley Scott *et al.* (2021) The genomic origins of the Bronze Age Tarim Basin mummies. *Nature* 599: 256–261.

Zhang, Hanzhi, Ting Ji, Mark Pagel, *et al.* (2020) Dated phylogeny suggests early Neolithic origin of Sino-Tibetan languages. *Scientific Reports* 10: 20792.

Zhang, Menghan, Shi Yan, Wuyun Pan & Li Jin (2019) Phylogenetic evidence for Sino-Tibetan origin in northern China in the Late Neolithic. *Nature* 569: 112–115.

Zhu, Yan-ping (2007) Seeking the roots of the design of Hongshan culture painted pottery. *Research of China's Frontier Archaeology* 6: 78–87.

5

Ancient Agricultural Borrowings between Sino-Tibetan and Transeurasian Languages

Bingcong Deng

Abstract: This chapter looks into agricultural lexical borrowings between Sino-Tibetan and Transeurasian languages during the Neolithic and Bronze Age, in order to investigate the prehistoric contact situations among the speech communities. It attempts to integrate archaeological and linguistic evidence to understand the nature of their early interactions. Six potential lexical borrowings are identified. The results based on linguistic research are consistent with previous archaeological findings: there are rare instances of agricultural lexical borrowings between the proto-languages during the Neolithic, but more cases of borrowings among lower taxonomic languages during the Bronze Age. Future research on other semantic domains involved in borrowing is needed to reveal more information about early contact dynamics.

Keywords: archeolinguistics, agriculture, lexical borrowings, prehistoric language contact, Sino-Tibetan, Transeurasian

5.1. Introduction

Sino-Tibetan (ST), also referred to as Trans-Himalayan, is a language family with at present the second largest number of speakers next to Indo-European, and stretching from Northeast Asia to the Tibetan Plateau. Chinese, Tibetan, and Burmese are amongst the most well-documented members of this language family. The classification of the family is disputed; but it is generally agreed upon that the earliest split occurred between Sinitic and Tibeto-Burman languages (Zhang et al. 2019, Sagart et al. 2019, Zhang et al. 2020). One of the recent phylogenetic studies based on shared lexical items in Sino-Tibetan supports an initial split of the Sinitic languages from the ancestral language approximately 8000 years BP (Zhang et al. 2020)[1]. There are also archaeological findings and linguistic data linking the Sino-Tibetan population to the Yangshao culture in the Yellow River valley, which is most likely to be the *Urheimat* of Sino-Tibetan people (see also David Bradley's proposal of a potential linkage of ST, Yenesian and Na-Dene, and his discussion of their homeland in this volume).

Transeuraisan (TEA) is a term referring to the collectiveness of five language groups: Mongolic, Turkic, Tungusic, Japonic and Koreanic[2]. The relationship amongst these languages has been debated extensively, forming the controversy around the grouping of 'Altaic languages'. Scholars against the unity suggest that the similarities among the languages could be explained by long-lasting areal contact and chance resemblance (e.g., Doerfer 1988; Vovin 2005). By contrast, researchers have also proposed sound correspondences and shared morphological features among Transeurasian languages that can be evidence for a genetic relationship (e.g., Starostin et al. 2003; Robbeets 2005, 2015). Phylogenetic analysis and multidisciplinary research also suggest that these languages share a common ancestor, and the language communities are genetically related (Robbeets et al. 2021). This research locates the homeland of the language family in the area of the West Liao River, with a proposed time-depth around 9000 years BP.

The geographical and chronological adjacency of Sino-Tibetan and Transeurasian homelands in prehistory (Fig. 5.1) implies that there was an early population interaction between the two speech communities during the Neolithic and Bronze Age. Furthermore, the two groups of populations were both agriculture-centered, hence the cultural interaction likely involved lexical exchange of agricultural vocabulary.

With this premise, this chapter will look into the ancient lexical borrowings related to agriculture from Sino-Tibetan and Transeurasian. 'Ancient' borrowings refer to the time frame of the Neolithic and Bronze Age. This includes borrowings at a proto-language level, as well as those after the initial split of the ancestral languages of ST and TEA. For these ancient loanwords, I will also look for a potential borrowing direction and a date of borrowing. The borrowings under investigation here will cover the vocabulary related to plant cultivation. It would also help our understanding of East Asian prehistory to scrutinise other fields of agricultural vocabulary, such as livestock and pastoral terms. Although beyond the scope of this chapter, this remains a topic of interest in future research.

[1] I follow the result of this research because of its expansive number of languages (131 ST languages) in their database, which includes, for example, Naga languages that lacked representatives in previous studies. At the same time, it is noteworthy that there are a great number of merits regarding Sagart et al. (2019), one of them being their attention to differentiate borrowings from cognates.

[2] This chapter follows the term 'Transeurasian' in accordance with its usage in Robbeets (2005, this volume), while the term 'Altaic' refers to the subgroup including Turkic, Mongolic and Tungusic languages.

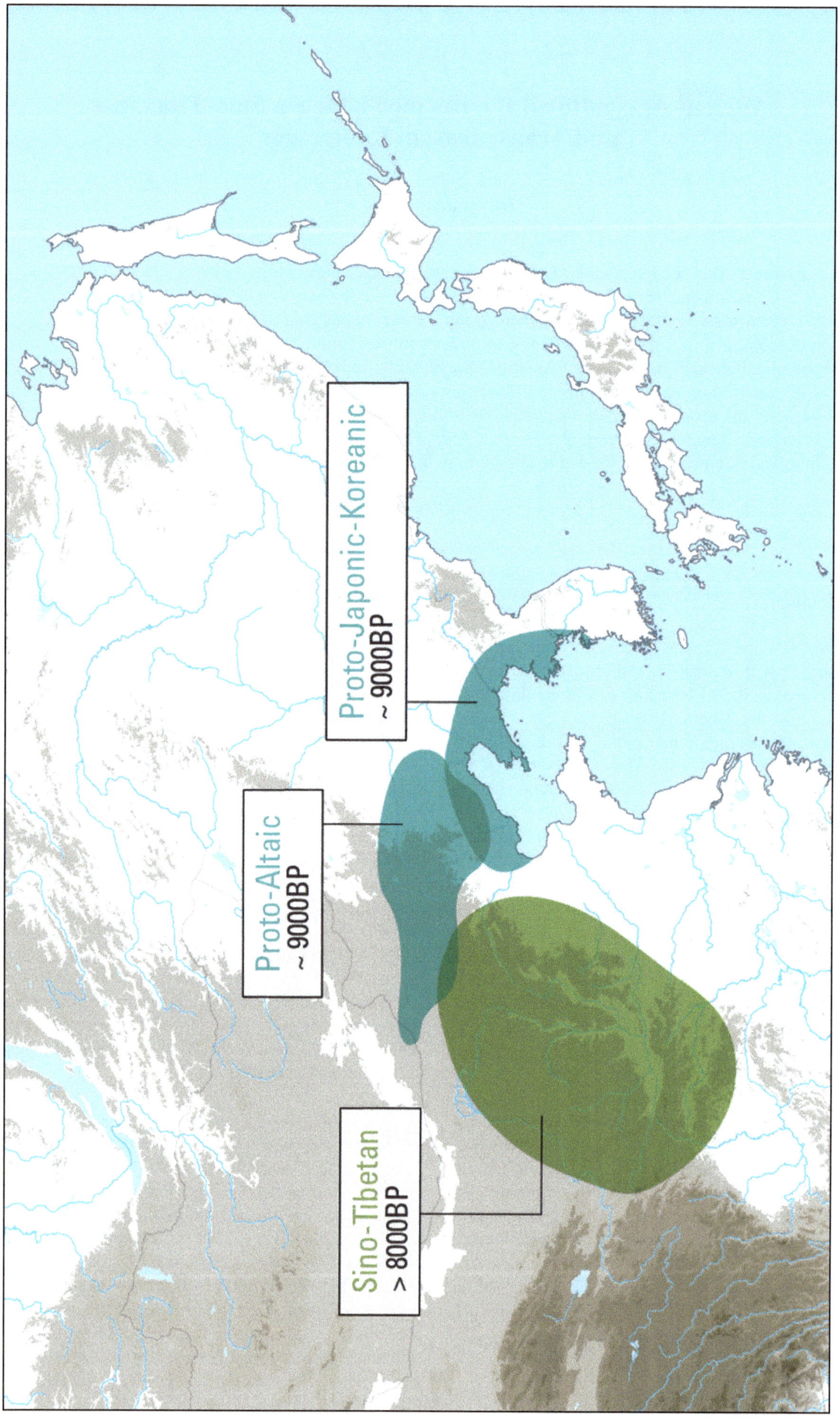

Figure 5.1. The estimated locations of Sino-Tibetan, Proto-Altaic and Proto-Japano-Koreanic at around 8000–9000 years ago. Source: map by M. O'Reilly, MPI-GEA.

In section 5.2, I will list the detailed research question. Section 5.3 will deal with methods I applied during this research and the literature I consulted for the data collection. Section 5.4 includes six possible lexical borrowings and reasons why I propose them to be borrowed. In section 5.5, I discuss the findings and conclude the chapter in section 5.6.

5.2. Research questions

This chapter aims to investigate the following questions:

1. What are the lexical items that are potentially borrowed?
2. What are the borrowing directions and borrowing dates of these lexical items? Is there evidence for language contact between Proto-Sino-Tibetan and Proto-Transeurasian? Is there any evidence of language contact after their initial splits?
3. On the basis of these lexical borrowings, can we infer what types of crops were transmitted across Eurasia in prehistoric times? In which direction did the crops move?
4. Is the linguistic evidence congruent with archaeological findings and historical records? (For instance, the multiple cultivation centres of millet in Northeast Asia.)

5.3. Methods and sources of data

5.3.1. Methods

The borrowings are identified through the rules of elimination. There are four possible explanations to similarities in lexical units in different languages: universal principles of linguistic structure, chance resemblance, inheritance, and borrowing. Universal principles of linguistic structure refer to the common patterns that can be observed cross-linguistically, such as nursery words and onomatopoeia. Chinese *māma* and English *mama* could both be used to refer to one's mother, but this similarity is due to children's phonetic limitation. Therefore, these forms will not be considered in the process of identifying borrowings. Similarities in forms could simply be due to coincidence, which are also referred to as chance resemblance. One example is that Japanese *koya* 'little house, hut' resembles Swedish *koja* 'little makeshift house, often in a tree' both in form and semantics, but the similar forms are only coincidental. This mostly happens between two languages, therefore introducing a third language would eliminate the possibility of similarities caused by coincidence. Lexical similarities can also be the result of genetic relationship between two or more languages; in that case we speak of 'cognates'. For instance, English *school* and German *Schule* share the same origin of proto-Germanic **skulō* 'crowd'. The process of separating inherited forms from borrowings is more complex in comparison to that of universals or chance resemblance. The principles to differentiate cognates and loanwords are based on phonological, morphological and semantic development that could be predicted from comparative historical linguistics. It can be expected that the development among genetically related languages is regular, and forms that cannot be categorised as regularities have the potential to be borrowed from another language. If similar forms cannot be explained by the former three possibilities, then it is highly likely to be a borrowed form.

However, in practice, when dealing with ancient borrowings there are a number of precarious factors concerning the two language families in discussion. For instance, on the Sino-Tibetan side, there is a lack of sufficient data of certain languages that is remote in terms of geographical locations of their linguistic communities, and as a consequence it is difficult to establish a consistent sound correspondence throughout the family and propose reconstructions when forms are missing. Meanwhile for Transeurasian languages, not all scholars agree upon the unity of the languages in the family. It is also difficult to deal with loanwords that could go back to prehistoric times. Therefore, it is of great importance to consult archaeological and genetic evidence to support proposals of such loanwords.

Similar to the protocols we follow to identify loanwords, principles of morphology, phonology and semantics are applied when proposing the date and the direction of the borrowing. For phonology, irregularities that cannot be explained by regular sound correspondences and sound changes within one language family are likely to be a result of borrowing. The date of a certain sound change can also help us propose a date when a lexical item was borrowed. Morphologically, a segmentable morpheme that can be found in the donor language is likely to be not segmentable in the recipient language; by comparison, if the morpheme can be easily separated from the root, then the possibility of it being a loan would be slim. Semantic shifts would help us in both identifying a loanword as well as suggesting a borrowing date. If an item is associated with a meaning that is a secondary semantic development in its counterpart languages, then it is potentially a loanword. To avoid presumptions of borrowing direction and borrowing date and circular reasoning, the most important factors for the identification process are that (1) the proposed loanwords show irregularity synchronically (e.g., do not comply with regular sound correspondences with its own language family) or diachronically (e.g., do not abide by certain sound changes), or (2) there is archaeological or genetic evidence to support a proposal of the borrowing direction and/or date.

5.3.2. Sources of data

Previous research on loanwords between Sino-Tibetan and Transeurasian is usually conducted on a lower taxonomic level on the family tree. The data collection for the following proposed loanwords started with Starostin (2008), in which he listed about two dozen lexical items that could be borrowed from Transeurasian into Sino-Tibetan. Starostin's work is one of the rare studies, if not the only one, that dealt with ancient borrowing at the level of proto-languages and involving macro-linguistic comparison. Another main source of the potential borrowings we consulted is Miyake (1997). He proposed a list of loanwords from Chinese to, in his terms, 'Pre-Sino-Japanese', and 'Pre-Sino-Korean'. These lexical items form a more expansive list of around 100 entries. Similarly, agricultural items were selected and investigated.

Other studies concerning the topic of lexical borrowings between the two language families started as early as Laufer (1916), who compiled a rather large number of loanwords between Transeurasian donor languages and Tibetan recipient languages. Some items on Laufer's list of borrowings were reprised in his 1919 book (Laufer 1919), a work which focuses more on archaeobotany than on linguistics.

Later works on the topic of loans include Menges (1995), a book in which he dedicated a few pages to lexical items that could be loanwords from Chinese into Turkic languages. Yakup (2005) also listed several 'early copies' from Chinese to the Turfan dialect of Uyghur, which is a Turkic language. More recent research from Shimunek (2017) discussed a small number of lexicons that could be borrowed from (Old) Chinese to Serbi-Mongolic.

Apart from the literature mentioned above, we also collected scattered borrowed lexical items discussed in other works; however, the main sources of the data collected for this research were obtained from the studies mentioned above. Amongst these works, except Starostin (2008), there is an especially limited number of studies on loanwords between Sino-Tibetan and Transeurasian at the top taxonomic level as well as the ancient strata from the Neolithic and Bronze Age. There are a number of reasons for the lack of research in this field. One of the most important factors is the issues related to the proto-languages of ST and TEA. The dispute on the unity of Transeurasian languages suggests that not all scholars would agree on a possible common ancestor for the languages and consequently the reconstructability of the proto-language. ST, on the other hand, still lacks a widely accepted reconstruction. This chapter will nevertheless hopefully move forward the discussion on ancient loanwords between the two speech communities.

5.4. Lexical borrowings

5.4.1. MILLET

5.4.1.1. Setaria italica

Linguistically, there is inherited vocabulary related to millet cultivation in both Sino-Tibetan and Transeurasian languages (cf. Sagart et al. 2019: SI 5.3 and 5.4 for Sino-Tibetan; Robbeets et al. 2021: SI 5 items (7) and (8) for Transeurasian). With the Yellow River farmers' focus on *Setaria italica* and the West Liao River population's familiarity with *Panicum miliaceum*, if there was a lexical exchange, it would suggest the introduction of a different species of millet to the speakers. However, this research detected no linguistic evidence of lexical borrowings concerning the names of millets at the early stage of the two language families. There are two potential loanwords detected in Transeurasian languages, on lower taxonomic levels, one for *Setaria italica*, another one for a derived product related to millet.

稷, Old Chinese (OC) **[ts]ək* 'millet (*Setaria italica*)' is the possible donor word for the proto-Korean (pK) form, **coh* 'millet'. Although there are no other attested forms in TEA that could relate pK **coh* to a more ancient protoform, the Chinese form is a reflex of a potential proto-Sino-Tibetan (pST) form, which probably takes the form of **tsək* '*Setaria italica*'. Cognates include Trung *tɕaʔ*[55] 'millet', and Hlopku *cək* '*Setaria italica*', as suggested by Sagart et al. (2019). There are also other forms that could suggest a pST velar or glottal ending: Lisu *tshøʔ* 'millet', Azha *tshɔ22* 'millet', Naxi *tshy55* 'millet'

In comparison, some similar forms in ST have a different final than pST **tsək,* which are listed as follows: Achang *sap55* 'millet', Burmese (Written) *chap* 'millet', Atsi *tsap55* 'millet', Nusu (Central) *tsa53* 'millet'.

These forms suggest a proto-Tibeto-Burman (pTB) *tsap 'millet'; however, it is questionable that they can go back to the same origin as **tsək* '*Setaria italica*'. The mismatch in the final of the two forms is where the question lays for their cognacy. It appears that the *-p final is only detected in non-Sinitic languages, hence suggesting a pTB ancestry other than pST. It is doubtful that these two similar forms are related; before more linguistic data are obtained, we tentatively suggest the existence of two different etyma concerning the *-p and *-k situation.

For borrowings, Starostin (2008) proposes a candidate for a potential donor word from Transeurasian languages, whose protoform is **ǯi̯úgi* for *P.miliaceum*, into ST. For this reconstruction, he lists the following reconstructed forms: proto-Turkic * **jügür-* 'millet', proto-Tungusic * **ǯija-/*ǯije-*(~-g-) 'millet', Proto-Koreanic * **čòh* 'millet' (Starostin et al. 2003:1547-8).

The unlikeliness of Starostin's theory is well explained by Sagart (2011). One problem is the irregular sound correspondences amongst Transeurasian languages. The correspondence is unexpected mainly because of the vowel mismatch, as pTEA *-*iu*- should be reflected as pTk *-*ö*-, pTg *-*ü*-, and pK *-*ə*- (Robbeets et al. 2021: SI 5), none of which could be found in the forms presented by Starostin. This suggests that these forms may not share a common origin and consequently do not have a pTEA ancestry.

However, it could be the case that there was a borrowing on a lower taxonomic level for this lexical item, namely from OC to pK. The OC **[ts]ək* 'millet (*Setaria italica*)' form serves as an ideal model for pK **coh* 'foxtail millet (*S. italica*)'. The borrowing direction was proposed with the support of archaeological findings; it is likely that this lexical item was spread to the Korean Peninsula together with the crop.

As discussed above, there are no existent Transeurasian cognates for the Korean form, suggesting a later either innovation or borrowing after the branches of pTEA split. In comparison to the broomcorn millet native to TEA speakers, it is possible that Korean speakers adopted the ST lexical item for *S. italica* with the introduction of foxtail millet. As the estimated divergence time of pK from pJ is around 5458 BP (Robbeets et al. 2021), the borrowing time of this item is likely to postdate the split. If the pK form truly has its origin in OC, then with the dating of reconstructable OC being 1250 BC-221 BC (Baxter and Sagart 2014:1), it is possible to narrow the borrowing date accordingly. Therefore, the tentative date of borrowing revealed by this research is 3270–2240 BP.

5.4.1.2. (MILLET) GRUEL

A lexical item related to processing cereals is potentially borrowed from ST to TEA, namely the ST word for 'to eat/drink (liquid cereal food)' borrowed into Transeurasian as '(millet) gruel'. In ST, cognates for the verb for 'to drink/eat' are widespread and well established. Several reconstructions for pTB 'eat/drink' have been proposed; for example, Benedict (1972) suggests the pTB form **am* 'eat/drink', and Matisoff (2003) proposes **ʔam 'eat/drink'*. The cognates were listed as follows in Matisoff (2015):

> Dhimal *am* 'drink', Thulung *um-* 'eat', Anong *am, εm*^{55} 'eat', Bwe *a*^{33} 'eat', Kayah *ʔε*^{11} 'eat', Kayan *ʔaŋ*^{11} 'eat', Kayaw *ʔɔ*^{11} 'eat', Pa-O (Southern) *ʔam*^{33} 'eat', Pa-O (Northern) *ʔam*^{55} 'eat', Pwo (Northern) *ʔε̃*^{33} 'eat', Pwo *ʔɔ̃*^{55} 'eat', Pwo (Bassein) *ʔàn* 'eat', Pwo (Moulmein) *ʔánʔ* 'eat', Sgaw (Northern) *ʔɔ*^{31} 'eat', Sgaw (Southern) *ʔɔ*^{45} 'eat'.

Their Chinese counterpart would be 飲 OC **q(r)[u]mʔ* 'drink (v.)'. The regular sound correspondence points to an inherited vocabulary. Sinitic cognates include proto-Min **əm*^{B1} 'rice water, the water of half-cooked rice', and Y-Guangzhou, ^{35}*(j)iɐm*^{B1}, 'rice water; to drink' (Schuessler 2007). Relating this etymon with the rice plant is found in other ST languages. Sagart (2018) listed examples of Bengni (Tani) *am* 'rice plant', Dulong (Nungish) *am*^{55} 'rice (paddy)', and Sak (Sal) *aŋ* 'rice plant'. These forms suggest a pST meaning of 'to eat/drink (rice as food)'. With the lack of evidence of rice being present prior to 4000 BC in the Yellow River area (Stevens and Fuller 2017), it is more likely that the original meaning for this etymon is 'to eat/drink (liquid cereal food)'. Therefore, a pST form is proposed here to be **ʔam*, with the meaning 'to eat/drink (liquid cereal food)'.

The pTEA **amʊ* 'cooked cereal, millet gruel' is likely to have a connection with pST **ʔam* 'to eat/drink (liquid cereal food)'. The pTEA reconstruction was proposed with shared vocabulary in three groups in Transeurasian, Proto-Japonic **amai* 'cereal starch', proto-Koreanic **amʌ* 'cooked cereal', and Proto-Mongolic **amu-n* 'cooked cereals; millet', as well as a borrowed form in Manchu *amsun* 'offerings of wine and food to a deity' from Mongolic (Robbeets et al. 2021). The widespread cognates in both language families suggest an ancient interaction related to cooking techniques between the speech communities.

The borrowing direction is proposed due to the initial present in pST but not in pTEA, as it could be lost as no glottal stop could be found in pTEA. Therefore, we propose an early dating for this borrowing, which would be between 9000-8000 BP.

5.4.2. AGRICULTURAL TOOLS

The word for the cultivation tool spade/pestle is another potential borrowing that could go back to a proto level for both ST and Transeurasian. Miyake (1997) proposes that Koreanic borrowed the word for spade from Chinese, quoting the following forms in his article: MK *salp, stapwu*, MJ *sarapi,* OJ *sapi*$_1$, *sape*$_1$, borrowed either from MC *chrap* or OC *tshrap.

The character for the Chinese form he refers to is 臿 *chā*, which is reconstructed by Baxter and Sagart (2014) as OC **[ts*^{h}*]<r>op* > MC *tsrheap*, meaning 'pestle'. There are several potential cognates for the Chinese form in other Sino-Tibetan languages, including Lahu ts*hʌ*^{54} 'spade' and Pumi *tʃhε̃*^{33}*tsə*^{55} 'spade'. These forms suggest the possibility of a shared ST vocabulary. However, the bisyllabic form could also suggest a later borrowing, probably from Chinese, considering its dominant status in mainland China. We also observed a set of lexical items in TB with some difference in semantics which can possibly be connected to Chinese **[ts*^{h}*]<r>op* 'pestle'. Examples include Bantawa (Southern Kiranti) *sep* 'sharpen (by grinding)', Bahing (Western Kiranti) *sεp-* 'sharpen (a blade)', Hayu (Western Kiranti) *tshop* 'sharpen a blade', and Thulung *sep-* 'sharpen'.

Based on the forms above, it is likely that they descended from the same pST origin, which is likely to resemble the form pST **ts(h)rop* 'spade'. As the TB languages share the meaning with relation to 'spade', it is possible that Chinese developed the meaning 'pestle' after the split. For the semantics of the previously mentioned reconstruction, I propose the meaning 'spade' instead of 'pestle', which is only found in Sinitic. If the donor word for the Japonic and Koreanic forms was truly found in Chinese as Miyake proposed, it appears that there is a less satisfactory semantic match.

It is more probable that this is a borrowing between pST and pTEA. The pTEA form is proposed as **sarpa* 'spade' by Robbeets et al. (2021). This form is inherited in pK **salp(V)* 'spadė' and proto-Turkic (pT) **sarpan* 'wooden plough (breast), spade'. OJ *sapi*$_1$ 'spade, hoe' lacks parallel attestation in Ryukyuan languages, suggesting the form cannot go back to a pJ origin and is likely to be a borrowing. However, by contrast to Miyake's proposal, it is a more probable scenario that OJ *sapi*$_1$ 'spade, hoe' is a loan from Koreanic, as proposed by Robbeets et al. (2021: SI 5), specifically from Paekche Old Korean **sap* 'spade' with its loss of liquid from the pK **salp(V)* 'spade'.

It is difficult to determine the borrowing direction for this item, one of the reasons being the similar agricultural tools shared by the ancestors of ST and TEA communities in the Neolithic, as shown by archaeological evidence (Stevens et al. 2020). Another factor is the vast time-depth in this case which is possibly around 8000–9000 years BP. With the absence of a plausible proposal for the borrowing direction of this lexical item, one can nonetheless infer that there was interaction between ST and TEA speech communities in the Neolithic, and that this interaction involved the introduction of a novel farming tool.

5.4.3. RICE/GRAINS

Rice is another important subsistence crop across Eurasia and the world in general. Two species are found in Asia, *Oryza indica* and *Oryza japonica*. The former was lowland rice grown throughout tropical Asia, while the latter was grown in temperate East Asia, upland areas of Southeast Asia, and high elevations in South Asia (Sagart 2008). The origin of East Asian cultivation of rice is attributed to southern China, particularly three areas: the lower Yangtze and lower Huaihe River, the lower Hanshui and upper Huai River Basin, and the Middle Yangtze-Lishui valleys (Stevens and Fuller 2017). There is no native vocabulary for rice in Transeurasian (Robbeets et al. 2021) or Sino-Tibetan (Sagart et al. 2019), which means that the rice vocabulary in both families is either a later innovation or loans from other sources.

5.4.3.1. RICE 1

Miyake (1997) proposed two distinct forms related to this crop in Sinitic which could be probable donor words into Transeurasian languages. The first one is 米 OC *(C.) *mˤ[e]j?* 'millet or rice grains, dehusked and polished.' The semantics of the word in Chinese appear to place emphasis on the form of cereals, regardless of whether it is millet or rice, giving the primary meaning 'grains'. Sagart (1999) proposes that the meaning related to 'rice' is secondary. Compared to other forms in ST, there was indeed a possible semantic narrowing process for this lexical item. Sagart (2018) included a few examples with the same item referring to 'millet': Garo (Burling) *mi-si-mi* 'millet' and Gyarong (Ma'erkang) *smər khri* 'millet'. These inherited forms that share the same origin as in Chinese point to the original meaning of 'grains'. Coblin (1986) proposed a similar semantic reconstruction for the item, which would be **miod* 'rice; grain sprouts'. Matisoff (2003) reconstructed the proto-Tibeto-Burman (pTB)[3] form for this item to be **may~mey*, 'rice/paddy'. This suggests that the meaning connected to 'rice' or 'rice plant' could go back to a pTB origin, with the possibility of being related to a pST form for 'grains'.

Considering internal phonological correspondences, the pST form is likely to resemble the form *[C.]mej, 'grains'. The rationale for this reconstructed form is that OC mostly preserved the vowel *e after diverging from pST, which might be changed in other languages in different phonological environments, whereas a *-y* final in TB could reflect an inherited *-j. It is unclear if there is an initial or a pre-initial (following the term coined by Baxter and Sagart) in the pST form, since it is absent in TB forms but is present in Baxter and Sagart's OC reconstruction as *C. The cognates in Sinitic languages could validate a pre-initial in OC. Early loanwords in other languages could also confirm the presence of a pre-initial. Vietnamese *mẻ* 'fermented rice', if borrowed, has a high register tone that could imply a nasal and final glottal stop, further meaning that there is a possibility for OC **hm-* (Miyake 1997). Proto-Mien, **hmei* B, 'rice, dehusked', has a voiceless resonant initial, which is a remnant of Chinese pre-initial (Baxter and Sagart 2014). It is safe to say that the pre-initial existed in the OC form, even though we cannot say what the consonant is exactly. For pST, it is possible that the Sinitic preserved the *C initial, when other ST languages lost it in progress, or evolved into a tonal variation. Comparing the Gyarong form *smər khri,* it is likely that there is a pre-initial going back to pST, like in Old Chinese. However, without more data we cannot decide what the consonant is exactly. Therefore, the tentative reconstruction for this etymon in pST is proposed here as **[C.]mej.* Here, following the Baxter-Sagart tradition, the *[C] refers to a consonant (pre-)initial that is phonologically possible for pST but unknown yet; it does not necessarily coincide with the pre-initials in OC or the Gyarong form.

For 米 **(C.)mˤ[e]j?*, a potential loanword in Old Japanese, *ko*$_2$*me*$_2$ is suggested by Miyake (1997). This could go back to proto-Japonic (pJ) **kəmai*, reflected by Old Shuri *kumi*, Yamatoma *khumï* 'rice', Asama (Tokunoshima) *kumɨɨ* 'rice', Mugiya-higashi (Yoron Island) *humi/mai*, 'rice,' Yonamine (Nakijin-son, Northern Okinawa) *humii* 'rice', Sesoko *kumi* 'rice' (Robbeets et al. 2021: SI 23). Miyake's proposal was followed with discussion by Sagart (2011), confirming that the word does not appear to be native Japanese because 'it would violate Arisaka's laws on cooccurrence of vowels,' and 'one would expect to find komo- alternating with *kome* in modern compound words'.

For the possible donor word, Miyake referred to an Old Chinese form **(h)meq*; he did not specify the source of this particular reconstruction. He further discusses the possibility of proto-Japono-Koreanic using **hm-* or **hVm-* to resemble the sound in Chinese, even though he admits that the fact that there are no cognates in Koreanic was brought to his attention. It is likely that the lexical item entered only the Japonic group in Transeurasian, suggesting that the borrowing date was late in history, possibly after the split of individual branches of TEA in the Bronze Age.

[3] PTB started to diverge about 6000 BP (Zhang et al. 2020) with its homeland approximately around present-day Gansu Province, China.

There are other proposals of potential donor words for Japanese *kome*. Vovin (1998) proposed an Austroasiatic origin of Japanese *kome*, quoting Nobuhiro Matsumoto on Vietnamese *mẻ* 'fermented rice'. The Vietnamese form is also likely to be a loanword from Sinitic. Sagart (2011) proposed that *kome* was borrowed from proto-Austronesian and possibly the proto language of Austronesian and Sino-Tibetan provided more genealogical evidence which could support his STAN hypothesis that Sino-Tibetan and Austronesian languages are genetically related, hence share a common ancestor. Related to this suggestion, Robbeets (2017b) proposes 糜 OC *C.m(r)aj 'rice gruel; destroy, crush' to be the donor word for para-Austronesian[4] (pAN) *hemay* 'cooked rice', which entered pJ and was borrowed as pJ **kəmai* 'dehusked rice'. The para-Austronesian form serves as a plausible model for the pJ form; if the para-Austronesian speakers were indeed in contact when Japonic speakers resided in the Liaodong-Shandong region (ca. 3000—2000 BC), then this hypothesis will be a possible scenario. However, OC *(C.)mˁ[e]j? 'millet or rice grains, dehusked and polished' remains a better candidate as the ultimate donor word in this borrowing chain, due to the probable existence of medial *-r- in OC *C.m(r)aj 'rice gruel; destroy, crush', whose omission in pAN or pJ requires further explanation.

Archaeological information suggests the ST/OC >> Japonic route of the lexical item is likely to be the case. In Japanese prehistory, the Jōmon and Yayoi cultures are differentiated by the introduction of cereal agriculture, including paddy rice (Nasu and Momohara 2016; Robbeets et al. 2021). The former dates to ca. 14,000—1000 BC and the latter 1000 BC—300 AD. Recent study on the spread of rice supports the earliest cultivation of rice in Japan by about 3000 years ago, driven by economic and political developments in China (Leipe et al. 2021). A possible earlier date can be proposed if more archaeological evidence is collected. Rice agriculture appeared in the Korean peninsula from the late second millennium BC as a result of contact with the Liaodong Bronze Age culture (Ahn 2010). Korea acquired rice from China, then the crop spread to Japan through the peninsula, hence rice cultivation in Korea is around 500 years earlier than that of Japan. There is no linguistic evidence for probable loans for rice vocabulary in Koreanic from Sinitic, leaving Japonic the only TEA group with a possible borrowed lexical item.

5.4.3.2. RICE 2

Another potential donor word in ST for Transeurasian rice vocabulary proposed by Miyake (1997) is 秈 OC **sa[n]*, as reconstructed by Schuessler (2007). Semantically, Sagart (2011) mentioned that the word denotes *indica* rice in modern standard Chinese; however, it might refer to a different type of rice in OC. It is recorded in a dictionary from the Hàn dynasty (202 BC – 9 AD) that this character is used to refer to rice by the people south of the Yangtze River. In 農政全書 *Nóng Zhèng Quán Shū*, 'A Complete Guide to Agriculture', written by 徐光啓 Xú Guāngqǐ (1562–1633), the author addresses the details of the cultivation of this type of rice:

> 南方水稻，其名不一。大槩為類有三：早熟而緊細者，曰秈[…]。
>
> Nánfāng shuǐdào, qí míng bùyī. Dàgài wéi lèi yǒu sān: zǎoshū ér jǐnxì zhě, yuē xiān [...].
>
> 'The rice cultivated in the south [of China] has different names. Approximately, there are three types: the type that matures early and looks firm and fine is called *Xiān.*'

The area of 'southern China' in the quote presumably refers to the south of the Yangtze River. It also appears that *xiān* is assigned for the *indica* type, which would match the southern domestication centre of *indica* rice.

For 秈 * *sa[n],* Miyake (1997) points out that the Old Japanese form *sine* might be reanalysed as the second half of an earlier compound **X-tu-ine*, as the initial s- may be secondary via the genitive marker *-*tu-*. Reconstructions for the Japanese form were attempted by scholars. Vovin (1998) proposed proto-Japonic **zina-Ci* or **hina-Ci*, following Unger's suggestion. Whitman (2011) reconstructed **jinaC*, arguing that there is weak evidence for voiced obstruent series in proto-Japanese. They uniformly glossed the item as 'rice plant', in alignment with the Chinese *xiān* for some type of rice. The main argument against it being a loanword in Japanese is that it is strange for Japanese to imitate the Chinese [a] with an [i], when [a] is also available in pJ, and the alternation from an open vowel to a close vowel would be a rare phenomenon.

Regarding the Chinese word, it could also ultimately be a borrowing, as suggested by Sagart (2011). The form *sa[n] would match the proto-Tai form **sa:l A*, suggesting that the irrigated rice field in Tai-Kadai is of Austronesian origin and dry field is Austroasiatic (Sagart 2011). Vovin (1998) proposed that Japanese *sine* is of Austroasiatic etymology. It could be that both Chinese *sa[n] and Japanese *sine* are loanwords, but regarding the Japanese form, it is unlikely that this could be a Chinese or ST loan when the linguistic forms are compared.

The investigation on J *kome* and J *sine* presented here suggests that it would appear that OC *(C.)mˁ[e]j? could be the donor words for pJ **kəmai*, and possibly an Austronesian form of *hemay* could be involved in the comparison. However, the Japanese form *sine* might be of another origin than Chinese or ST, which would be beyond the discussion scope of this article.

5.4.4. WHEAT/BARLEY

For the vocabulary related to wheat or barley, Starostin (2008) proposed the Chinese form, 麥 OC *m-rˁək 'a kind of wheat', to be a loanword from Transeurasian

[4] A sister language of proto-Austronesian.

languages, whose proto-form was reconstructed by him as **mi̯urgu* 'wheat'. The late introduction of barley and wheat to eastern China around 4500 to 3500 BP suggests that the lexical items for the crops cannot go back to either pST or pTEA. In this case, the lexical loan is not likely to be at a proto level, as Starostin proposed, and the direction of borrowing appears to be from Sinitic to Transeurasian.

The Chinese written characters for the names of wheat and barley are attested in 詩經 *Shījīng* (a poetry work dating from the 11th to 7th centuries BC). Barley is referred to as 牟 *móu*, OC *mu 'barley' (reconstruction suggested by Zhengzhang 2003), as a part of a compound with 來 OC *mə.rˤək 'a kind of wheat'. 來 and 牟 both served as the phonetic element in the later, more widely used characters 麥 *mài* and barley 麰 *móu*, which would further correlate the written characters to the crops. This attestation suggests that co-cultivation of wheat and barley could go back to the 11th to 7th centuries BC.

Words for wheat or barley in non-Sinitic languages in ST appear to have no relation to the Sinitic forms mentioned above, but there is a possibility of a common ancestral form for TB. Written Tibetan refers to wheat as གྲོ *gro*, which is also different from its counterparts in other TB languages. The distribution of different etymons for wheat/barley in ST suggests a later introduction of the crops after the early splits of the language family. Hill (2019) proposed that Tibetan *gro* 'wheat' is a reflex of the proto-Tibetan form *ɢroy 'wheat'. Another similar form in ST is found in Newari, a Tibeto-Burman language spoken in Nepal, the form is *cho* 'wheat', going back to proto-Newari *chro, after the language lost its consonant clusters (Tamot 2002). These forms could possibly go back to a common origin, meaning there is a pTB form for wheat.

A similar Chinese form could be compared to the TB forms; that would be 禾 OC *[ɢ]ˤoj 'growing grain'. The correspondence between the Chinese form and the Tibeto-Burman forms appears to be regular, but being an introduced plant, it is unlikely for ST languages to share a common ancestral form. There is a possibility that the Chinese form, *[ɢ]ˤoj 'growing grain', is only a look-alike, or that there was a semantic shift in the TB branch after the initial split. This would need further research to determine the nature of their relation.

Going back to Chinese *m-rˤək 'wheat', Schuessler (2007) draws the connection between *m-rˤək and WT *bra-bo* 'buckwheat', in which the *-bo* component is a nominal suffix. This is an unlikely connection and one of the reasons is the mismatch in their finals: it is strange to have Tibetan zero matching Chinese *-k final. The semantic match is also less satisfactory as buckwheat is native to temperate East Asia and not a cereal.

Other suggested etymologies for Chinese *m-rˤək include Mair's (1996) proposal that it is a borrowing from Indo-European **melhχ* 'to grind'. This appears to be a less likely scenario, with the semantics being a dubious match. However, it was indeed through IE speakers that ST populations acquired wheat and barley, therefore it is possible for the vocabulary for wheat and barley in ST to be under Indo-European (IE) influence. Another candidate for the source of the borrowing is proposed by Blažek (2019), namely proto-Indo-European **mr̥k*, with the gloss 'seed of barley, products derived from barley'. Blažek reconstructed the form based on proto-Celtic **mraki-* 'corn or seed of barley', Luvian *marwali-* 'barley-stem', Hittite *marnuwa(nt)-* 'a kind of beer' and Old Indic *markaṭaka-* 'a kind of corn'. However, the form **mr̥k* itself is irregular for Indo-European. As proposed by Rasmus Bjørn, the proto-form that entered OC is likely to be **mr̥(n)k*, possibly going back to a language originating in southwest Asia (*per litteras* 5 January 2022; see also Bjørn, this volume). This suggests a probability that there is a borrowing chain for the lexicon to enter Sinitic languages through IE speakers, then into Tungusic and then into Japonic and Koreanic after their split.

Archaeological evidence can offer us more information about the dispersal of wheat and barley in Eurasia. Wheat and barley were domesticated in the Fertile Crescent as winter crops, and they were introduced to Sinitic speaking people through the Tibetan Plateau and Hexi corridor, as these plants were found in these regions around 2000 BC. From there, wheat was transported to central and eastern China around 1500 BC but barley is not recorded until 900 BC (Liu et al. 2017). It was through Shandong that the Korean Peninsula acquired wheat and barley. Wheat was grown in Korea approximately by at least 2300 BP (Crawford 1992), and it is likely that wheat was transmitted to the Japanese Archipelago through Korea (Robbeets et al. 2021).

The transmission route of wheat and barley also eliminates the possibility that they were known to ancestors of either Sino-Tibetan speakers or Transeurasian; there is also a lack of consistent shared vocabulary for the plants in ST. As archaeological evidence suggests, the crops went through the TB speaking areas first, and it was through these regions that Sinitic people acquired wheat and barley. If there were loanwords (instead of purely innovation) for the crops in ST, it is likely for the donor words to enter TB first; then through TB, the lexicon entered Sinitic.

As the potential donor word for wheat in ST is proposed to be a proto form that looks like **mr̥(n)k*, it is possible for the lexical item to have entered Tocharian according to its route suggested by archaeological findings. However, there is no Tocharian form attested in this case, and the expected pTB form—my expectation would be *mrə(k)—is also absent. OC m-rˤək 'wheat' entered proto-Tungusic (pTg) as **mirgi* 'barley'. Then, following the spread of barley, the pTg form was borrowed separately as pK **milk* 'wheat', and pJ **munki* 'wheat'.

5.4.5. CULTIVATED FIELD

The Chinese word for agriculture is a potential donor word for 'cultivated field' in Koreanic and Japonic, and possibly their ancestor pJK. Chinese 農 *nóng*, OC *nˁ[o]ŋ could refer to either the industry of agriculture but also the action of farming. A later derivation also related this item to the people who engage in the action of farming, hence farmers. It lacks an established set of cognates in ST, with most of the similar forms being possible borrowings from Chinese. Similar forms in other ST languages include: Achang *nuŋ*^{31}ɳε^{31}/ *nɔŋ*^{31}ɳε^{31}'agriculture', Nusu *nɔ̃*^{31}*nie*^{31} 'agriculture', Hani (Gelanghe) *nuŋ*^{31}*jε*^{31} 'agriculture', Jinuo *noŋ*^{42}ɳə^{42} 'agriculture', Lisu (Northern) *nɔ*^{21}*je*^{21} 'agriculture', Gurung (Chachok) *nɔ̃* 'farm labourer', Dulong (Nujiang) *luŋ*^{31}ɳ*e*^{31} 'agriculture', Jingpo *nuŋ*^{31}*je*^{33} 'agriculture', Tujia (Northern) *lũ*^{21}*ne*^{21}'agriculture', Tujia (Southern) *nɔ̃*^{13}*ne*^{33}'agriculture', and Bai (Dali) *nu iε* 'agriculture'. The disyllabic construction resembles the modern Chinese form of 农业/農業 *nóng yè*, literally 'agricultural industry', implying a more probable situation of being loanwords from Chinese instead of actual cognates.

The similar forms in Japanese and Korean are pJ **no* 'field' and pK **non* 'irrigated field, rice paddy field'. With its semantic relation to rice, it is probable that this lexical item was borrowed from Sinitic to Japonic and Koreanic, as suggested by the dispersal of rice mentioned above. The other meaning 'irrigated field' suggests the introduction of a new cultivation technology, which could explain the motivation behind borrowing this item. Robbeets et al. (2021) proposed the pJK form to be **non*, based on the attested forms in Japonic and Koreanic languages. It is more likely for the ancestor language of Japonic and Koreanic to borrow the form from Old Chinese instead of Middle Chinese, because the component *ow* in the MC form *nowng* would leave a trace in the loans, producing loaned forms similar to *nou*, compared to later borrowings such as 濃 *nóng* MC *mrjowng* 'thick (esp. as a liquid); strong (sc drink)' >> Sino-Japanese 濃 *nou*, Kan-on *jou*, Go-on *nyuu* (Mair 2016). However, this is not found in Japonic or Koreanic forms, therefore we would expect a direct borrowing from Old Chinese, prior to the pJK split.

5.5. Discussion

The lexical items discussed in this chapter are listed in Table 5.1. There are two items that are possible borrowings on a proto-language level, namely SPADE and (MILLET) GRUEL. This is suggestive of the existence of interaction among the ancestors of these language communities which led to lexical exchange regarding agricultural vocabulary; the prehistoric interaction between the two groups of speakers can be better understood with more domains under investigation. However, the agricultural items borrowed between the proto-languages are few, which can be explained by the immense time-depth in question, so there might not be a large agricultural vocabulary for either Transeurasian or Sino-Tibetan. The two borrowing items on a proto-language level suggest an introduction of a new agricultural tool (SPADE) and possibly the introduction of a new cooking and food-processing method (GRUEL). This finding suggests that the speech communities had no need to borrow a large amount of agricultural lexical items and were farmers with sufficient vocabulary for cultivation essentials.

There is also a potential borrowing between OC and pJK, which is CULTIVATED FIELD. This could suggest the introduction of irrigation for field cultivation. On lower taxonomic levels, more intense lexical borrowings happened among Sinitic speakers and pJK (pJ and pK) speakers during the Bronze Age. The lexical borrowings amongst these speakers involve MILLET (*S. italica*), WHEAT/BARLEY, and RICE. These borrowings suggest either a later incorporation of new crops into their agricultural system or new vocabulary replacing the previously existing words.

Table 5.1. Agricultural lexical borrowings from ST to TEA discussed in this chapter in the order of proposed borrowing date.

Item	Binomial	ST form(s)	Transeurasian form	Borrowing direction	Time of borrowing
(millet) Gruel	N/A	pST **ʔam* 'to eat/drink (liquid cereal food)'	pTEA **amʊ* '(millet) gruel)	pST >> pTranseurasian	9000—8000 BP
Spade	N/A	pST **ts(h)rop* 'spade'	pTEA **sarpa* 'spade'	?	9000—8000 BP
Cultivated field	N/A	OC **nˁ[o]ŋ* 'agriculture'	pJK **non* 'cultivated field'	OC >> pJK	8000—5458 BP
Rice	*Oryza japonica*	(pST **[C.]mej* 'grains' >) OC *(C.)mˁ[e]jʔ 'millet or rice grains, dehusked and polished'	J *kome* 'rice'	(pST ? >) OC >> pJ	5458 BP—-2200 BP
Millet	*Setaria italica*	OC **[ts]ək* 'millet (*Setaria italica*)	pK *coh* 'millet (*S. italica*)	OC >> pK	3270—978 BP
Wheat/barley	*Triticum aestivum/ Hordeum vulgare*	OC **m-rˁək* 'wheat'	pK **milk* 'wheat', pJ **munki* 'wheat', pTg **mirgi* 'barley (*Hordeum vulgare*)	(pTB? >>) OC >> pTg >> K /pTG >> J	2600—978 BP (pK), 2600—2200 BP (pJ)

The findings presented here are consistent with the archaeobotanic findings in previous studies. The independent millet vocabulary in ST and TEA is consistent with the multiple domestication centres of millet. The Xinglongwa culture, associated with the TEA speech community, is linked with predominately broomcorn millet, dating from 6200—5400 BC; foxtail millet was found in archaeological sites in the Yellow River Valley, dating 6500—6000 BC (Stevens and Fuller 2017). The dates and millet cultivation centres are in line with the lack of lexical borrowings between the two families at the proto-language stage. The lack of inherited rice vocabulary suggests that the ancestors of ST or TEA did not cultivate rice as a native crop, which is supported by the southern China domestication centres. The co-cultivation of rice and millet started around 4000—3500 BC, with millet farmers adopting rice supported by findings in Yangshao sites. Rice dispersed to the Korean peninsula around 1500 BC and to Japan after 700—600 BC. The later borrowings for rice vocabulary from OC to Japonic also followed this timeline, as the lexical item for 'rice' only entered Japonic after its break-up from Koreanic around 5458 BP. Wheat and barley were brought to China around 2000 BC through Central Asia; archaeological sites in the Southern Primorye suggest the arrival of barley is associated with the Krounovska culture around 2600—1800 BP. The borrowed item in Chinese could have an origin from TB languages, then moved into pTg, and eventually entered Koreanic and Japonic separately.

5.6. Conclusion

This chapter investigated six lexical items that were possibly borrowed from ST to TEA. There are rare cases of lexical borrowings between the two proto-language communities, but more borrowed items between Sinitic and separately into Japonic and Koreanic. This finding provides linguistic support of prehistoric interaction between the two speech communities, updating previous research on probable borrowing dates and directions proposed by Starostin (2008) and Miyake (1997). The potential borrowings discussed in this chapter also contribute to the reconstruction of prehistoric Northeast Asia, and the possible reasons behind people's interaction. Combined with archaeological and genetic information, the linguistic evidence presented here has the potential to help understand the nature of the interaction between speech communities. With this in mind, investigating ancient lexical loans covering more semantic domains will illustrate a more complete picture of Northeast Asia in prehistoric times.

References

Ahn, S-M. (2010) The emergence of rice agriculture in Korea: Archaeobotanical perspectives. *Archaeological and Anthropological Sciences* 2(2): 89–98.

Baxter, W.H. and L. Sagart (2014) *Old Chinese: A New Reconstruction*. Oxford University Press.

Benedict, P.K. (1972) *Sino-Tibetan: A Conspectus*. Cambridge University Press.

Blažek, V., M. Schwarz and O. Srba (2019) *Altaic Languages: History of Research, Survey, Classification and a Sketch of Comparative Grammar*. Brno: Masaryk University Press.

Coblin, Weldon S. (1986) *A Sinologist's Handlist of Sino-Tibetan Lexical Comparisons*. Routledge.

Doerfer, G. (1988) *Grundwort und Sprachmischung: Eine Untersuchung an Hand von Körperteilbezeichnungen*. Franz Steiner: Wiesbaden.

Hill, N. W. (2019) *The Historical Phonology of Tibetan, Burman and Chinese*. Cambridge University Press.

Laufer, B. (1916) Loan-Words in Tibetan. *T'oung Pao* 17(4/5): 403–552.

Laufer, B. (1919) *Sino-Iranica, Chinese contributions to the history of civilization in ancient Iran*.

Leipe, Christian, E. Endo, S. Kuramochi, M. Wagner and P.E. Tarasov (2021) Crop cultivation of Middle Yayoi culture communities (fourth century BCE-first century CE) in the Kanto region, eastern Japan, inferred from a radiocarbon-dated archaeobotanical record. *Vegetation History & Archaeobotany* 30: 409-421.

Li, T., C. Ning, I. Zhushchikhovskaya, M. Hudson and M. Robbeets (2020) Millet agriculture dispersed from Northeast China to the Russian Far East: Integrating archaeology, genetics, and linguistics. *Archaeological Research in Asia* 22: 100177.

Liu, X., D.L. Lister, Z. Zhao, C.A. Petrie, X. Zeng, P.J. Jones, R.A. Staff, A.K. Pokharia, J. Bates, R.N. Singh, S.A. Weber, G. Motuzaite Matuzeviciute, G. Dong, H. Li, H. Lü, H. Jiang, J. Wang, J. Ma, D. Tian ... Jones, M.K. (2017) Journey to the east: Diverse routes and variable flowering times for wheat and barley en route to prehistoric China. PLOS ONE 12: e0187405.

Mair, V.H. (1996) Language and script: biology, archaeology, and (pre)history. International Review of Chinese Linguistics 1: 31–41.

Mair, V.H. (2016) ABC Dictionary of Sino-Japanese Readings. University of Hawai'i Press.

Matisoff, J.A. (2003) *Handbook of Proto-Tibeto-Burman: System and Philosophy of Sino-Tibetan Reconstruction*. University of California Press.

Matisoff, J.A. (2015) *The Sino-Tibetan Etymological Dictionary and Thesaurus Project (STEDT)*. University of California, Berkeley.

Menges, K. H. (1995) *The Turkic Languages and Peoples: An Introduction to Turkic Studies* (2nd, rev. ed.). Harrassowitz.

Miyake, M. (1997) Pre-Sino-Korean and Pre-Sino-Japanese: Reexamining an old problem from a modern perspective. *Japanese/Korean Linguistics* 6: 179–211.

Nasu, H. and A. Momohara (2016) The beginnings of rice and millet agriculture in prehistoric Japan. *Quaternary International* 397: 504–512.

Norman, J. (1988) *Chinese*. Cambridge University Press.

Robbeets, M. (2005) *Is Japanese Related to Korean, Tungusic, Mongolic and Turkic?* Harrassowitz.

Robbeets, M. (2015) *Diachrony of Verb Morphology: Japanese and the Transeurasian Languages*. Mouton de Gruyter.

Robbeets, M. (2017a) The Transeurasian languages. In R. Hickey (ed), The Cambridge Handbook of Areal Linguistics, pp. 586–626. Cambridge University Press.

Robbeets, M. (2017b) Austronesian influence and Transeurasian ancestry in Japanese: A case of farming/language dispersal. *Language Dynamics and Change* 7(2): 210–251.

Robbeets, M., R. Bouckaert, M. Conte et al. (2021) Triangulation supports agricultural spread of the Transeurasian languages. *Nature* 599: 616–621.

Sagart, L. (1999) *The Roots of Old Chinese*. Amsterdam: John Benjamins.

Sagart, L. (2008) The expansion of Setaria farmers in East Asia: a linguistic and archaeological model. In *Past Human Migrations in East Asia*, pp. 165–190. Routledge.

Sagart, L. (2011) How many independent rice vocabularies in Asia? *Rice* 4: 121–133.

Sagart, L. (2018) The names of the rice plant. II 'Tibeto-Burman' (continued). Sino-Tibetan-Austronesian. Online source: https://stan.hypotheses.org/248

Sagart, L., G. Jacques, Y. Lai, R.J. Ryder, V. Thouzeau, S.J. Greenhill and J-M. List (2019) Dated language phylogenies shed light on the ancestry of Sino-Tibetan. *Proceedings of the National Academy of Sciences USA* 116 (21): 10317–10322.

Schuessler, A. (2007) *ABC Etymological Dictionary of Old Chinese*. University of Hawai'i Press.

Sergusheva, E. A. and Y.E. Vostretsov (2009) The advance of agriculture in the coastal zone of East Asia. In A. S. Fairbairn and E. Weiss (eds.), *From Foragers to Farmers*, pp. 205–219. Oxford: Oxbow Books.

Shimunek, A. (2017) An Old Chinese word for 'Silk' in Mongolic, Turkic, Tungusic, and Persian. *Eurasian Studies* 15(1): 142–151.

Starostin, S.A., A.V. Dybo and O.A. Mudrak (2003) *Etymological Dictionary of the Altaic Languages*. Leiden: Brill.

Starostin, S. (2008) Altaic loans in old Chinese. In *Past Human Migrations in East Asia: Matching Archaeology, Linguistics and Genetics*, pp. 254–262. Routledge.

Stevens, C. J. and D.Q. Fuller (2017) The spread of agriculture in eastern Asia: archaeological bases for hypothetical farmer/language dispersals. *Language Dynamics and Change* 7: 152–186.

Stevens, C.J., G. Shelach-Lavi, H. Zhang, M. Teng and D.Q. Fuller (2021) A model for the domestication of *Panicum miliaceum* (common, proso or broomcorn millet) in China. *Vegetation History and Archaeobotany* 30: 21–33.

Tamot, K. (2002) Some characteristics of the Tibeto-Burman stock of Early Classical Newari. In *Medieval Tibeto-Burman Languages*, pp. 13–26). Brill.

Thomason, S. G., & Kaufman, T. (1988) *Language Contact, Creolization, and Genetic Linguistics*. University of California Press.

VanBik, K. (2009) *Proto-Kuki-Chin: A Reconstructed Ancestor of the Kuki-Chin Languages*. Sino-Tibetan Etymological Dictionary and Thesaurus Project, Dept. of Linguistics Research Unit, University of California, Berkeley.

Vovin, A. (1998) Japanese rice agriculture terminology and linguistic affiliation of Yayoi culture. In R. Blench and M. Spriggs (eds.), *Archaeology and Language II: Archaeological Data and Linguistic Hypotheses*, pp. 366–378. Routledge.

Vovin, A. (2005) The end of the Altaic controversy: In memory of Gerhard Doerfer. *Central Asiatic Journal* 49: 71–132.

Whitman, J. (2011) Northeast Asian linguistic ecology and the advent of rice agriculture in Korea and Japan. *Rice* 4: 149–158.

Yakup, A. (2005) *The Turfan Dialect of Uyghur*. Harrassowitz.

Zhang, H., T. Ji, M. Pagel and R. Mace (2020) Dated phylogeny suggests early Neolithic origin of Sino-Tibetan languages. *Scientific Reports* 10: e20792.

Zhang, J., H. Lu, W. Gu, N. Wu, K. Zhou, Y. Hu, Y. Xin and C. Wang (2012) Early mixed farming of millet and rice 7800 years ago in the middle Yellow River region, China. *PLoS ONE* 7: e52146.

Zhang, M., S. Yan, W. Pan and L. Jin (2019) Phylogenetic evidence for Sino-Tibetan origin in northern China in the Late Neolithic. *Nature* 569(7754): 112–115.

Zhengzhang, S. (2003) 上古音系 Shanggu Yinxi [Phonological System of Old Chinese]. Shanghai: Shanghai Jiaoyu Chubanshe [Shanghai Education Press].

6

Amuric-Tungusic language contact and the Amuric homeland

Martijn G.T.M. Knapen

Abstract: Amuric (or Nivkh) is a shallow language family spoken on the Lower Amur and Sakhalin and with no apparent affiliation to other language families. As a consequence, reconstructing its more distant past has so far proven difficult. Conversely, extensive and prolonged borrowing between the Amuric and Tungusic languages has long been established. This chapter provides an overview of Amuric and Tungusic historical phonology and shows on the basis of Amuric-internal sound changes that the vocabulary shared between the two language families results from several periods of contact, with some words dating back to Pre-Proto-Amuric and Proto-Tungusic. The geographically specific terms among the material suggest that the territory of the Amuric ancestor language was spoken in a coastal region and that its territory bordered the Tungusic ultimate homeland. While the chapter advances the study of Amuric and the linguistic history of Northeast Asia more broadly, it acknowledges that its conclusions require additional support.

Keywords: Amuric, Nivkh, Tungusic, Manchu, Paleo-Siberian, language contact, loanwords, Okhotsk culture

6.1. Introduction[1]

At the mouth of the Amur River, the Asian mainland is separated from Sakhalin Island by the Strait of Tartary. The lower course of the Amur and much of Sakhalin form the lands historically and presently inhabited by the Nivkh people, one of the indigenous peoples of the Russian Far East. Their language (or languages, as discussed below) has gradually decreased in number of speakers due to the socioeconomic upheavals of the last century. Nivkh is currently classified as moribund, although efforts have been set up in recent years to reverse this trend (Gruzdeva 2015; Gruzdeva and Janhunen 2018).

The language of the Nivkh people is most often described as a single language, simply referred to as Nivkh or, historically and nowadays sporadically, as Gilyak. In this interpretation, Nivkh is made up of two major dialect groups. The differences between these two are, however, sufficiently great to merit analysing them as two branches of a language family, bearing the name 'Amuric' (Janhunen 1996: 73–74). The two branches are referred to as *Nivkh* (proper), spoken on the mainland, the Amur Liman on the western coast of Sakhalin, and *Nighvng*, spoken on the eastern coast of the island and formerly on its southern half. The North Sakhalin variety, spoken on the Schmidt peninsula, is intermediate between the two, sharing features with both branches (Gruzdeva 1998: 7; Gruzdeva and Janhunen 2020: 2–3).

The breakup of Proto-Amuric – the last common ancestor of Nivkh and Nighvng – appears to have happened relatively recently. While the two branches are distinct enough to suppose such a split, they are also similar enough to suggest that this event occurred merely 500 years before the present, at most (Janhunen 2010: 290). As a result, Proto-Amuric, as it has been reconstructed (for which, see Fortescue 2016; Halm 2017; 2019; Halm and Slater 2018; 2020), differs very little from its descendants.

Greater time depths are difficult to reach, as Amuric does not appear to form a node in a larger family tree. Although regarded as one of the last remnants of an ancient 'Palaeo-Asiatic' (or 'Palaeo-Siberian') linguistic lineage from the times when von Schrenck (1881) first coined the term, Amuric appears not to be related to any of the languages to which this same label has been applied (see, however, Fortescue 2011b for a recent attempt to establish a relationship with Chukotko-Kamchatkan).

With its shallow time depth and lack of any relatives, the means for reconstructing the more distant past of the Amuric language family appear limited. Nevertheless, attempts have been made to create deeper reconstructions (to which I return in Section 2). The most promising approach appears to be exploiting the fact that the Amuric languages, for most of their history, seem to have been in contact with the apparent newcomers to the Amur region and Sakhalin: the Tungusic languages. Although later surrounding the Amuric language family completely (see Figure 6.1), the ultimate homeland of the Tungusic

[1] The present chapter builds on my MA thesis, Knapen (2021).

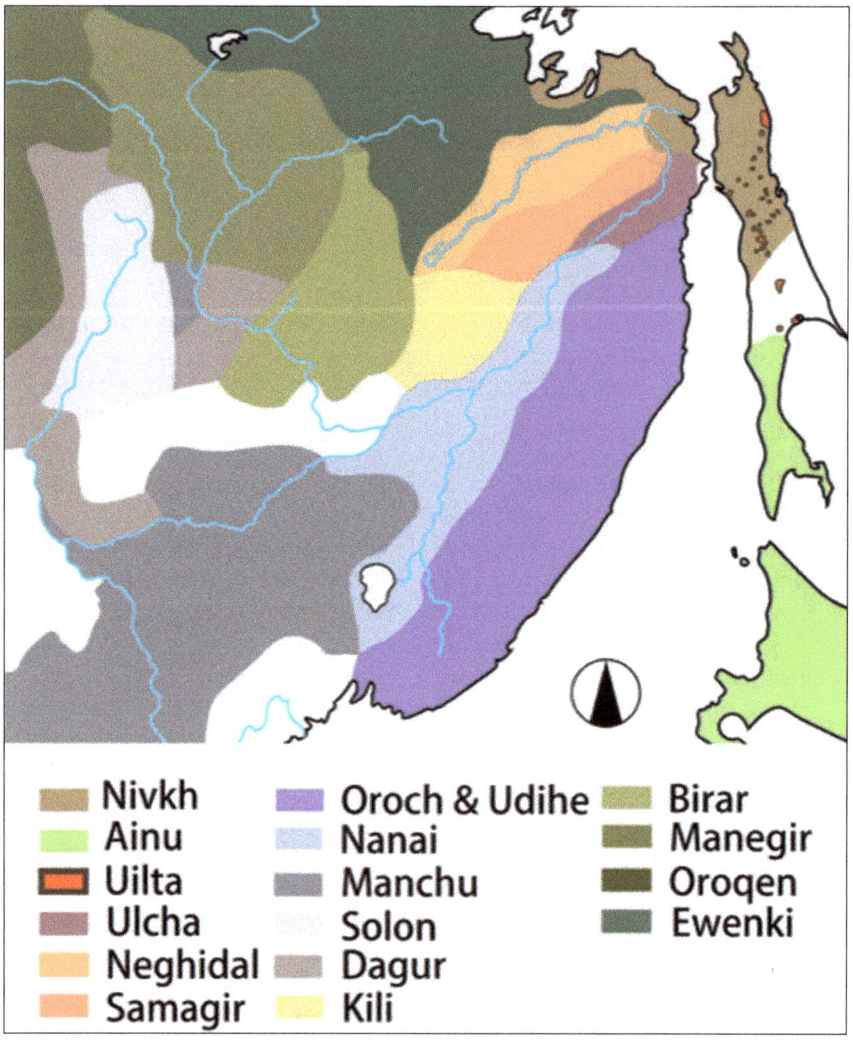

Figure 6.1. The ethnolinguistic groups of the Lower Amur in the second half of the nineteenth century. Source: Map by J. Uchiyama, based on information in von Schrenck (1881).

languages was to the south, in the Lake Khanka area. There, their ultimate ancestor, Proto-Tungusic, must have been spoken somewhere between 600 BC and 700 AD (Wang and Robbeets 2020). From the south, the Tungusic languages spread northward, to areas already occupied by Amuric, which was then pushed towards its current location (Janhunen 2010: 294; 2012: 10–11; Robbeets et al. 2020: 761–762; Wang and Robbeets 2020: 5–7). Whether this means a displacement of the Amuric-speaking population or simply replacement of their languages (i.e., language shift) remains to be investigated.

The Amuric homeland in this scenario was also originally located further south. Evidence that points to this southern origin includes metallurgical terminology, toponyms (e.g. the final element *-ri* in *Sungari* and *Ussuri* has been identified as Am (*)*eri* 'river') and loanwords between Amuric and Tungusic (Janhunen 1996: 231; 2010: 294; 2016: 5–6; Gruzdeva and Temina 2020: 173). Evidence of Amuric loanwords in Proto-Tungusic would provide strong evidence that the ultimate ancestor of the Amuric language family was contemporaneous with, as well as geographically adjacent to, Proto-Tungusic. Therefore, investigating these loanwords in more detail may be able to remove the veil covering the prehistory of the Amuric language family.

On lexical parallels between Amuric and Tungusic, there is a substantial body of research, tracing its inception to the first descriptive work on the Amuric languages: Grube (1892). This line of research has been continued by von Möllendorff (1894), Laufer (1902), Krejnovič (1937; 1955a), Austerlitz (1956; 1977; 1983a; 1983b; 1984a; 1984b; 1985; 1986b), Bouda (1960), Jakobson (1971), Panfilov (1973), Cincius (SSTM1; SSTM2), Doerfer (1985a; 1985b), Janhunen (1986; 1997; 2005; 2008; 2011; 2014; 2016), Burykin (1987; 1988; 1989; 2005; 2012), Pevnov (1992; 2016; 2018), Blažek (1997), Shternberg (1999), Abondolo (2005), Burykin and Popov (2012), Robbeets (2017), Tangiku (2015; n. d.), Shiraishi and Botma (2016), Fortescue (2016), and Li et al. (2020). Krejnovič (1955a) is of particular importance, since he was cited by most scholars after him. The full breadth of this research field is difficult to determine, as articles vary from extensive lists to brief mentions of single words. The totality of the research listed here provides however ample material to investigate the history of Amuric-Tungusic contact in the lexical domain.

The main objective of this chapter is therefore to critically assess the available data and what it may reveal about the earliest interactions between Amuric and Tungusic and where they would have occurred. Due to the complex nature of this problem, I provide a short disposition of the topics that are to be addressed here. First follows a brief discussion of the phonology of the ancestors of the two lineages in question. Next follows an outline of the methodology used. Lastly, the data is analysed further, examined for clues to periodise the borrowings as well as their implications for the homeland question.

6.2. Overview of Tungusic and Amuric historical phonology

A discussion of borrowing requires a sufficient understanding of the phonological systems that belong to the languages in contact. Loanwords tend to undergo a process referred to as 'phonological adaptation', due to differences in phonological makeup. If a foreign word contains a structure that violates native constraints, then this structure is repaired so that it conforms to them instead. Among these structures are counted segments but also combinations of segments (e.g., syllables and roots) (Kang 2011). If donor and recipient language have identical constraints, then there is little need to apply phonological adaptation, since no structures of one would be illicit in the other. For that reason, the phonologies and phonotactics of the ultimate ancestors to the Amuric and Tungusic language families need to be compared.

Because phonological adaptation is to be understood as a synchronic process, the two languages to be compared need to have been contemporaries. Proto-Amuric and Proto-Tungusic, two protolanguages reconstructed through the Comparative Method, are however not to be regarded as contemporaneous. Reconstruction on the basis of the Comparative Method departs from the extant members of a language family. Therefore, a protolanguage produced through this method is essentially a rendering of what an ancestor to a linguistic lineage looked like just moments before it split into branches. The time depth associated with a protolanguage is thus a metaphorical time of death: after this point in time, it ceased to exist, being no longer whole. The time depths proposed for Proto-Amuric and Proto-Tungusic discussed in the introduction to this chapter are widely apart, and therefore their respective breakups represent different points on the linguistic timeline of Northeast Asia.

Yet without evidence to the contrary, it could be assumed that the structure of Proto-Amuric remained stable for an indefinite amount of time until it fractured into Nivkh and Nighvng. The difference between late and early Proto-Amuric would then be largely irrelevant when comparing them to Proto-Tungusic, since late and early Proto-Amuric would be exactly the same language. However, it is a well-established fact that all languages change and thus it must be assumed that Proto-Amuric was also susceptible to change. The research cited below, has indeed shown that there was a stage preceding the immediate ancestor of Nivkh and Nighvng. As the language that brought forth these two branches is termed 'Proto-Amuric' any stages preceding this language could be termed 'Pre-Proto-Amuric'. However, in the context of this chapter, Pre-Proto-Amuric should be understood in more narrow terms: the stage of Proto-Amuric that existed at the same point in time as Proto-Tungusic did. That such a language did in fact exist at that point in time, is of course to be proven on the basis of Amuric loans in Proto-Tungusic.

Synchronising the Amuric ancestor language with Proto-Tungusic requires the historical linguist's toolkit to be expanded. The methodology devised by Janhunen (2016: 6) differentiates, in addition to the Comparative Method, two other techniques: internal and external reconstruction. Internal reconstruction makes use of synchronic data from within a single language (in this case Proto-Amuric) to reconstruct earlier stages. For that same purpose, external reconstruction employs data external to the lineage, i.e., loanwords from unrelated languages. The reconstruction of Pre-Proto-Amuric phonology and phonotactics discussed in this section is an amalgamation of results from these two methods, while previous reconstructions on the basis of the Comparative Method have also been corrected with more up-to-date evidence. Whether the phonology and phonotactics reconstructed in this manner represent that of an Amuric contemporary of Proto-Tungusic is difficult to ascertain without any branching events that could serve as reference points as typically relied on in more conventional approaches. Nevertheless, this reconstruction should be representative of a language closer to it than the immediate ancestor of the two branches of Amuric.

6.2.1. Proto-Tungusic and (Pre-)Proto-Amuric vowel inventories

Both Amuric and Tungusic languages have vowel inventories organised around vowel harmony. These systems are however different with respect to the direction in which the harmony operates. In Amuric, it is back to front, with prefixes harmonising with the vowels of the root. In Tungusic, it is front to back, with suffixes harmonising with the vowels of the root (Li 1996: 123–131; Gruzdeva 1998: 11). Aside from the direction, the feature that is harmonised with differs between the two families. Vowel harmony in most Tungusic languages is based on the feature RTR (retracted tongue root), which was most likely the case in Proto-Tungusic as well (Ko et al. 2014; Joseph et al. 2020). In this system, vowels are organised in two sets: RTR and nonRTR (see Table 6.1).

Commonly, this system is also reconstructed as palatal harmony, where the harmonic sets are front versus back vowels (Benzing 1956: 19–25; Janhunen 1981: 136). For that reason and to ease conversion between the two interpretations, Table 6.1 also presents in brackets the

Table 6.1. Proto-Tungusic vowel phonemes.

	non-back		back	
	nonRTR	**RTR**	**nonRTR**	**RTR**
High	*i (*i)	*ị (*ï)	*u (*ü)	*ụ (*u)
Low	*ə (*ä)	*a (*a)	*o (*ö)	*ọ (*o)

Table 6.2. Amuric vowel phonemes (adapted from Hattori 1962a: 1; 1962b: 68).

	Front	Central	Back
High	(*)i	(*)ə	(*)u
Low	(*)e	(*)a	(*)o

equivalent characters used for palatal harmony. Arguments in favour of palatal harmony have been provided by Barrere and Janhunen (2019: 59–60), though at present, these arguments do not seem sufficient to favour palatal harmony over RTR-harmony. Therefore, the latter system is maintained here.

Instead of RTR-harmony, the Amuric languages have two harmonic sets based on vowel height (Table 6.2). Additionally, the total inventory is slightly smaller, numbering six instead of eight vowels. While in Tungusic vowel harmony also puts a restriction on the vowels that may occur in the root (Li 1996: 123–125), in Amuric vowel harmony affects prefixes only (Gruzdeva 1998: 11). Without restrictions on vowel co-occurrence within the root, the Amuric system is best described as rudimentary.

There is evidence that, in the past, co-occurrence restrictions did also apply to the root. The system appears functionally defunct as high vowels may occur in the second syllable of any root, no matter the height of the vowel of the preceding syllable. Yet, in contrast, the distribution of low vowels is more restricted. If the first syllable is headed by a high vowel, no low vowel generally occurs in the second syllable. This restriction on distribution is apparently also retained in borrowings from Tungusic and Ainu. In these loanwords, vowels in the first syllable are raised or lowered depending on the height of the vowel in the second syllable (Hattori 1962a: 78–82; Shiraishi and Botma 2016).

The collapse of root-internal vowel harmony may be explained in several ways. It could be the result of reduction, which results in raising (Shiraishi and Botma 2017: 208–210). Alternatively, non-harmonious roots could go back to compounds. In this scenario, the vowels did originally not belong to the same root and hence were not required to harmonise. Another explanation may be that some instances of PAm **i* go back to PPAm **Vj*, as evidenced by certain loanwords, which means that in certain instances **i* in the second syllable may not descend from a vowel at all (Janhunen 2016: 22).

With or without root-internal harmony, the antecedent of the current system may have been characterised by RTR-harmony. In this scenario, high vowels descend from non-RTR vowels while low vowels descend from RTR vowels (Janhunen 1981: 139–140; Ko et al. 2014: 167; Joseph et al. 2020: 505). In North Sakhalin Nivkh/Nighvng, there are still remnants of such a system, as evidenced by an additional vowel /ʊ/, which contrasts with /u/ and only occurs after uvulars (Shiraishi and Botma 2017: 210–211). Postulating **ụ* for (Pre-)Proto-Amuric would explain the irregular correspondence between AN *laʁo-* 'to be frightened' and ESN *laʁu-* 'to be frightened' (cf. Savel'jeva and Taksami 1970: 153). AN /o/ generally corresponds to ESN /o/ and therefore this correspondence is irregular. In such a situation, positing an additional segment could provide a solution and it so happens that this same word is attested with /ʊ/ in North Sakhalin Nivkh/Nighvng: *laʁʊ-* (Shirashi and Botma 2017: 211). It would therefore seem safe to reconstruct PAm/PPAm **ụ*. To make the system symmetrical there should also be a non-RTR counterpart to (*)*o* < **ọ*, for which direct evidence is yet to be found. Despite the lack of data, it would appear that Pre-Proto-Amuric may have had the same vowel inventory as Proto-Tungusic as given in Table 6.1. Its system of vowel harmony may have been characterised by the same RTR vs. non-RTR contrast as well.

6.2.2. Proto-Tungusic and (Pre-)Proto-Amuric consonant inventories

Janhunen (2016) and Gruzdeva and Janhunen (2020: 14) have argued that the consonant inventory of Pre-Proto-Amuric was nearly identical to that of Proto-Tungusic. A comparison of Tables 6.3 and 6.4 shows that Proto-Amuric is quite different however.

A detail, though an important detail nonetheless, is that Proto-Tungusic had a voicing contrast in obstruents, while in Proto-Amuric there was a contrast here indicated with the terms *fortis* and *lenis*. Fortis plosives surface as voiceless aspirated plosives, while fortis fricatives surface as voiceless fricatives. Lenis plosives surface as voiceless unaspirated plosives, while lenis fricatives surface as voiced fricatives. Fortis plosives pattern with fortis fricatives and lenis plosives pattern with lenis fricatives in a morphophonological process termed 'Consonant Alternation'. At morpheme boundaries, initial plosives become fricatives and vice versa, depending on the segment that precedes (for a more detailed discussion, see Shiraishi

Table 6.3. Proto-Tungusic consonant phonemes (adapted from Benzing 1956: 27; Whaley and Oskolskaya 2020: 84).

	Labial	Dental	Palatal	Velar
Voiceless plosive	*p	*t	*č	*k
Voiced Plosive	*b	*d	*ʤ	*g
Voiceless fricative		*s		*x
Nasal	*m	*n	(*ɲ)	*ŋ
Liquid		*r, *l		
Glide	*w		*j	

Table 6.4. Proto-Amuric consonant phonemes.

	Labial	Dental	Palatal	Velar	Uvular	Glottal
Fortis plosive	*pʰ	*tʰ	*cʰ	*kʰ	*qʰ	
Lenis plosive	*p	*t	*c	*k	*q	
Fortis fricative	*f	*r̥	*s	*x	*χ	
Lenis fricative	*v	*r	*z	*ɣ	*ʁ	
Nasal	*m	*n	*ɲ	*ŋ		
Liquid		*l				
Glide	*w		*j			*h

2006). Since the opposition surfaces as aspiration vs. non-aspiration and voicing vs. voicelessness, the feature [± spread glottis] may provide a more apt description (Shiraishi 2006: 47–49). However, given that the use of the terms fortis and lenis is more well-established in historical linguistic literature on the language family (e.g. Fortescue 2016), they are more suitable in this context.

The number of fricatives also sets Proto-Amuric apart from Proto-Tungusic. These are equally numerous in the attested Amuric languages (cf. Gruzdeva 1998: 10; Hattori 1962a: 2; 1962b: 68; Shiraishi 2006: 24). In final position, the contrast between fortis and lenis fricatives disappears in most varieties, with the exception of South Sakhalin Nighvng (Shiraishi 2006: 70–74). Fortescue (2016: 4–5) only reconstructs a voiced series of fricatives and derives all voiceless fricatives from Consonant Alternation, word-final devoicing or devoicing caused by adjacency to voiceless plosives. However, this does not properly account for the behaviour of fricatives attested in South Sakhalin Nighvng. Therefore, a voiceless series should be reconstructed as well.

This number of fricatives may have been much smaller in Pre-Proto-Amuric or there may have been no fricatives at all. Currently, certain nouns in Amuric have non-alternating initial fricatives. These nouns may reveal the loss of an initial segment that left spirantisation as its only trace. For instance, AN *ruv* 'sibling' has ESN *tuvŋ* 'sibling' as its cognate. The fricative /r/ in Amur Nivkh can be explained as the result of spirantisation caused by a preceding third person possessive prefix **i-*, and therefore AN *ruv* translates more literally to'his/her sibling' (Jakobson 1971: 93–94). Further evidence of this change can be found in borrowings from Amuric. For example, the antecedent of PAm **ralŋ* 'frog' can be reconstructed with an initial vowel on the basis of Uilta *udala(n)* 'frog', which must have been borrowed from Amuric (Austerlitz 1977: 18–19; Pevnov 2016: 58; 2018: 449). All initial fricatives in the inherited Amuric lexicon can then be reconstructed as plosives.

Austerlitz (1982; 1990: 20; 1994) has further postulated that not just initial fricatives descend from plosives but medial and final fricatives as well. All are the product of intervocalic spirantisation. This sound change was blocked only by gemination, hence why there are still plosives remaining in Proto-Amuric. Glottal **h* is the only fricative like segment not eliminated in this reconstruction, although it is better analysed as a glide in the extant Amuric languages, given that it does not partake in consonant alternation (Shiraishi 2006: 28–29; Gruzdeva 1998: 13). If Pre-Proto-Amuric is to be reconstructed without any fricatives, then that moves its phonology both closer to and further from Proto-Tungusic. In this reconstruction, the contrasts **cʰ* vs. **s* and **t* vs. **r* are absent, as PAm **s* descends from PPAm **cʰ* and PAm **r* descends from PPAm **t*. In Proto-Tungusic, **s* and **r*, were most certainly separate phonemes from **č* and **t*. Janhunen (2016: 17–19) and Gruzdeva and Janhunen (2020: 14–16) propose that Pre-Proto-Amuric also had these consonants, yet given that their evidence is limited, this hypothesis requires additional support.

Less problematic is Gruzdeva and Janhunen's (2020: 14) choice of equating Am (*)*h* and Tg (*)*x*. They are also perceived as equivalents in many loans. Although Krejnovič (1955a: 138) found other categorisations of Tg (*)*x* in addition to Am (*)*h*, those that are discussed in the present chapter exclusively show Am (*)*h*. Consequently, the feature [+glottal] is largely irrelevant in differentiating the two inventories. The velar-uvular opposition is however a place contrast that requires further discussion. The opposition is possibly an innovation, since uvulars are mostly found adjacent to /a, o/ and also after /e/. The previous subsection has proposed that RTR vowels are the ancestors to Amuric low vowels. Given that uvulars are consonants specified for [+RTR] and velars are not, uvularisation can be seen as assimilation of a dorsal consonant to an RTR vowel (Hattori 1962b: 69; Jakobson 1971: 83–84; Austerlitz 1990: 20–21; Shiraishi 2006: 25–26; Shiraishi and Botma 2017: 204). Yet Fortescue (2011b: 1362–1363) has shown that it is possible to reconstruct roots with both **a* and **k* for Proto-Amuric, thereby contradicting these claims. However, these could also be explained. Some roots in which velars co-occur with low vowels may go back to compounds that are no longer transparent. The velar would then not be part of the same root as the low (originally RTR) vowel, thereby avoiding the its assimilatory effects. Alternatively, borrowing may be an explanation. A number of roots in which low vowels and velars co-occur are borrowed from Tungusic, which then suggests that the velar quality they had in the donor language was differentially maintained (Austerlitz 1956: 262; Jakobson 1971: 84). Consequently, the velar-uvular opposition may be discarded from the reconstruction of

Pre-Proto-Amuric, thereby making its consonantism completely identical to Proto-Tungusic.

6.2.3. Proto-Tungusic and (Pre-)Proto-Amuric root and syllable structure

One of the more salient characteristics that sets Amuric apart from Tungusic is its complex roots and syllables. Tungusic syllables follow a relatively simple structure of (C)V(C) (Li 1996: 19–20). This was not much different in Proto-Tungusic and a survey of Benzing (1956) finds only roots of up to three syllables in length. Clusters, which are maximally CC, only occur at the border of syllables. In contrast, Amuric roots allow for consonants clusters in both onset and coda of a syllable, yielding the maximal structure (C)(C)V(C)(C)(C). Amuric roots are usually only one syllable in length, although disyllabic roots also occur (Shiraishi 2009b: 86–87). Yet words may have been longer in the past and consisted of more simple syllables.

By proposing that most fricatives descend from intervocalic plosives, Austerlitz (1982; 1990: 20–22; 1994) effectively expands the number of syllables that a Pre-Proto-Amuric root consisted of. Clusters must then be broken up by reconstructing vowels in between them, which yields simple syllables of the type (C)V. However, there are relatively few roots attested that reflect these original vowels, as Shiraishi (2009a: 196) points out. The main etymon proposed by Austerlitz (1983a: 78, 1986a: 188; 1994) for which the extant Amuric languages still show evidence of vowels, is PPAm **ci-ka-tV* ‘tree’. The vowel **i* is attested in AN *ciɣ̥r* and the vowel **a* is attested in ESN *cʰχar̥* (cf. Fortescue 2016: 52). Although there is relatively little intra-Amuric evidence for these vowels, loanwords generally suggest that vowels were lost. A parallel in Tungusic or Ainu often retains vowels absent in Amuric. This thereby provides support for reconstructing the same type of roots and syllables for Pre-Proto-Amuric as are found in Proto-Tungusic (Austerlitz 1985; Janhunen 2016).

6.3. Data selection and methodology

The choice of methodology in approaching the present topic is in large part dictated by the type of data that is being used. For the purpose of reconstructing human prehistory, the method best suited for a dataset made up exclusively of lexemes, is cultural reconstruction. The lexicon of a language may contain words connected with a certain area in the world, such as terms for particular flora and fauna. This can also include the vocabulary associated with the culturally specific use of these natural resources. Thus, such words are diagnostic of the particular place where the speakers of this language lived (Wang and Robbeets 2020: 3–4). Coupled with the study of diachronic contact, there is a possibility of determining where one people was located relative to the other. If there are loanwords from Pre-Proto-Amuric in Proto-Tungusic, then it can be concluded that there was interaction between the people that spoke them. The geographical and ecological association of the lexical items borrowed from Pre-Proto-Amuric would then show where this language was spoken relative to Proto-Tungusic.

While cultural reconstruction is relatively straightforward, reconstructing the vocabulary that can be used for this purpose is more complicated. Among the large number of Amuric-Tungusic lexical parallels, there appear to be those belonging to different periods of interaction between the two linguistic lineages. This then requires certain criteria and techniques to tell apart recent from ancient loans.

The diachronic analysis of loanwords follows a procedure that has been termed ‘stratigraphy’ by Kallio (2015). Loanword strata are analogous to layers of sediment where every stratum has its own particular characteristics. In the history of languages, these characteristics are brought about by sound change. Sound laws are only operational at a certain point in time and all lexemes borrowed before that point are affected by them. When they cease to be productive, no new loanwords will partake in these sound changes. These developments effectively alter the phonological constraints that loanwords need to obey. To illustrate, consider Krejnovič’s (1937: 52–54; 1955a) observation that initial fricatives in certain Tungusic loans were categorised as plosives in Amuric, while in others they were preserved as fricatives. He suggested that their transfer to Amuric had occurred at different points in time. This assessment appears to be correct. As discussed in the previous section, initial fricatives seemed to have emerged later. Without any roots that can be reconstructed with an initial fricative, only those with plosives are left over, suggesting that fricatives were originally illicit in initial position. Before the emergence of initial fricatives, loanwords were amended to conform with this constraint, while afterwards this was no longer necessary. In this manner, Krejnovič’s examples can be put on a timeline. He compared Am (*)*cʰafq* ‘chopsticks’ and (*)*setaŋ* ‘sugar’ to Ma *sabka* ‘chopsticks’ and *šatan* ‘sugar’ respectively. Because a fricative is categorised as a plosive in (*)*cʰafq*, it must have been borrowed before (*)*setaŋ* as the latter preserves the initial fricative.

While this procedure aids in dating loans in Amuric and Tungusic, it is not sufficient for determining whether the parallelism is the result of interaction between the forebears of these language families. The two items that make up these parallels should also be reconstructable to these ancestors. For this purpose, the Comparative Method is most commonly used. The method’s first four steps in the description provided by Millar and Trask (2015: 196) require that the comparanda have similar meanings and exhibit regular sound correspondences. When these requirements can be met, it is possible to reconstruct protoforms.

The accuracy of such reconstructions can be hindered by extensive borrowing, which is particularly relevant to the case under study here. As noted by Doerfer (1985a: 254–261), Jurchenic borrowings appear in the Nanaic and Orochic branches of Tungusic as well as certain Ewenic languages such as Solon and southern and eastern varieties of Ewenki.[2] Only the western varieties of Ewenki as well as Ewen fell outside of the Jurchenic sphere. Consequently, words found in both the idioms affected by Jurchenic and those that were unaffected by Jurchenic can be reconstructed to Proto-Tungusic with complete confidence. On the Amuric side, differing degrees of Tungusic influence on the members of the family have also been noted. The Nivkh branch mostly came about due to influence of Tungusic (Gruzdeva 1998: 7). Therefore, lexemes exclusively found in this branch are more likely to have been borrowed from Tungusic at a later time. Those also found in the Nighvng branch, which was less heavily affected by Tungusic influence, are more likely to have been inherited from Proto-Amuric or even Pre-Proto-Amuric. For this reason, items without a safe range of attestation have been excluded from the dataset.

The great influence of Jurchenic on the surrounding Tungusic languages also shows the importance of considering the historical conditions under which language interact. The influence of this branch can largely be attributed to the powerful position its speakers held: the branch's main representatives, Jurchen and Manchu, were the languages of the founders of the Jin (1115–1234) and Qing (1644–1911) empires respectively (Janhunen 1996: 93–99, 132, 138). Therefore, the influence of these languages can largely be linked to influence emanating from these states. Considering the historical context also affects how individual words should be treated. Technically, the words **jaakọọ* 'Yakut' and **lụụča* 'Russian' can be reconstructed for Proto-Tungusic, as both have reflexes widely attested in all Tungusic languages. Since there are also reconstructable Amuric correspondents, **joqo(n)* and **loci~*lụcj* respectively (cf. Grube 1892: 74; Bouda 1960: 405; Savel'jeva and Taksami 1970: 82, 164; Janhunen 2016: 13–14; Sangi and Gašilova 2016: 22, 43; Shiraishi and Botma 2016: 44), it would appear that these were also subject to early contact between the two lineages. Yet it would be wholly anachronistic to assume that Yakuts and Russians lived close to the speakers of Proto-Tungusic and Pre-Proto-Amuric, as these ethnic groups started to migrate this far east no earlier than the 17th century (Forsyth 1992: 56, 104). In addition, there are also such *Wanderwörter* for which it is best to assume that they spread through trade. For example, there are sufficient cognates to reconstruct PTg **čaị* 'tea' and PAm **cʰai* 'tea' (Savel'jeva and Taksami 1970: 440; SSTM2: 377; Tangiku et al. 2008: 69; Tangiku 2015: 2; n.d. 1). Probably, these two are ultimately borrowed from Chinese 茶, which is read *chá* in modern Mandarin. Yet a more contemporaneous pronunciation of the character is †*draɨ~drɛː*, its reading for Early Middle Chinese (Pulleyblank 1991: 46). Therefore, it is more likely to have been borrowed from Mandarin or a close predecessor to Mandarin and thus at a later date. This also pertains to PAm *(*)arak* 'liquor'. Again, there is a Tungusic counterpart: PTg *(*)arakịị* 'liquor', which is found throughout the family (Grube 1892: 44; von Möllendorff 1894: 143; Krejnovič 1955a: 158; 1955b: 23; Savel'jeva and Taksami 1970: 34; SSTM1: 48; Tangiku et al. 2008: 29; Tangiku 2015: 2; n.d.: 1). Equivalents are in fact found all over Eurasia, with Arabic most likely its ultimate source. It entered into Tungusic through Mongolic, which it cannot have done so at an early date, as it is attested relatively late in Mongolic: the word does not appear in Middle Mongol (Doerfer 1985a: 37; Nugteren 2011: 271). Amuric might form the last link in this chain, possibly acquiring it from Manchu, as the historical supplier of liquor to the Nivkh were Manchu and Chinese traders (Schrenck 1891: 608).

The selection based on all these criteria and considerations only includes those items that can be confidently reconstructed for both Proto-Amuric and Proto-Tungusic. This then leaves the formulation of what criteria the validity of the comparison is judged on. First and foremost, they should be nearly identical in structure. This decreases the chance that the observed similarity is the result of coincidence, as relatively long matching strings of (nearly) identical segments have been shown to be a strong indicator of a shared source of two comparanda, either shared through inheritance or borrowing (Nichols 1996: 50–54). Additionally, the etyma can also be expected to have been identical, given that Pre-Proto-Amuric and Proto-Tungusic seem to have had the same phoneme inventory and phonotactics, leaving few incentives for applying phonological adaptation. In this methodological framework, I have so far relied on the Comparative Method, even though it cannot be applied successfully to the reconstruction of Pre-Proto-Amuric. Thus, when comparing the Proto-Amuric and Proto-Tungusic etyma, they are judged on the basis that they are potentially similar if the proposed developments preceding Proto-Amuric can be proven correct.

In addition to comparisons regarded as doubtful on this particular structural basis, otherwise doubtful parallels have been left out of the dataset. This includes PAm **cʰi* 'you (sg.)' which appears similar to PTg **sii, *sin-* 'you (sg.)' (cf. Krejnovič 1955a: 163; SSTM2: 72–73; Burykin 1988: 148; Fortescue 2016: 32). This comparison is phonologically problematic (see Knapen 2021: 87–88). Although this issue also requires further attention, its discussion is outside of the scope of the present chapter. Other words that have been dismissed deviated too greatly in function and meaning from each other. For instance, PAm **wa-* 'to fight' could be compared to PTg **waa-* 'to kill' (cf. Krejnovič 1955a: 165; Bouda 1960: 400; SSTM1: 127–129; Fortescue 2016: 158). In Tungusic, *(*)waa-* is a transitive verb, while in Amuric *(*)wa-* is functionally intransitive, requiring either a plural subject or a patient in the allative or comitative case (Otaina and Nedjalkov 2007: 1737–1738). Similarly problematic comparisons have been left out as well.

[2] There are various classifications of Ewenki varieties. By eastern, southern and western varieties of Ewenki, I mean those varieties that Doerfer (1985a: 11–12) classifies as such.

The items that met all these criteria can be found in the Supplement to this chapter. It provides first an etymon for Proto-Amuric and then its Proto-Tungusic correspondent. Since it includes no reconstruction of Pre-Proto-Amuric yet, it allows room for the method behind stratigraphy to be applied and thereby determine their place on the timeline. The strongest evidence for sound changes in Amuric are those words borrowed from Tungusic in Amuric. The field of Tungusic historical phonology has advanced far enough to trust that the etyma reconstructed for Proto-Tungusic accurately reflect the structure they had at the time when this language was spoken. Therefore, these protoforms can also be regarded as the model or the ultimate ancestor to the model on which the Proto-Amuric etyma were based. Nevertheless, it must first be proven of course that they are indeed borrowed from Tungusic. For that purpose and also to establish the opposite direction, I use two criteria derived from Clauson (1959: 175–178), Rozycki (1994: 4–5), Haspelmath (2009: 45) and Campbell (2013: 62–66). These criteria are morphological segmentability and adaptational patterns.

1. Morphological segmentation can reveal which language is the donor language and which language is the recipient language, if the lexeme being investigated was borrowed as an unsegmented block of morphemes. Only in one language are these morphemes productive. Therefore, if in one language they can be identified successfully as certain derivational or inflectional morphemes, then that language must be the donor language.
2. Adaptational patterns are useful as a criterion when their results are correspondences that only work in one direction. Krejnovič (1955a: 137–138) has shown that initial voiceless and voiced obstruents found in Tungusic loans are categorised in Amuric as fortis and lenis obstruents, respectively. Thus, PTg (*)*t*- is interpreted as PAm (*)*tʰ*- and PTg (*)*d*- is interpreted as PAm (*)*t*-. There are however also instances of parallels in which a Tungusic voiceless plosive corresponds to both fortis and lenis plosives in Amuric. This correspondence is suspicious, since languages tend to settle on a single strategy for adapting illicit structures (Kang 2011: 2259). Therefore, it is improbable that Amuric would be any different. The most parsimonious explanation appears to be the following. Both fortis and lenis obstruents are realised as voiceless in Amuric (the only distinction being aspiration), both types were most likely categorised as voiceless in Tungusic, due to the absence of audible voicing that would motivate otherwise. Thus both PAm (*)*tʰ*- and (*)*t*- are interpreted as PTg (*)*t*-. If a Tungusic voiceless plosive corresponds to an Amuric lenis plosive, this strongly suggests that the word was borrowed from Amuric, as per the previous reasoning. Fortis plosives are however silent on this matter since they always correspond to voiceless plosives, obscuring their origin. Other patterns that may be employed include segments found in environments where they are or were otherwise illicit. An example of this is the observation that words with velars in the environment of low vowels tend to have a Tungusic counterpart. As this appears to be disallowed in native Amuric words, that would suggest they are borrowed from Tungusic (Austerlitz 1956: 262; Jakobson 1971: 84; Janhunen 2016: 13). This is essentially an exception to normal restrictions and is referred to as 'differential importation' (Kang 2011: 2260).

6.4. Stratigraphy of Amuric-Tungusic lexical parallels

In the Supplement are listed the 36 parallels that were selected after careful consideration. Every Proto-Amuric comparandum in this list is made up of segments that each have an identical or nearly identical correspondent in the same position in their respective Proto-Tungusic counterpart, the only exception of course being those words in which nothing corresponds to a Proto-Tungusic vowel or cases of metathesis, e.g. PAm **amχa-* 'to be a father in law', cf. PTg **amaa-kAA* 'grandfather, father-in-law'. Although the correspondences are consistent between two words they are not regular when the set is taken as a whole. A plosive in a Proto-Tungusic word sometimes has a fricative as its counterpart in Proto-Amuric and sometimes a plosive. It also bears repeating that Proto-Tungusic vowels are only sporadically reflected in Proto-Amuric. This unevenness in parallelism provides fertile ground for the purposes of stratigraphy.

The first characteristic to be considered is the vowel-zero correspondence. Four words easily segmented in Tungusic and for which this correspondence is found in non-initial syllables are PAm **tʰoχ* 'elk (*Alces alces*)' (from PTg **tǫǫ-kII* 'elk (*Alces alces*)'), **pʰuvŋ* 'saw' (from PTg **pụụ-pOOn* 'saw') and **əryŋ* 'breath, life' (from PTg **ərii-gAAn* 'breath, life'). Therefore, it can be concluded that these vowels had been lost by the Proto-Amuric stage. There are also PAm **aka-* 'to be older' (cf. PTg **akaa* 'older brother'), **aχma-* 'to be a father-in-law' (cf. PTg **amaa-kAA* 'grandfather, father-in-law') that seem to preserve vowels. Both are plausible loans from Tungusic. PAm **aka-,* contains a velar between two low vowels. As differential importation of velars in an environment where uvulars are generally found in native words, it is most likely a borrowing from Tungusic. As for PAm *aχma-*, it is sufficient to observe that it consists of two morphemes in Tungusic.

Either as a phenomenon of phonological adaptation or sound change, vowel deletion would not have been applied haphazardly but systematically instead if these words entered the Amuric lexicon at the same time. Thus, these words must have been borrowed into Amuric at different points in time. Since it would be impossible to find words like **aka-* or **aχma-* if they were borrowed before vowel deletion in non-initial syllables, they must have been borrowed later. That makes the preservation of vowels in non-initial syllables a typical characteristic of younger loans.

To some extent, the same can be demonstrated with spirantised plosives. As discussed in Section 3, non-alternating initial fricatives can be used for that purpose. Thus, PAm **χoɲ* 'sheep' appears to have been adopted from Tungusic more recently, most likely from a Tungusic language in which initial **k* shifted to (*)*x*. Although it may have passed through other Tungusic languages first before reaching Amuric, Ma *χonin* is most likely its ultimate source. Only Jurchenic and Solon underwent this change (Benzing 1956: 28), while Solon also has numerous loanwords from Jurchenic (Doerfer 1985a: 37, 254–261). In cases where the fricative goes back to a fricative in Proto-Tungusic, there is a split between the two fricatives of Proto-Tungusic. PTg **x* corresponds to PAm **h* and PTg **s* corresponds to PAm **cʰ* in initial position, cf. PAm *hal(ŋ)* 'collar' and PAm **cʰamŋ* 'eagle, shaman'. For the latter, it is not clear which language was the donor language. Possibly, PAm **cʰamŋ* 'shaman, eagle' is ultimately a derivation of **cʰma-* 'to visit', which would then suggest Pre-Proto-Amuric had roots beginning with **s*. However, this comparison is not without issue (see Knapen 2021: 85–87). The matter of PPAm **s* requires additional attention, yet is outside of the scope of the present chapter. Since the contrast between **s* and **cʰ* was either lost in Proto-Amuric or Pre-Proto-Amuric never possessed it, it can safely be assumed that the correspondence PTg **s* = PAm **cʰ* suggest the item has been part of the Amuric lexicon before the Proto-Amuric stage.

The effectiveness of using initial fricatives in this manner is contrasted by the ineffectiveness of using medial fricatives. In PAm **aka-* 'to be older' a plosive is retained, in the same position where it has become a fricative in **pʰuvŋ* 'saw' and **tʰoχ* 'elk (*Alces alces*)'. As shown above, the retention of the final vowel in **aka-* suggests that it was borrowed recently, while the deletion of vowels in **pʰuvŋ* and **tʰoχ* suggest that they were borrowed earlier. As the latter two also have fricatives, this may be a characteristic of earlier adoption as well. The opposition between fricatives and plosives in Proto-Amuric seems to be the result of the degemination of long plosives and the spirantisation of short plosives in medial position. It would therefore appear that Tungusic medial plosives were always categorised as short plosives in Pre-Proto-Amuric, which then became fricatives in Proto-Amuric. Plosives in words borrowed after this period, were then simply categorised as plosives as the contrast was maintained by shortening geminates. While plausible, this scenario runs into complications. PAm **pʰuvŋ* 'saw', **tʰoχ* 'elk' and **uγɲ~*uγŋ* 'loon' may also be compared to Ma *fufun* 'saw' < PTg **pụpOOn*, *toχo* 'elk' < PTg **tọọ-kII* and Neg. **uxəəŋi* 'loon' < PTg **ookəəŋi* respectively (SSTM1: 34–35; SSTM2: 191–192, 257, 336). For these examples, it is difficult to tell spirantisation in Amuric apart from spirantisation in Tungusic. Thus, it becomes difficult to tell whether these were borrowed from Proto-Tungusic or one of the languages it was the ancestor to.

In sum, on the basis of the current state of reconstruction of Pre-Proto-Amuric, only vowel deletion and the categorisation of fricatives as plosives appear to provide some indication to a loanword's antiquity. It should however be noted that it is also found in PAm **χoɲ* 'sheep', a word that must be a newer addition to the Amuric lexicon on account of its initial fricative. Despite that, it does not reflect the vowel in the final syllable of Ma *χonin* 'sheep'. This suggests that vowel elision is a more recent development than the emergence of initial fricatives. When the former sound change ceased to be operational is difficult to estimate, but certainly before the time when **aka-* 'to be older' and *aχma-* 'to be a father-in-law' were borrowed.

On the basis of these findings, four strata can be identified, which are listed below from oldest to newest.

1. PTg **s* and **x* correspond to PAm **cʰ* and **h* in initial position.
2. An initial fricative in Proto-Amuric corresponds to a Proto-Tungusic plosive or a post-Proto-Tungusic fricative.
3. Proto-Tungusic vowels in non-initial syllables are absent in their Proto-Amuric counterpart. The Proto-Tungusic syllables correspond to Proto-Amuric clusters instead.
4. Proto-Tungusic vowels in non-initial syllables are present in their Proto-Amuric counterpart.

The numbers found in the Supplement refer to the characteristic correspondences defined and numbered here. Every subsequent stratum retains the feature introduced in the preceding stratum. Of course, if the Proto-Tungusic etymon in the parallel starts with neither **s* nor **x*, there is no basis to assign the parallel to either (1) or (2). These characteristics mainly serve to separate those words that have them. Otherwise, parallels that fall in stratum (3) can be regarded as old, but not necessarily ancient. It should be noted that these are only correspondences and not a statement on the direction in which these words were borrowed, although in the future they could be used to differentiate recipient from donor language.

6.4.2. The Amuric homeland question

With such limited means of determining the age of these loanwords, it would appear that Pre-Proto-Amuric cannot be dated more precisely than its name already suggests: before Proto-Amuric. Nonetheless, this list provides some support to Janhunen's (2010: 292–296; 2016: 5–6, 23) hypothesis. The fact remains that the etyma have reflexes widely attested in their respective language families. Those Amuric etyma that also lack vowels in non-initial syllables and fricatives in initial syllables (excepting PTg **x* = PAm **h*) appear reasonably old. If these words are clearly borrowed from Amuric in Proto-Tungusic, the parallel is most likely the product of contact between Proto-Tungusic and Pre-Proto-Amuric. Yet in the case of borrowing in the opposite direction, it is still difficult to establish how long a given lexeme was part of the Amuric lexicon, without any other means of dating them. Therefore, the hypothesis must still be regarded as tentative.

This fact notwithstanding, there are plausible loans from Amuric in Tungusic whose examination may reveal where Pre-Proto-Amuric was located. It has been noted previously that Amuric borrowings in Tungusic tend to refer to fish, marine mammals and other aquatic life (Doerfer 1985a: 261; 1985b). Those that are found in only a few Tungusic languages are most likely later borrowings, but those that are widely attested may have been inherited from Proto-Tungusic. Since these animals are bound to a certain environment and grant to humans the possibility of a certain mode of subsistence, they are suitable evidence for assessing where Pre-Proto-Amuric was located.

Two terms for cetaceans that can be reconstructed for Proto-Tungusic have Amuric counterparts. These are PTg *kalịmụ and *kịaŋa, both of which translate to 'whale'. These correspond to PAm *q^halm 'whale of small size' and *keŋ 'whale of intermediate size', respectively. PAm *q^halm cannot be analysed further and therefore it would be difficult to establish whether it was borrowed from Amuric into Tungusic or the other way around. In contrast, PTg *kịaŋa appears to have an Amuric origin. Its initial voiceless plosive corresponds to a lenis plosive in Amuric. Thus, it would seem that the ancestor to PAm *keŋ was its model.

This same argument may be applied to the word for 'stroganina', which is a dish made of strips of raw fish. For Proto-Tungusic, this can be reconstructed as *tala or *tala(k)ka, which corresponds to PAm *talq 'stroganina'. Again, a voiceless plosive in Tungusic corresponds to a lenis plosive in Amuric. As PAm *talq appears to be a derivation from *talqa-, *talva- 'to be raw' (cf. Fortescue 2016: 38), there is sufficient reason to argue that it was borrowed from Amuric into Tungusic. Then, even the dish itself could have been adopted from speakers of Pre-Proto-Amuric.

PAm *keŋ 'whale of intermediate size' and *talq 'stroganina' suggest that familiarity with sea life and dependency on fishing was more strongly associated with speakers of Pre-Proto-Amuric than speakers of Proto-Tungusic. One could then postulate that speakers of Pre-Proto-Amuric lived in the vicinity of a body of water, while speakers of Proto-Tungusic did not. Pevnov (2012: 31) has suggested that the Proto-Tungusic homeland was located near the sea, as the word *laamọ 'sea' can be reconstructed for the Tungusic protolanguage. This appears to have a parallel in PAm *lam 'very small wave with foam on top', which could be reconstructed on the basis of SSN *lam* 'very small wave with foam on top'. Unfortunately, within Amuric it is only attested in this variety (Yamaguchi and Izutsu 2004: 48). Nevertheless, internal reconstruction appears to support reconstructing this etymon and even that it was morphologically complex, that is, a compound.

A segmentation has been proposed by Janhunen (2008: 97–100; 2016: 12–13). Although he does not cite SSN *lam*, he finds a similar root *la-mo in PAm *lams < PPAm *la-mo-s, which he compares to PTg *laamos 'east wind', reflexes of which are only attested in Ewenki and Ewen (cf. SSTM1: 491). Wind coming from the east of where the Ewenki and Ewen are located would come from the sea. Its association with the sea could be suggested further by the fact that the element *la is found in various words related to aquatic life and bodies of water. However, the meaning 'east wind' is only attested in Amur Nivkh, while it is attested as 'north (wind)' in the Nighvng varieties (cf. Fortescue 2016: 92). Yet it should be noted that the cardinal directions are of little importance to the Amuric system of spatial reference. Rather, one orientates oneself on the basis of the local environment, the direction that a river flows or position relative to the shore (Fortescue 2011a: 88–92). It is therefore not inconceivable that the original meaning of PAm *lams is indeed related to the wind's direction relative to the sea. With this root *la-mo also present in PAm *lam 'very small wave with foam on top', it should be possible to reconstruct PPAm *la-mo.

PAm *keŋ 'whale of intermediate size', *talq 'stroganina', *lam 'very small wave with foam on top' (<*la-mo) and possibly *q^halm suggest that speakers of Pre-Proto-Amuric had knowledge of the marine environment and possibly exploited aquatic resources. Fishing and hunting of sea-mammals as subsistence activities of the Nivkh are known from the anthropological literature (see, for instance, Taksami 1975: 12–40). These activities were also common among prehistoric communities that were present around the Strait of Tartary. The earliest archaeological evidence of the processing of fish on the Lower Amur dates to the Late Glacial (ca. 16,000–10,000 BP) (Shoda et al. 2020: 6). That is, however, not to imply that communities of the Late Glacial already spoke an Amuric language. A plausible candidate among later archaeological cultures may be the Okhotsk culture (AD 550–1200) of Sakhalin and Northern Hokkaido, which is often described as a precursor of the Nivkh and which was characterised by sea-mammal hunting (Hudson 2004: 295–298; Zgusta 2015: 83–96). The vocabulary related to cetaceans would be consistent with the view that the language of the Okhotsk people was Amuric in origin, which also has been suggested by Vovin (2016) on the basis of evidence of early contact between Amuric and Ainu. Although his data still require explanation, Sakhalin, Hokkaido and the Kurils, as per Vovin's proposal, cannot be regarded as the ultimate homeland of Amuric on account of its early contact with Tungusic as well as toponymic evidence.

As argued by Wang and Robbeets (2020), the ultimate homeland of the Tungusic languages was located in the vicinity of Lake Khanka. The body of old lexical parallels that these languages share, supports the geographical adjacency of their homelands. Pre-Proto-Amuric must have been spoken to the northeast of Proto-Tungusic, as evidenced by its vocabulary that suggests closeness to the sea. Although this is relatively little evidence to base a homeland localisation on, it does argue against placing Proto-Tungusic anywhere near the coast. Instead, it suggests that direct access to the ocean was blocked off to them by speakers of Pre-Proto-Amuric. That places the homeland further to the east than implied by Janhunen (1996: 231; 2010: 292-296; 2016: 5–6, 23). The same counterproposal was brought forward by Li et al. (2020: 16), though exclusively on the basis of loanwords. On the basis of toponymic data, the sea appears to be only the eastern boundary of the homeland. The western boundary must have encompassed parts of the Sungari and Ussuri, if these hydronyms are indeed Amuric in origin. Combining these findings then yields the roughly demarcated area pictured in Figure 6.2. On this map, Pre-Proto-Amuric occupies only a section of the Sungari, while Tungusic occupies the remainder. Quite possibly, their territories may have overlapped.

6.5. Concluding remarks

In sum, the Amuric ancestor language appears to have been a contemporary to Proto-Tungusic. This is evidenced by Amuric loanwords in Proto-Tungusic. Since these loans include maritime terminology, it would suggest that Proto-Tungusic was not spoken along the sea shore, which instead must have been Amuric-speaking territory. The phonological developments reflected in these parallels as well as others suggest at least four different periods of borrowing that occurred between Proto-Amuric and its preceding stage, Pre-Proto-Amuric. Whether the oldest layer of these was borrowed directly from Proto-Tungusic or a descendant of it, is difficult to ascertain. Other than that these developments preceded the Proto-Amuric level, there are currently no means for proving when exactly they occurred. Therefore, these borrowings could have entered the Amuric inherited lexicon at the time before the breakup of Proto-Tungusic or well after it.

This issue encapsulates the current obstacles facing Amuric historical comparative linguistics. Loanwords from Amuric in Tungusic and vice versa clearly show great potential for reconstructing Amuric prehistory. Yet too little diachronic analysis has been performed on the Amuric language

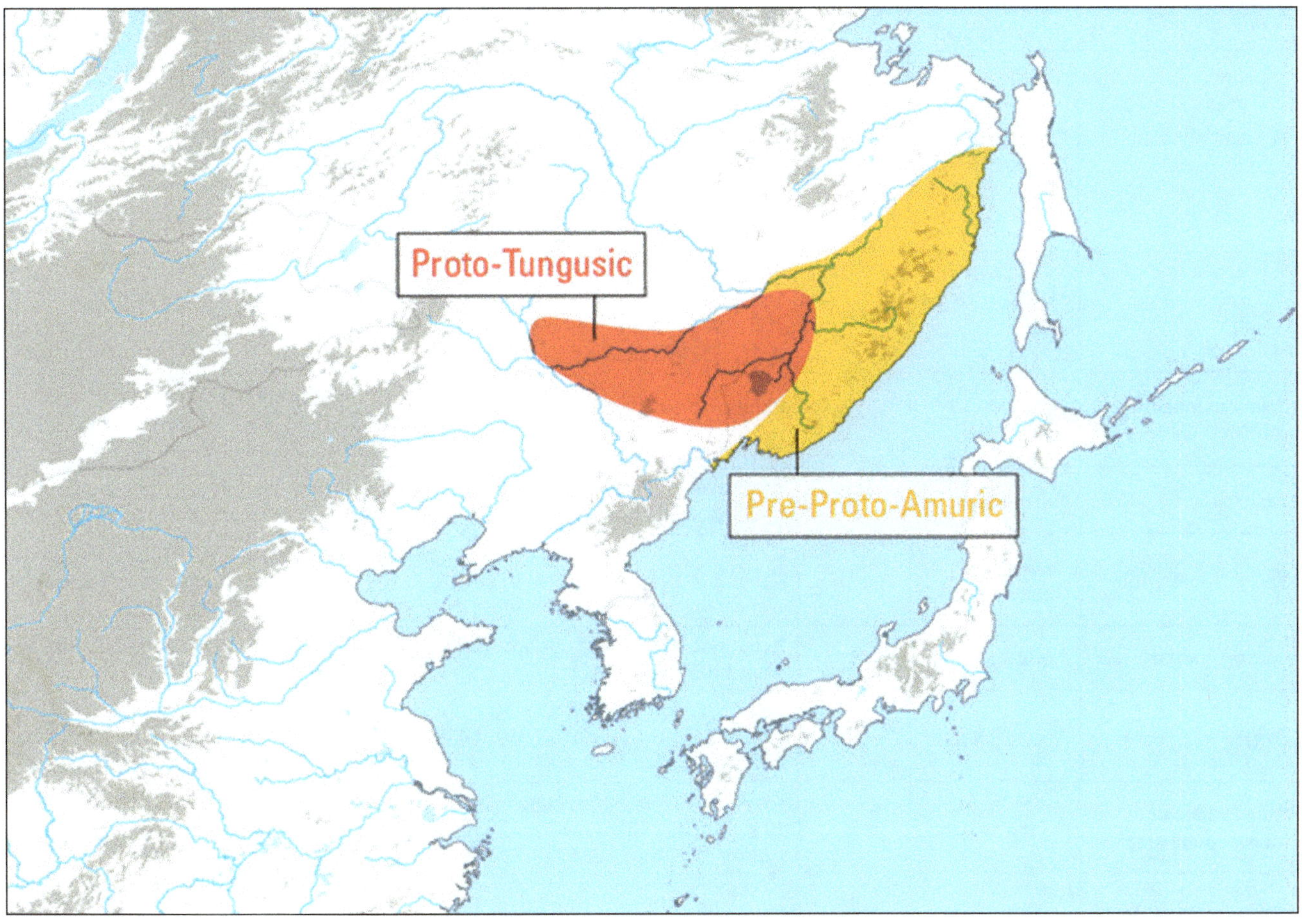

Figure 6.2. The Tungusic and Amuric homelands. Source: Map by M. O'Reilly, MPI-GEA, based on information by Martine Robbeets.

family to regard any conclusions that can be drawn from loanwords as proven facts. I would therefore suggest that the following lines of inquiry must be taken up.

Firstly, the reconstruction of Pre-Proto-Amuric has to be improved. The posited sound laws need to be confirmed and more sound changes need to be identified for the purposes of classifying loanword strata. Additionally, these sound changes can also aid in discovering hereto unidentified parallels. Secondly, though loanwords certainly form complementary data, an attempt should be made to identify the Amuric homeland using more conventional cultural reconstruction. Until that time, the conclusions presented in this chapter are tentative.

Supplement

Proto-Amuric	Proto-Tungusic	Stratum	Proposed by	Lexicographic references
**aka-* 'to be older'	**akaa* 'older brother'	(4)	Grube 1892: 53; von Möllendroff 1894: 144; Krejnovič 1955a: 158; Panfilov 1973: 10; Bouda 1960: 403; Shternberg 1999: 34; Burykin & Popov 2012: 81; Fortescue 2016: 10	SSTM1: 23-24; Fortescue 2016: 10
**amr̥a-* 'to taste'	**amta(-)* 'taste'	(4)	Grube 1892: 45; von Möllendorff 1894: 144; Krejnovič 1955a: 158; Austerlitz 1983a: 86; Fortescue 2016: 13	SSTM1: 39; Fortescue 2016: 13
**amɣ~*amx* 'mouth'	**amŋa* 'mouth'	(3)	Von Möllendorff 1894: 144; Krejnovič 1955a: 158; Panfilov 1973: 10; Burykin 1987: 52; 1988: 147	Grube 1892: 45, 46, 53; SSTM1: 38-39; Fortescue 2016: 12
**ar(a)(-)* 'almost'	**aran* 'barely'	(3)/(4)	Von Möllendorff 1894: 144	SSTM1: 48; Grube 1892: 44; Krejnovič 1955b: 23, 48; Fortescue 2016: 15
**aχma-* 'to be a father in law'	**ama-kAA* 'grandfather, father-in-law'	(4)	Krejnovič 1955a: 158; Burykin & Popov 2012: 84	Grube 1892: 44-45; SSTM1: 34-35; Fortescue 2016: 8
**capŋ* 'pincers'	**ʤapa-pOOn* 'grabbing tool'	(3)	Krejnovič 1955a: 162	SSTM1: 240-241; Savel'jeva & Taksami 1970: 375
**cʰamŋ* 'shaman, eagle'	**samaan* 'shaman'	(1)	Grube 1892: 78, 81; von Möllendorff 1894: 145; Krejnovič 1937: 53; 1955a: 163; Bouda 1960: 402; Jakobson 1971: 93; Austerlitz 1984b: 233; 1986b: 143–144; Janhunen 1986: 110–111; 2005: 25–26; 2016: 18-19; Abondolo 2005: 58–59; Fortescue 2016: 143	Grube 1892: 78, 81; SSTM2: 59; Fortescue 2016: 30,143
**cʰoxt(u̡)-* 'to get drunk'	**so̡kto̡o̡-* 'to get drunk'	(1)	Grube 1892: 79; von Möllendorff 1894: 145; Krejnovič 1955a: 163; Jakobson 1971: 90; Panfilov 1973: 9; Fortescue 2016: 35	Grube 1892: 79; SSTM2: 106; Fortescue 2016: 35
**əlv-* 'to make a roof'	**əlbə-* 'to cover'	(3)	Krejnovič 1955a: 159; Bouda 1960: 405; Burykin 1987: 47	Savel'jeva & Taksami 1970: 463; SSTM2: 445; Sangi & Gašilova 2016: 103; Fortescue 2016: 166
**ərŋ* 'time'	**əruun* 'time'	(3)	Krejnovič 1955a: 159; Bouda 1960: 402; SSTM2: 464; Burykin 1987: 47	SSTM2: 463-464; Fortescue 2016: 167
**ərɣŋ* 'breath, life'	**ərii-gAAn* 'breathe, life'	(3)	Krejnovič 1955a: 159; SSTM2: 464	Takahashi 1942: 227; Krejnovič 1955a: 159; Savel'jeva & Taksami 1970: 466; SSTM2: 464
**hal(ŋ)* 'collar'	**xala* 'tug'	(1)	Krejnovič 1955a: 165; Bouda 1960: 405; Tangiku 2015 4; n.d. 1	SSTM1: 28–29; Fortescue 2016: 70
**hilɣ~*hilx* 'tongue'	**xilŋu* 'tongue'	(1)	Von Möllendorff 1894: 145; Krejnovič 1955a: 165; Burykin 1987: 50; 1988: 149; 2012: 281	SSTM1: 316–317; Fortescue 2016: 74
**keŋ* 'whale of intermediate size'	**ki̡aŋa* 'whale'	(3)	Krejnovič 1955a: 164; Bouda 1960: 403; SSTM1: 388; Doerfer 1985a: 261; 1985b: 198; Burykin 1989: 45	SSTM1: 388; Fortescue 2016: 58
**lam* 'small wave'	**laamo̡* 'sea'	(3)	Janhunen 2008: 97–100; 2016: 12–13	SSTM1: 490–491; Yamaguchi & Izutsu 2004: 48
**luk* 'shaggy'	**lo(k)ko* 'shaggy'	(3)	Krejnovič 1955a: 160; Bouda 1960: 405; SSTM1: 509	Savel'jeva & Taksami 1970: 165; SSTM1: 508–509; Sangi & Gašilova 2016: 43

Proto-Amuric	Proto-Tungusic	Stratum	Proposed by	Lexicographic references
**mam-* 'to squeeze'	**nama-* 'to hold firmly'	(3)	SSTM1: 581	SSTM1: 581; Fortescue 2016: 101
**maŋq-* 'to be strong'	**ma(-)ŋ(g)a* 'hard, strong, difficult'	(3)	Grube 1892: 108; von Möllendorff 1894: 146; Krejnovič 1955a: 160; Bouda 1960: 401; Panfilov 1973: 9; SSTM1: 530; Fortescue 2016: 101–102	SSTM1: 529-530; Fortescue 2016: 101–102
**me(-)q* 'musk deer'	**mik(-)čaan* 'musk deer'	(3)	Burykin 1987: 49–50	SSTM1: 536; Fortescue 2016: 103–104
**murŋ* 'horse'	**mụrịn* 'horse'	(3)	Grube 1892: 111; von Möllendorff 1894: 146; Krejnovič 1955a: 160; Bouda 1960: 401; Panfilov 1973: 8; Tangiku 2015: 24; n.d.: 7; Robbeets 2015: 34–35; Janhunen 2016: 15–16; Fortescue 2016: 109	SSTM1: 558–559; Fortescue 2016: 109
**ɲoɲ(-)j* 'goose'	**ɲụŋna(-kII)* 'goose'	(3)	Grube 1892: 87; von Möllendorff 1894: 145; Krejnovič 1955a: 161; Bouda 1960: 405; SSTM1: 647; Burykin 1987: 49	Grube 1892: 87; Savel'jeva & Taksami 1970: 209; SSTM1: 646–647; Sangi & Gašilova 2016: 52
**pʰisk* 'kite'	**pigəəčuən* 'kite'	(3)	Grube 1892: 101–102; von Möllendorff 1894: 146; Krejnovič 1955a: 161; Bouda 1960: 399	Grube 1892: 101–102; Savel'jeva & Taksami 1970: 287; SSTM2: 322–323; Sangi & Gašilova 2016: 69
**pʰutm* 'midge'	**pormii-g-tA* 'midge'	(3)	Krejnovič 1955a: 161; Burykin 1987: 48	SSTM2: 348; Fortescue 2016: 137
**pʰuv(-)ŋ* 'saw'	**pụụ-pOOn* 'saw'	(3)	Grube 1892: 103; von Möllendorff 1894: 146; Krejnovič 1955a: 161; Bouda 1960: 400; Burykin 1987: 49; Tangiku 2015: 14, 21; n.d.: 3, 5	SSTM2: 336; Fortescue 2016: 138
**pur-* 'to spread'	**boruu-* 'to drop'	(3)		SSTM1: 113; Fortescue 2016: 27
**qʰalm* 'whale of small size'	**kalịmụ* 'whale'	(3)	Krejnovič 1955a: 164; Doerfer 1985a: 261; 1985b: 197–198; Janhunen 2011: 206–207; 2016: 13	Grube 1892: 54; Savel'jeva & Taksami 1970: 147; SSTM1: 366–367; Sangi & Gašilova 2016: 36
**qʰot(-)r* 'bear'	**kut(ị)* 'bear'	(3)	Von Möllendorff 1894: 144; Burykin 1987: 49; Pevnov 1992: 28	Grube 1892: 59; Savel'jeva & Taksami 1970: 151; SSTM1: 440
**qoŋqŋ* 'bell, jaw harp'	**kọọŋ(g)aa(n) (-g-tA)* 'bell'	(3)	Krejnovič 1955a: 164; Tangiku 2015: 21, 22; n.d.: 5; Fortescue 2016: 67	SSTM1: 410; Fortescue 2016: 67–68
**tʰat-u-* 'teach'	**tatị-* 'learn, get used to'	(3)	Grube 1892: 97; von Möllendorff 1894: 146; Krejnovič 1955a: 162	Grube 1892: 97; Krejnovič 1955a: 162; 1955b: 590; Savel'jeva & Taksami 1970: 304, 345; SSTM2: 170–171
**tapsŋ* 'salt'	**dabụsụn* 'salt'	(3)	Grube 1892: 89; von Möllendorff 1894: 143; Krejnovič 1955a: 162; Bouda 1960: 401; Panfilov 1973: 8–9; Abondolo 2005: 58; Tangiku 2015: 14, 21; n.d. 3, 5; Fortescue 2016: 41	SSTM1: 186; Fortescue 2016: 41
**talq* 'stroganina'	**tala((k)ka)* 'stroganina'	(3)	Krejnovič 1955a: 162	Grube 1892: 88; Krejnovič 1955a: 162; 1955b: 618; SSTM2: 156; Fortescue 2016: 38
**taʁ* 'epidemic'	**da(x)a-* 'to be infected'	(3)	Krejnovič 1955a: 162	SSTM1: 184; Fortescue 2016: 40
**tʰə(-)l(-)ŋkVw-* 'to tell a tale'	**təəloŋo(-)* 'legend, tale'	(3)	Krejnovič 1955a: 162; Bouda 1960: 401; Panfilov 1973: 11; SSTM2: 233; Janhunen 2016: 22	SSTM2: 233; Fortescue 2016: 154
**tʰoχ* 'elk (*Alces alces*)'	**tọọ-kII* 'elk (*Alces alces*)'	(3)	Grube 1892: 90; von Möllendorff 1894: 145; Krejnovič 1955a: 162; Bouda 1960: 402; SSTM2: 192	SSTM2: 191–192; Fortescue 2016: 150
uɣɲ*~uɣŋ* 'loon'	**ookəəŋi* 'loon'	(3)	Burykin 1987: 48	SSTM2: 257; Fortescue 2016: 155
**χoɲ* 'sheep'	**kọnịn* 'sheep'	(2)	Krejnovič 1955a: 165; Bouda 1960: 404; Panfilov 1973: 8; Fortescue 2016: 165	SSTM1: 409–410; Fortescue 2016: 165

Abbreviations of languages

Am	Amuric
AN	Amur Nivkh
ESN	East Sakhalin Nighvng
NSN	North Sakhalin Nivkh/Nighvng
Ma	Manchu
PAm	Proto-Amuric
PPAm	Pre-Proto-Amuric
PTg	Proto-Tungusic
SSN	South Sakhalin Nighvng
Tg	Tungusic

Abbreviations of references

SSTM1	Cincius, V. I. 1975. *Sravnitel'nyj slovar' tunguso-man'čžurskix jazykov: Tom I.* Leningrad: Nauka.
SSSTM2	Cincius, V. I. 1977. *Sravnitel'nyj slovar' tunguso-man'čžurskix jazykov: Tom II.* Leningrad: Nauka.

References

Abondolo, Daniel (2005) Invariance in variance: Language, prominence, and silence in the verbal art of ritual. In Juha Pentikäinen and Péter Simoncsics (eds.), *Shamanhood: An Endangered Language*, pp. 49–63. Oslo: Novus Forlag.

Austerlitz, Robert (1956) Gilyak nursery words. *WORD* 12 (2): 260–279.

Austerlitz, Robert (1977) The study of the Paleosiberian languages. In Daniel Armstrong and C. H. van Schooneveld (eds.), *Roman Jakobson: Echoes of his Scholarship*, pp. 13–20. Lisse: The Peter de Ridder Press.

Austerlitz, Robert (1982) Gilyak internal reconstruction, 1: Seven etyma. *Folia Slavica* 5 (1–3): 81–88.

Austerlitz, Robert (1983a) On the penumbra of questions surrounding the internal reconstruction of Gilyak. *Research on Language & Social Interaction* 16 (3–4): 75–87.

Austerlitz, Robert (1983b) Préhistoire des Guiliaks: Métaux. Gilyak Prehistory: Metals. *Colloque international Sibérie 1582–1982* (Paris 24–27 mai 1983). Résumés/Abstracts/Rezjume: 1–2. Paris: C.R.N.S., Laboratoire de Slavistique.

Austerlitz, Robert (1984a) Gilyak internal reconstruction, 2: Iron and questions related to metallurgy. *Folia Slavica* 7 (1–2): 38–48.

Austerlitz, Robert (1984b) On the vocabulary of Nivkh shamanism: The etymon of *qas* ('drum') and related questions. In Mihály Hoppál (ed.), *Shamanism in Eurasia,* pp. 231–239. Göttingen: Herodot.

Austerlitz, Robert (1985) Etymological frustrations (Gilyak). *International Journal of American Linguistics* 51 (4): 336–339.

Austerlitz, Robert (1986a) Contrasting fact with fiction: The common denominator in internal reconstruction, with a bibliography. In Dieter Kastovsky and Aleksandr Szwedek (eds.), *Linguistics Across Historical and Geographical Boudaries. In Honour of Jacek Frisiak on the Occasion of his Fiftieth Birthday*, pp. 183–192. Berlin: Mouton de Gruyter.

Austerlitz, Robert (1986b) Shaman. *Ural-Altaische Jahrbücher* 58: 143–144.

Austerlitz, Robert (1990) Typology in service of internal reconstruction: Saxalin Nivx. In Winfred Lehmann (ed.), *Language Typology 1987*, pp. 17–34. Amsterdam: John Benjamins Publishing Company.

Austerlitz, Robert (1994) Gilyak internal reconstruction, 3: Ligneous matter. In Howard I. Anderson (ed.), *Non-Slavic Languages of the USSR: Papers from the Fourth Conference [1985],* pp. 229–223. Columbus, Ohio: Slavica Publishers.

Barrere, Ian G. and Juha A. Janhunen (2019) Mongolian vowel harmony in a Eurasian context. *International Journal of Eurasian Linguistics* 1: 46–77.

Benzing, Johannes (1956) *Die Tungusischen Sprachen: Versuch einer Vergleichenden Grammatik.* Wiesbaden: Franz Steiner Verlag.

Blažek, Václav (1997) Altaic numerals. *Studia Etymologica Cracoviensia* 2: 33–75.

Bouda, Karl (1960) Die Verwandtschaftsverhältnisse des Giljakischen. *Anthropos* 55 (3/4): 355–415.

Burykin, A. A. (1987) Tunguso-man'čžursko-nivxskie leksičeskie paralleli i istorija nivxskogo vokalizma. *Lingvističeskie Issledovanija* (1987): 46–54.

Burykin, A. A. (1988) Tunguso-man'čžursko-nivxskie svjazi i problema genetičeskoj prinadležnosti nivxskogo jazyka. In G. A. Čemodanov (ed.), *Voprosy leksiki i sintaksisa jazykov narodov Krajnego Severa SSSR*, pp. 136–150. Leningrad: Leningradskij gosudarstvennyj pedagogičeskij institut.

Burykin, A. A. (1989) Tunguso-man'čžursko-nivxskie leksičeskie paralleli i istorija nivxskogo vokalizma II. *Lingvističeskie issledovanija* (1989): 23–25.

Burykin, A. A. (2005) Nazvanija metallov v nivxskom jazyke. In *Kjunerovskie čtenija (2001–2004): Kratkoe soderžanie dokladov*, pp. 177–185. Saint-Petersburg: MAÈ RAN.

Burykin, A. A. (2012) Ètnogenez nivxov, mežètničeskie svjazi v Priamursko-Saxalinskom regione i problema genetičeskoj prinadležnosti nivxskogo jazyka. In E. A. Rezvan (ed.), *Lev Šternberg – graždanin, učenyj, pedagog: K 150-letiju so dnja roždenija*, pp. 275–286. Saint-Petersburg: MAÈ RAN.

Burykin, A. A. and V.A. Popov. (2012) Problemy izučenija sistemy terminov rodstva i svojstva nivxov.

In V. A. Popov (ed.), *Algebra rodstva*, pp. 74–94. Saint-Petersburg: MAÈ RAN.

Campbell, Lyle (2013) *Historical Linguistics: An Introduction.* Edinburgh University Press.

Clauson, Gerard (1959) The earliest Turkish loan words in Mongolian. *Central Asiatic Journal* 4 (3): 174–187.

Doerfer, Gerhard (1985a) *Mongolo-Tungusica.* Wiesbaden: Harrassowitz.

Doerfer, Gerhard (1985b) Terms for aquatic animals in the Wu T'i Ch'ing Wên Chien. In *Proceedings of the International Symposium on B. Piłsudski's Phonographic Records and the Ainu Culture,* pp. 190–202.

Forsyth, James (1992) *A History of the Peoples of Siberia.* Cambridge University Press.

Fortescue, Michael (2011a) *Orientation Systems of the North Pacific Rim.* Copenhagen: Museum Tusculanum Press.

Fortescue, Michael (2011b) The relationship of Nivkh to Chukotko-Kamchatkan revisited. *Lingua* 121: 1359–1376.

Fortescue, Michael (2016) *Comparative Nivkh Dictionary.* München: LINCOM.

Grube, Wilhelm (1892) *I. Giljakisches Wörterverzeichniss nach den Originalaufzeichnungen von L. v. Schrenck und P. v. Glehn, Zusammengestellt und mit Grammatischen Bemerkungen Versehen.* St. Petersburg: Buchdruckerei der Kaiserlichen Akademie der Wissenschaften.

Gruzdeva, Ekaterina (1998) *Nivkh.* München: LINCOM.

Gruzdeva, Ekaterina (2015) Explaining language loss: The Sakhalin Nivkh case. In Heiko F. Marten, Michael Rießler, Janne Saarikivi, and Reetta Toivanen (eds.), *Cultural and Linguistic Minorities in the Russian Federation and the European Union*, pp. 233–252. Springer.

Gruzdeva, Ekaterina and Juha Janhunen (2018) The revitalization of Nivkh on Sakhalin. In Leanne Hilton, Leena Huss, and Gerald Roche (eds.), *The Routledge Handbook of Language Revitalization*, pp. 464–472. Routledge.

Gruzdeva, Ekaterina and Juha Janhunen (2020) Notes on the typological prehistory of Ghilyak. *International Journal of Eurasian Linguistics* 2: 1–28.

Gruzdeva, Ekaterina and Marina Temina (2020) Nivkh toponyms in the Amur-Sakhalin region of the Russian Far East. *Altai Hakpo* 30: 167–204.

Halm, Robert (2017) Application of the comparative method to vocoid sequences in Nivkh. *Kansas Working Papers in Linguistics* 38: 12–22.

Halm, Robert (2019) Initial clusters in pre-Proto-Nivkh: Internal reconstruction from Proto-Nivkh. *Kansas Working Papers in Linguistics* 40: 1–3.

Halm, Robert and Jay Slater (2018) Application of the comparative method to Nivkh: Other regular sound correspondences. *Kansas Working Papers in Linguistics* 39: 21–36.

Halm, Robert and Jay Slater (2020) Application of the comparative method to morpheme-final nasals in Nivkh. *Kansas Working Papers in Linguistics* 40: 1–21.

Haspelmath, M. (2009) Lexical borrowing: Concepts and issues. In M. Haspelmath and U. Tadmor (eds.), *Loanwords in the World's Languages: A Comparative Handbook*, pp. 35–54. Berlin: De Gruyter Mouton.

Hattori, Takesi (1962a) Bemerkungen zur Phonologie des Südostgiljakischen. *Studia Phonologica* 2: 1–26.

Hattori, Takesi (1962b) Versuch einer Phonologie des Südostgiljakischen (I): Phonembestand und Verteilung. *Journal of Hokkaido Gakugei University, Section 1 (A)* 13(1): 67–130.

Hudson, Mark J. (2004) The perverse realities of change: world system incorporation and the Okhotsk culture of Hokkaido. *Journal of Anthropological Archaeology* 23: 290–308.

Jakobson, Roman (1971) Notes on Gilyak. In *Roman Jakobson: Selected Writings II*, 72–97. The Hague: Mouton.

Janhunen, Juha (1981) Korean vowel system in North Asian perspective. *Hangeul* 172: 129–146.

Janhunen, Juha (1986) Siberian shamanistic terminology. *Mémoires de la Société Finno-Ougrienne* 194: 97–177.

Janhunen, Juha (1996) *Manchuria: An Ethnic History.* Helsinki: Mémoires de la Société Finno-Ougrienne.

Janhunen, Juha (1997) The Russian monsters: On the etymology of an ethnonymic complex. *Studia Etymologica Cracoviensia* 2: 159–165.

Janhunen, Juha (2005) The Eurasian shaman: Linguistic perspectives. In Juha Pentikäinen and Péter Simoncsics (eds.), *Shamanhood: An Endangered Language*, pp. 17–29. Oslo: Novus.

Janhunen, Juha (2008) Liao: A Manchurian hydronym and its ethnohistorical context. *Studia Etymologica Cracoviensia* 13 (1): 89–102.

Janhunen, Juha (2010) Reconstructing the language map of prehistorical Northeast Asia. *Studia Orientalia 108: 281–303.*

Janhunen, Juha (2011) Unicorn, mammoth, whale: Mythological connections of zoonyms in North and East Asia. *Linguistics, Archaeology and the Human Past* 12: 189–222.

Janhunen, Juha (2012) The expansion of Tungusic as an ethnic and linguistic process. In Andrej L. Malchukov and Lindsay J. Whaley (eds.), *Recent Advances in Tungusic Linguistics,* pp. 5–16. Wiesbaden: Harrassowitz.

Janhunen, Juha (2014) On the ethnonyms Orok and Uryankhai. *Studia Etymologica Cracoviensia* 19: 71–81

Janhunen, Juha (2016) Reconstructio Externa Linguae Ghiliacorum. *Studia Orientalia* 117: 3–27.

Joseph, Andrew, Seongyeon Ko and John Whitman (2020) A comparative approach to vowel systems and harmonies in the Transeurasian languages and beyond. In Martine Robbeets and Alexander Savelyev (eds.), *The Oxford Guide to the Transeurasian Languages*, pp. 486–508. Oxford University Press.

Kallio, Petri (2015) The stratigraphy of the Germanic loanwords in Finnic. In John Ole Askedal and Hans Frede Nielsen (eds.), *Early Germanic Languages in Contact*, pp. 23–38. Amsterdam: John Benjamins Publishing Company.

Kang, Yoonjung (2011) Loanword phonology. In Marc van Oostendorp, Colin Erwen, Elizabeth Hume and Keren Rice (eds.), *Companion to Phonology*, pp. 2258–2282. Malden: Blackwell.

Knapen, Martijn G.T.M. (2021) *The Oldest Layer of Amuric-Tungusic Lexical Contacts.* Master's thesis, University of Helsinki.

Ko, Seongyeon, Andrew Joseph and John Whitman (2014) Comparative consequences of the tongue root harmony analysis for Proto-Tungusic, Proto-Mongolic, and Proto-Korean. In Martine Robbeets and Walter Bisang (eds.), *Paradigm Change in the Transeurasian Languages and Beyond*, pp. 141–176. Amsterdam: John Benjamins Publishing Company.

Krejnovič, E. A. (1937) *Fonetika nivxskogo (giljackogo) jazyka.* Moscow: Gosudarstvennoe učebno-pedagogičeskoe izdatel'stvo.

Krejnovič, E. A. (1955a) Giljacko-tunguso-man'čžurskie jazykovye paralleli. *Doklady i soobščenija Instituta jazykoznanija AN SSSR* 8: 135–167.

Krejnovič, E. A. (1955b) *Slovar' nivxskogo jazyka.* Unpublished manuscript.

Laufer, Berthold (1902) The decorative art of the Amur tribes. *Memoirs of the American Museum of Natural History* VII: 1–86.

Li, Bing (1996) *Tungusic Vowel Harmony: Description and Analysis.* Doctoral dissertation, University of Amsterdam.

Li, Tao, Chao Ning, Irina S. Zhushchikhovskaya, Mark Hudson and Martine Robbeets (2020) Millet agriculture dispersed from Northeast China to the Russian Far East: Integrating archaeology, genetics, and linguistics. *Archaeological Research in Asia* 22: e100177.

Millar, Robert McColl and Larry Trask (2015) *Trask's Historical Linguistics*. Routledge.

Möllendorff, P. G. von. (1894) The Ghilyak language. *China Review* 21 (3): 141–146.

Nichols, Johanna (1996) The comparative method as heuristic. In Mark Durie and Malcolm Ross (eds.), *The Comparative Method Reviewed: Regularity and Irregularity in Language Change*, pp. 39–71. Oxford University Press.

Nugteren, Hans (2011) *Mongolic Phonology and the Qinghai-Gansu Languages.* Utrecht: LOT.

Otaina, Galina A. and Vladimir P. Nedjalkov (2007) Reciprocal constructions in Nivkh (Gilyak). In Vladimir P. Nedjalkov (ed.), *Reciprocal Constructions*, pp. 1715–1747. Amsterdam: John Benjamins Publishing Company.

Panfilov, Vladimir Zinov'evich (1973) Nivxsko-altajskie jazykovye svjazi. *Voprosy jazykoznanija* 6: 3–12.

Pevnov, Alexander M. (1992) Nivxskij i tunguso-man'čžurskie jazyki: Problemy kontaktov. In: *B. O. Pilsudskij - issledovatel' narodov Saxalina (materialy meždunarodnoj naučnoj konferencii. 31 oktjabrja – 2 nojabrja 1990g. g. Južno-Saxalinsk) – T. 2, 25–29.* Južno-Saxalinsk: Saxalinskij oblastnoj kraevedčeskij muzej.

Pevnov, Alexander M. (2009) On some features of Nivkh and Uilta (in connection with prospects of Russian-Japanese collaboration). In Toshiro Tsumagari (ed.), *Saharin no gengo sekai: Hokudai bungakukenkyūka kōkai symposium hōkokusho*, pp. 113–125. Sapporo: Hokkaido University.

Pevnov, Alexander M. (2012) The problem of localization of the Manchu-Tungusic homeland. In Andrej L. Malchukov and Lindsay J. Whaley (eds.), *Recent Advances in Tungusic Linguistics,* pp. 17–40. Wiesbaden: Harrassowitz.

Pevnov, Alexander M. (2016) On the specific features of Orok as compared to the other Tungusic languages. *Studia Orientalia* 117: 47–63.

Pevnov, Alexander M. (2018) Lingvističeskie svidetel'stva istoričeskix kontaktov orokov i oročej na Saxaline. *Acta Linguistica Petropolitana* XIV (1): 441–463.

Pulleyblank, Edwin G. (1991) *Lexicon of Reconstructed Pronunciation in Early Middle Chinese, Late Middle Chinese, and Early Mandarin.* Vancouver: UBC Press.

Robbeets, Martine (2017) Proto-Trans-eurasian: where and when? *Man in India* 97 (1): 19–46.

Robbeets, Martine, Juha Janhunen, Alexander Savelyev and Evgeniya Korovina (2020) The homelands of the individual Transeurasian proto-languages. In Martine Robbeets and Alexander Savelyev (eds.), *The Oxford Guide to the Transeurasian Languages*, pp. 753–771. Oxford University Press.

Rozycki, William (1994) *Mongol Elements in Manchu.* Bloomington, IN: Indiana University, Research Institute for Inner Asian Studies.

Sangi, Vladimir Mikhailovich and L. B. Gašilova (2016) *Slovar' nivxsko-russkij (saxalinskij dialekt)*. Saint Petersburg: Posveščenie.

Savel'jeva, V. N., and Č. M. Taksami (1970) *Nivxsko-russkij slovar'*. Moscow: Sovetskaja ènciklopedija.

Schrenck, Leopold von (1881) *Reisen und Forschungen im Amur-Lande in den Jahren 1854–1856*. Vol. 3 (1). *Die Völker des Amur-Landes: Georaphisch-Historischer u. Anthropologisch-Etnologischer Theil*. St. Petersburg: Buchdruckerei der Kaiserl. Akademie der Wissenschaften.

Schrenck, Leopold von (1891) *Reisen und Forschungen im Amur-Lande in den Jahren 1854–1856*. Vol. 3 (2). *Die Völker des Amur-Landes: Ethnographischer Theil – Erste Hälfte*. St. Petersburg: Buchdruckerei der Kaiserl. Akademie der Wissenschaften.

Smoljak, Anna Vasil'evna (1984) *Tradicionnoe xozjajstvo i material'naja kul'tura narodov Nižnego Amura i Saxalina*. Moscow: Nauka.

Shiraishi, Hidetoshi (2006) *Topics in Nivkh Phonology*. PhD dissertation, Rijksuniversiteit Groningen.

Shiraishi, Hidetoshi (2009a) Modelling initial weakenings. In Kuniya Nasukawa and Phillip Backley (eds.), *Strength Relations in Phonology*, pp. 183–218. Berlin: De Gruyter.

Shiraishi, Hidetoshi (2009b) On the alternation between CVCVC and CVCC forms in Nivkh. In Toshiro Tsumagari (ed.), *Saharin no gengo sekai: Hokudai bungakukenkyūka kōkai symposium hōkokusho*, pp. 85–94. Sapporo: Hokkaido University.

Shiraishi, Hidetoshi and Bert Botma (2016) Asymmetric distribution of vowels in Nivkh. *Studia Orientalia* 117: 39–46.

Shiraishi, Hidetoshi and Bert Botma (2017) On the diachronic origin of Nivkh height restrictions. In Geoff Lindsey and Andrew Nevins (eds.), *Language Faculty and Beyond 14*, pp. 202–214. Amsterdam: John Benjamins Publishing Company.

Shoda, Shinya, Alexandre Lucquin, Oksana Yanshina, Yaroslav Kuzmin, Igor Shevkomud, Vitaly Medvedev, Evgeniya Derevianko, Zoya Lapshina, Oliver E. Craig and Peter Jordan (2020) Late Glacial hunter-gatherer pottery in the Russian Far East: Indications of diversity in origins and use. *Quaternary Science Reviews* 229: 1–11.

Shternberg, Lev Iakovlevich (1999) *The Social Organization of the Gilyak*. American Museum of Natural History.

Takahashi, Moritaka (1942) *Karafuto Giriyakugo*. Osaka: Asahi Shinbun.

Taksami, Č. M. (1975) *Osnovnye problemy ètnografii i istorii Nivxov*. Leningrad: Nauka.

Tangiku, I., N. Ja. Tanzina, and N. V. Nitkuk (2008) *Basic Vocabulary of the Sakhalin Dialect of Nivkh Language (Nogliki Dialect)*. Tokyo: ILCAA.

Tangiku, Itsuji (2015) *Ainugo to hoppō gengo no sesshoku: Ainugo Karafuto hōgen no kōekihin, dōshokubutsumei ni okeru shakuyō goi*. Unpublished lecture handout.

Tangiku, Itsuji (n.d.) *Common Words for Instruments among Nivkh, Uilta and Ainu*. Unpublished lecture handout.

Vovin, Alexander (2016) On the linguistic prehistory of Hokkaidō. *Studia Orientalia* 117: 29–38.

Wang, Chuan-Chao and Martine Robbeets (2020) The homeland of Proto-Tungusic inferred from contemporary words and ancient genomes. *Evolutionary Human Sciences* 2: e8.

Whaley, Lindsay J. and Sofia Oskolskaya (2020) The classification of the Tungusic languages. In Martine Robbeets and Alexander Savelyev (eds.), *The Oxford Guide to the Transeurasian Languages*, pp. 81–91. Oxford University Press.

Yamaguchi, Kazuhiko and Katsunobu Izutsu (Eds.) (2004) Publication Series of Hattori Bunko 2: Nivkh-English Lexical Materials (Poronaisk Dialect). *Bulletin of the Hokkaido Museum of Northern Peoples* 13: 36–58.

Zgusta, Richard (2015) *The Peoples of Northeast Asia through Time: Precolonial Ethnic and Cultural Processes along the coast between Hokkaido and the Bering Strait*. Leiden: Brill.

7

Sino-Tibetan, Yeniseian and Na-Dene

David Bradley

Abstract: While it is normally not considered possible to use standard comparative linguistic methodology to investigate linguistic relationships beyond about 6000 BP due to the usual processes of change in every language, it appears that there is a variety of linguistic evidence for an earlier linkage between the Sino-Tibetan languages of East Asia, the Yeniseian languages of Siberia, and the Na-Dene languages of northwestern North America. This paper summarises earlier proposals for the wider connections of these three well-established language families and presents a variety of new linguistic evidence for their linkage. This evidence is then briefly placed in the context of current archaeological and historical understanding, ancient and modern genome studies, and the structure of the societies of the speakers of these languages. Finally, a tentative phylogeny and chronology is suggested.

Keywords: Sino-Tibetan, Yeniseian, Na-Dene, remote linguistic linkage

7.1. Prior Comparative/Historical Linguistic Studies

Most of the established language families of the world have undergone extensive and rigorous linguistic comparison. Sino-Tibetan (ST) and Na-Dene (ND) are among those which have thorough reconstructions; available data on Yeniseian languages has also been well analysed.

In order to establish genetic relationship between languages, it is essential to demonstrate systematic sound correspondences in core vocabulary within the languages which has not been borrowed from other languages. For Proto-ST, the overall framework is provided by Benedict (1972), with numerous more recent updates by various scholars. Matisoff (2003) provides a reconstruction of Proto-Tibeto-Burman (TB), one major subgroup of Proto-ST; Baxter and Sagart (2014) provide a reconstruction of Proto-Sinitic, the other major subgroup. There is also a long tradition of comparative linguistic research within Proto-ND, culminating in Leer (2011) on Proto-Athapaskan, which is the major subgroup of Proto-ND. Until recently, most comparative work on Yeniseian languages was in Russian, but the work of Vajda (2004), Georg (2007) and Kotorova and Nefedov (2015) provide excellent and thorough documentation of Ket, the sole surviving Yenisean language, in English. Most scholars working on these three language families have not interacted with each other, and they use different phonetic symbols and terminologies, which obscure similarities.

Apart from regular sound correspondences in a substantial body of non-borrowed vocabulary, it is also normal to expect similarities of form and structure in morphology (word formation) and syntax (grammar) in related languages. Most comparative/historical linguists reject any proposed genetic linguistic relationships which are not based on systematic comparison and reconstruction of all three of these: sound correspondences, shared non-borrowed vocabulary and shared morphology. Wolfenden (1929) did solid early comparative work on Proto-ST morphology, which has been continued by many scholars since. From Sapir on, many linguists have worked on Proto-ND morphosyntax.

Vajda (2010, 2012, 2013, 2018) has documented a large number of sound, vocabulary and morphological correspondences between Proto-ND and Yeniseian, which have clearly and solidly demonstrated a linguistic relationship. The purpose of this article is to add further data showing how Proto-ST may be related to Proto-ND and Yeniseian, and how this linkage fits with what we know about the earlier societies from the relevant etyma (reconstructed vocabulary items) and from archaeology, history, ancient and modern genetics and social structure.

7.1.1. Problematic methodologies for establishing linguistic relationships

Most historical linguists agree that it is in principle impossible to reconstruct securely beyond about 6000 BP (see for example Campbell and Poser 2008). On the other hand, there are a few linguists such as Greenberg (2005) who have tried to do exactly that, calling their approach 'mass comparison' and proposing very remote and ancient linguistic relationships based on limited lexical evidence, with little or no attempt at showing sound correspondences or morphosyntactic similarities. Examples include Greenberg's claim that most languages of the Americas other than Eskimo-Aleut and ND form one macrofamily (which he called Amerind), or his claim that most of the languages of Eurasia can be linked in what he calls the Eurasiatic macrofamily. These and other hypotheses of

Greenberg have led to extreme controversy and most are rejected by nearly all historical linguists.

There are other linguists doing long-distance comparison, mainly Russians such as S. Starostin (1984/1991), who attempt both vocabulary and sound correspondences, but allow extreme latitude in meaning changes in the words compared and in the regularity of the sound changes required to link them. Shevoroshkin and Manaster Ramer (1991) summarise much of this work up to the end of the 1980s; Georgiy Starostin (2012) tries to add rigor by limiting the words compared to a small predetermined basic wordlist.

There are also linguists who do pairwise comparison of a well-established language family and a language isolate and find similarities; for example Čašule (2016, 2017 and so on) comparing Indo-European and the South Asian isolate Burusho (also known as Burushaski), or Holst (2017) comparing Kartvelian (South Caucasian) and Burusho. These approaches are problematic; most do not follow the rigorous requirements of the established comparative method. Also, comparing any two sets of linguistic data will certainly allow some chance similarities to be located in a large enough corpus of data, particularly if semantic and phonological rigor is relaxed.

7.1.2. Proposals concerning the linkages of ST, Yeniseian and ND

There were several early proposals linking the Proto-ST and Proto-ND languages. The eminent ND specialist Edward Sapir is said to have believed in this connection from the early 1920s, and wrote so in letters to colleagues, but he never published his results on this topic. Swadesh (1952, 1971), a student of Sapir, reported on Sapir's views much later, after the ST specialist Robert Shafer produced a series of articles giving examples of possible shared ST/ND etyma (1952, 1959, 1967); this was followed up by Bengtson (1994). Trombetti (1922/1923) also linked Proto-ST and Proto-ND within the context of his now-discredited attempt to show that all human languages are ultimately related.

Toporov (1971) compared the South Asian isolates Burusho and Kusunda with Yeniseian and ND. Another set of proposals emanating from the Russian school linked Yeniseian instead with ST and with northwest and northeast Caucasian, under the name Sino-Caucasian (S. Starostin 1984/1991, 1994). This was extended by Bengtson (1991) who added nearly all the language isolates further west in Eurasia: Basque, Etruscan, Sumerian, Burusho and so on, under the name Macro-Caucasian. Nikolaev (1991), following Toporov, further connected Sino-Caucasian with ND. Ruhlen (1994), working within Greenberg's version of mass comparison, accepted S. Starostin's Sino-Caucasian and its linkage with ST and ND as proposed by Sapir, Trombetti, Swadesh, Shafer and Nikolaev under the name Dene-Caucasian, as well as Bengtson's additions of Basque, Burusho and possibly Sumerian, and added the South Asian isolate Nahali. Conversely, van Driem (2021: 200–209) supported the linkage of Yeniseian and ND and the observations of Čašule concerning the linkage of Burusho and Indo-European, but not a connection of Yeniseian and ND with ST.

Benedict (1942) demonstrated a more solidly-based linkage between Proto-Austronesian (Proto-AN), another well-established language family whose current maximum diversity and probable point of dispersal is modern Taiwan, and Proto Tai-Kadai of Southern China and mainland Southeast Asia; this linkage has since been further documented by many others. Benedict (1975) provided extensive additional examples of cognates, but unwisely further extended this to include the unrelated Miao-Yao language family, also more recently known as Hmong-Mien (Ratliff 2010). Benedict (1990) added Japanese, a Transeurasian language related to Korean and the Altaic languages, which discredited his original insight for most scholars working within Proto-AN and Transeurasian. Sagart (1993, 2005) proposed instead to link Proto-ST with Proto-AN in what he calls STAN, with a hypothetical Urheimat at the mouth of the Yellow River on the coast of eastern China. However, STAN has been completely rejected by nearly all Proto-AN and Proto-ST historical linguists.

Elements of the proposal for the linkage made in this paper between ST, Yeniseian and ND, as well as competing proposals, have thus been in the linguistic literature for a long time in various forms, but no one has used the full range of comparative linguistic methodology to clarify and solidify the relationships. This may be because the time depth of the linkage is so remote that normal processes of internal change within every language as well as contact with adjacent languages have obscured the similarities. Here I compare basic vocabulary and morphosyntax in not two but three established families, which greatly reduces the possibility of chance similarities.

Figure 7.1 summarises the distribution of the ST languages in China and surrounding areas, the Yeniseian languages in Russia, and the ND languages in northwestern North America. The ND and particularly the Yenisiean languages have had their distribution and diversity reduced over recent centuries. A likely Urheimat for each family is indicated by dotted lines.

7.2. Linguistic Evidence for a Sino-Tibetan/Yeniseian/Na-Dene Linkage

The evidence will be presented in the order (1) syntax, (2) morphology, (3) phonology and (4) lexicon. Within (2), examples will be given from verb and noun morphology. Within (4), examples will be drawn from basic vocabulary, including three structured areas of the lexicon which may be more resistant to change: motion up and down, numerals and pronouns.

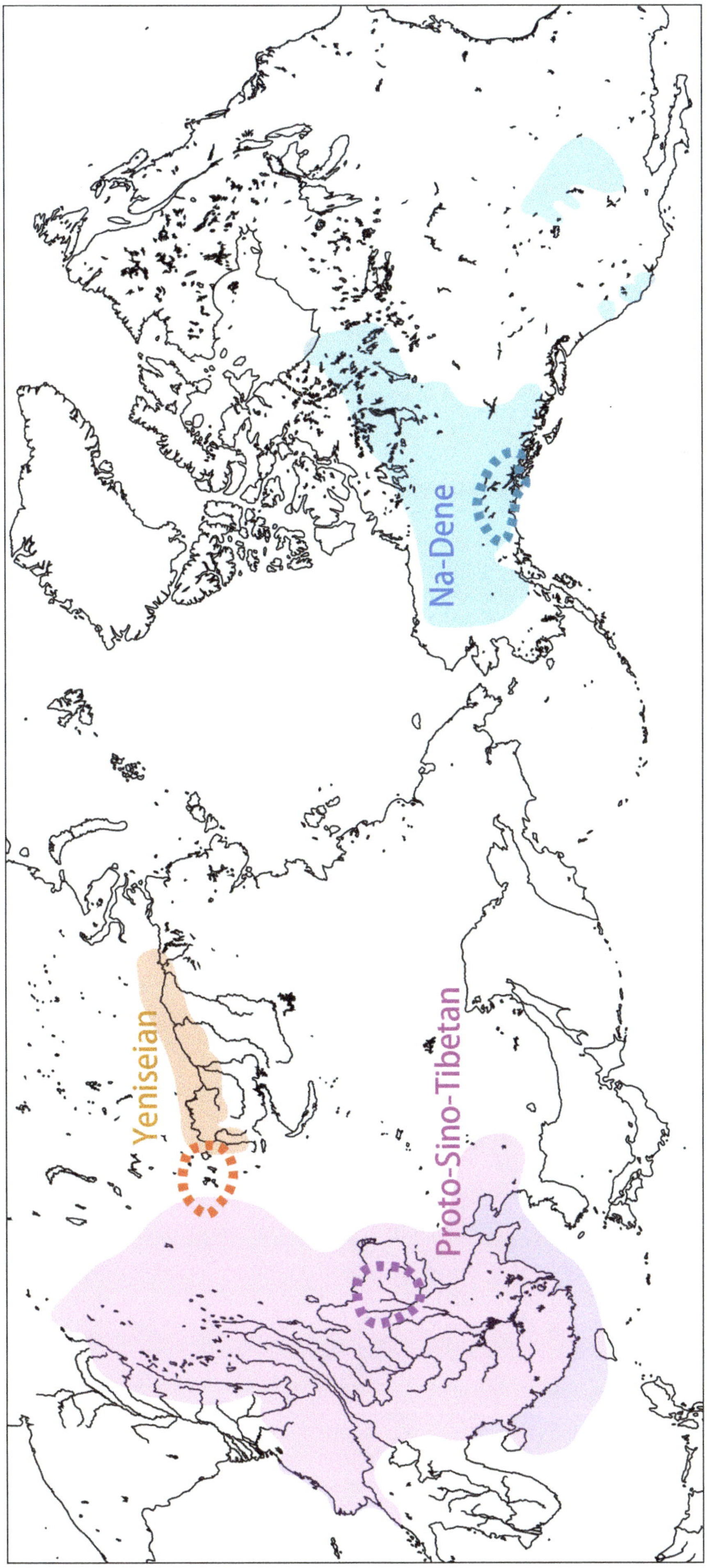

Figure 7.1. Distribution of the Sino-Tibetan, Yeniseian and Na-Dene languages (colour shading) with probable homelands for each family shown in dotted lines. Source: Map by J. Uchiyama.

7.2.1. Syntax

There are numerous syntactic similarities between ST, Yeniseian and ND languages. Nearly all of the many hundreds of languages are verb-final; the main exception is Chinese whose shifted from verb-final to verb-medial was not complete by the Classical Chinese period (2500 BP onward) as copula clauses were still copula-final. Of the TB languages, the only ones with verb-medial main clause word order are Bai, which has been under very strong verb-medial Chinese influence for two millennia, and the Karenic languages mainly in Burma, which were in contact with a variety of verb-medial Austroasiatic languages in the area for many millennia. All Sinitic and Karenic languages as well as Bai show a wide variety of conservative syntactic characteristics more typical of verb-final languages, which they share with other ST, Yeniseian and ND languages. These include various suffixes on nouns and on verbs; relative clauses marked with a clause-final relative marker followed by the head noun; nominalised clauses marked by a clause-final nominaliser; noun possessor preceding possessed noun, and so on. Having the verb in final position in the clause is one of the two dominant main clause orders in languages worldwide, so it is possible that these parallels are accidental and due to general linguistic typological tendencies.

There are also some less widespread parallels. There is a tendency to conflate relativisation and nominalisation strategies by using similar markers and often having no following head noun; adjectives are a subclass of verbs; nouns are often absent from sentences when they are already clear from the context, and do not need to be replaced by pronouns; and so on.

7.2.2. Morphology

It is typologically usual for verb-final languages to have suffixing morphology. However, all three language families show extensive evidence of prefixing morphology, both productive and fossilised. This tendency to prefixing is extreme on verbs in Yeniseian and ND languages, which have long sequences of highly productive prefixes for a wide variety of functions, many exactly parallel between the two; Vajda (2010) discusses these parallels in depth. There is also evidence for prefixing on noun and related forms.

All three language families also have some suffixing morphology, particularly on nouns. In Proto-ST, there is also more innovative prefixing and suffixing morphology; some categories expressed by prefixes in Yeniseian and in ND languages are expressed by suffixes in ST languages. These include verbal markings such as tense/aspect, agreement with nouns, and so on.

7.2.2.1. Verb morphology

There are very striking similarities in some highly frequent and productive preverbal and postverbal grammatical elements. These include the three prefixes and one suffix shown in Table 7.1. Note that all reconstructed forms in

Table 7.1. Verb morphology in Proto-ST, Ket and Proto-Athapaskan.

	Proto-ST	Ket	Proto-Athapaskan
Negative	***ma-**	bən-	***iʔ-/ʔi-** or ***s-**
Prohibitive	***ta-/da-**	ət-	***da-**
Valency Increase	***s-**	s-	***ɬ-**
Nominaliser	***-su**	-s	***-hi**

Table 7.1 and all other forms below have been adjusted to standard International Phonetic Alphabet symbols. Proto-ST etyma are from Benedict (1972) and Proto-Athapaskan etyma are from Leer (2011) except where noted. Proto-Athapaskan forms are cited instead of Proto-ND forms as the latter have few generally-agreed reconstructions.

As we will see in section 2.4.1 below, metathesis (reversal of segment order) in Ket as seen in the prohibitive and partly in the negative is very frequently observed in Ket cognates of forms reconstructed for Proto-ST and/or Proto-ND. The Proto-Athapaskan negative prefix appears not to be cognate with the Proto-ST and Ket forms, but it is still a prefix, which as noted in 2.1 above is not expected in a verb-final language. In some cases, the functions of the cognate forms differ; for example, the Proto-ST ***-su** suffix is an agentive nominaliser, but in Ket, the /-s/ suffix is an abstract nominaliser. Terminological differences have also obscured cognacy: the Proto-ST valency-increasing verb prefix ***s-** has since Wolfenden (1929) been called a causative prefix, but the Proto-Athapaskan valency-increasing prefix ***ɬ-** is called a classifier; Ket also has an innovative and more productive valency-increasing verb suffix /-q/; both are also sometimes called causatives.

7.2.2.2. Noun morphology

There is not much prefixed noun morphology which can be reconstructed across the three families, but noun stems, to a greater extent than verb stems, show a number of prefixes which can be reconstructed, particularly within Proto-ST; these may reflect earlier noun compounding and/or morphology. There is also some shared suffixing morphology, such as a female suffix ***-ma** which may be related to the 'mother' etymon (see 2.4.1 below).

Most of the Proto-ST noun morphology is innovative, often transparently grammaticalised. For example, there is a body-part prefix ***m** which is presumably derived from the Proto-ST noun ***mi** PERSON, as seen in LIVER in Table 7.3 below. Some of the apparent irregularities in phonological correspondences of initial consonants in noun forms may be due to differences in prefixation between the three families. See for example the discussion of numerals (2.4.3), which show various different prefixes and suffixes in Proto-ST, Yeniseian and Proto-ND.

In most ST languages other than Chinese, and in Yeniseian and ND, case, topic and other grammatical markings are postpositions after the noun, as expected in a verb-final language. Both Yeniseian and ND languages often incorporate nouns into the verb, usually in the middle of the prefixes which precede a verb stem; the origin of the verb agreement system of Proto-ST is pronouns which have been incorporated as suffixes on the verb stem. In Sinitic languages, since the change to verb-medial main clause word order, various verbs have been grammaticalised into prepositional constructions; the locational verb 'be at' becomes the nominal preposition 'at', the verb 'give' becomes the nominal preposition 'for' and so on.

7.2.3. Phonology

There are various strong parallels between the reconstructed sound systems of Proto-ST and Proto-ND, some of which are shared with Yeniseian Ket. The initial consonant systems of Proto-ST and Proto-ND are very complex, with three distinct manners of syllable-initial stops and coronal affricates, also some typologically unusual places of articulation including uvulars. Ket has merged nearly all manners of articulation, maintaining only a distinction between /t/ and /d/; otherwise, all Ket stops are underlyingly voiceless. There are also complex consonant clusters reconstructed in syllable-initial position, also actually attested in some modern languages; maximally one or two consonantal prefixes, an initial, then a medial liquid or glide. In many languages, and in the Proto-ND reconstruction, some such clusters are separated by weak vowels; this is also seen in a variety of modern ST languages.

By contrast, in Proto-ST the syllable-final consonants are very restricted: usually only one consonant, with only one manner of stop consonants, ***p *t *k *q**, three nasal consonants ***m *n *ŋ**, liquids ***r** or ***l**, or occasionally with a suffixed ***s** following a vowel or following another consonant. Proto-ND has somewhat fewer restrictions on final consonants, and Ket of course has a reduced consonant system overall.

The typologically unusual uvular stops are retained in Proto-ND and many modern ND languages and in Ket. They have been merged with other consonants in most ST languages, but can be reconstructed for Proto-ST, both initially (Baxter and Sagart 2014) and finally (Sagart 2017).

There is a widespread tendency for ST, Yeniseian and ND languages to develop systems of tones; these often include not only pitch differences, but also phonation differences (breathy, modal or creaky voice) and other characteristics. Most reconstructions suggest that these tones are secondary, conditioned by differences in the manner of initial consonants. Some unrelated languages in close contact with ST languages also develop tone systems in parallel ways; this includes all Tai-Kadai languages and all Miao-Yao languages, as well as some Austroasiatic languages spoken in areas strongly influenced by speakers of ST languages, such as Vietnamese.

In comparing cognate lexical material and reconstructing phonology in related languages, there are systematic

Table 7.2. Correspondences in Proto-ST, Yeniseian, Proto-ND and Proto-Athapaskan.

	Proto-ST	Ket	Proto-ND	Proto-Athapaskan
CRY	***ŋu**	qoʔd	***kʷəχ**	***tʃrəʁ**
EVIL	***ŋan**	qiliŋ	***k'ʷa:n**	***tʃ'ra:**
WOMB	***ŋal**	daqaj 'nest'	***kʷa:l**	***tʃra:tɬeʔ**
SUN	***ŋi**	i/iqot	***gʷe**	***dʒre:j**
WRINKLED	***ŋar**	kudaŋ	***k'ʷal**	***tʃ'rətɬ'**

sound correspondences showing the same sound change; sometimes the sound changes are very substantial, but consistent. Caveney (2014) has identified one such correspondence: in some etyma, reconstructed Proto-ST initial velar nasal ***ŋ** corresponds to reconstructed Proto-ND labialised velar stops and reconstructed Proto-Athapaskan palatoalveolar affricates. Table 7.2 shows that most Ket forms have a voiceless uvular or occasionally velar stop in their cognates. For another possible example, see FIVE in 2.4.3 and for a partial parallel see the 1st person pronoun in 2.4.4.

As more reconstructions for Proto-Yeniseian/ND become available, we will probably find many additional correspondence patterns.

7.2.4. Lexicon

Various sources have listed a large number of possible lexical cognates between ST and ND, including Shafer (1952, 1957, 1959), Nikolaev (1991), Bengtson (1991) and Ruhlen (1994). Vajda (2010, 2012) identifies a large number of likely Yeniseian/ND cognates, some contested by G. Starostin (2012) who cites ST, North Caucasian and Burusho alternatives. The extremely valuable Ket dictionary of Kotorova and Nefedov (2015) now allows much more extensive comparison with Yeniseian than has ever previously been possible.

7.2.4.1. Basic vocabulary

A few of the many apparent cognates across all three families are listed in Table 7.3. In some examples, forms in Ket, which is only attested over the last couple of centuries, have undergone meaning change.

Table 7.3. Proto-ST, Yeniseian and Proto-Athapaskan lexical similarities.

	Proto-ST	Ket	Proto-Athapaskan
HAND	***g-lak**	laʔŋ	***tlaʔ**
LIVER	***m-sin**	seŋ	***sənt'**
DOG	***kwi(n)**	kəqən 'fox'	***tlEɲ-k'je**
DRINK	***daŋ**	dóò	***ta:**
WATER	***twe**	tik 'snow'	***tu**

Table 7.4. Examples with metathesis in Yeniseian.

	Proto-ST	**Ket**	**Proto-Athapaskan**
MOTHER	***ma/*nu**	am	***ʔEne:**
FATHER	***ba**	op	***waʔ**
PERSON	***mi**	keʔd sg., deʔŋ pl.	***dEne:**
WOMAN	***Cmi**	qim	***tʃʰrêqe:**
GRANDFATHER	***puw**	qip	***Owe̱**

In some cases, the Yeniseian cognates show metathesis (reversal of segment order); this process is widespread within Ket. Table 7.4 shows some examples. PERSON is an example which shows that Yeniseian and ND show greater lexical similarity: the plural stem in Ket and the usual Proto-Athapaskan reconstructed form are cognate, but the Proto-ST etymon shows up in Ket only in the WOMAN etymon, where the metathesised Ket cognate has no cognate in Proto-Athapaskan. The MOTHER etymon ***ma** is attested in Proto-ST and in Ket; the alternative Proto-ST form BREAST/MOTHER ***nu** may be related to the Proto-Athapaskan form.

7.2.4.2. Up and down

The basic motion verbs for upward and downward motion reconstructed for Proto-ST appear to be cognate with the Ket forms for uphill and downhill motion, and with the Proto-ND directional forms for upward and downward motion. The forms are presented in Table 7.5. Leer (1999) discusses the Proto-ND forms, and Vajda (2013: 11) connects the Proto-ND and Ket forms.

Table 7.5. Up and Down.

Proto-ST	**GO UP**	***dak**	**GO DOWN**	***jak**
Ket	uphill	-aged-	downhill	-igd-
Proto-ND	UPWARD	***dəq**	DOWNWARD	***jəχ**

7.2.4.3. Numerals

It is often but not always the case that lower numerals tend to be stable cognates. In the East Asian linguistic area, even these have been borrowed, for example from Sinitic into Tai-Kadai for all but ONE and from various ST sources into Miao-Yao for all but TWO and THREE. However, as there has been no recent contact between ST, Yeniseian and ND languages, any similarities must be residual. There is also considerable internal diversity in the reconstructed form of ONE in Proto-ST. The Ket and Kott forms in Table 7.6 are from Werner (2004). Note that there is various innovative morphology: several alternative prefixes in Proto-ST, a /-k/ suffix in some forms in Yeniseian which voices to [g] when intervocalic in Kott, a distinct animate stem for ONE in Ket, an /-a/ suffix in Kott and so on. Proto-Athapaskan also has various animate and inanimate suffixes which are not included in the forms given here.

Once the affixes are stripped off, there are resemblances between Proto-ST and Yeniseian for all seven numeral stems listed, which extend to Proto-Athapaskan for ONE to FOUR.

The Ket form for TWO again shows metathesis compared to Proto-ST. In FIVE, we again see a velar nasal initial corresponding to a Ket uvular voiceless stop initial, but the Proto-Athapaskan forms for FIVE, SIX and SEVEN appear not to be cognate. Note also some parallels in compound forms, especially SEVEN (5+2) in Proto-ST and Kott. It is notable that without Kott data, it would be difficult to discern some of the links between Proto-ST and Yeniseian; with more data from the many Yeniseian languages which are no longer spoken, perhaps even more could have been done.

7.2.4.4. Pronouns

Pronouns form a small closed grammatical system and often have stable forms and meanings, especially for Speech Act Participant (SAP, 1st and 2nd person) forms. Table 7.7 shows the SAP pronouns. We see likely cognates for all three reconstructed forms. The Proto-ST 1st and 2nd person forms are from Benedict (1972) as usual; the 1st

Table 7.6. Numerals.

	ST		**Ket**	**Kott**	**Proto-Athapaskan**
ONE	***it/*kat/**	animate	qɔʔk		***tlÂq'**
	***g-tjik**	inanimate	qu·s	hu:tʃa	
TWO	***g-ni-s**		i·n	i:na	***nâ(dE)qe**
THREE	***g-sum**		dɔʔŋ	to:ŋa	***ta:q**
FOUR	***b-li**		si·k	(t)ʃe:ga	***denk**
FIVE	***l-/b-ŋa**		qa·k	kega	***nelaʔi**
SIX	***d-ruk**		aʔ	χelutʃa	***qusdEta:nI**
SEVEN	***s-/k-ni-s**		ɔʔn	χelina	***kjusk'jEdi:/**
					***qUnEts'Eʁeji:**

Table 7.7. SAP Pronouns in Proto-ST, Yeniseian and Proto-Athapaskan.

	Proto-ST	Ket/Kott	Proto-Athapaskan
1	***ŋa(i)**	ad/ŋai	***ʃwí˜**
2	***naŋ**	u	***ni~**
1+2	***ni**	ətn	***ne~**

inclusive form (you and I, 1+2) is reconstructed in Bradley (1979/1991 etymon 442–1). The 1st person form shows an initial velar nasal in Proto-ST corresponding in this case to a Proto-Athapaskan labialised pataltoalveolar affricate, similar to the correspondence we saw in Table 7.2, but the Yeniseian forms do not follow the expected pattern: Ket has an innovative form for both 1st and 2nd person, while Kott has a 1st person form identical to the Proto-ST form. We again see metathesis in the Ket 1st inclusive pronoun.

Pronouns are often cliticised onto other forms. In all three language families being compared, we see that SAP possessive pronouns, sometimes in shortened forms, are cliticised before the possessed noun. This Possessor + Possessed order is also found for possessed nouns when the possessor is not a pronoun. Note also that we see a likely cognate for the 2nd person pronoun in the Ket possessive 'your' form which has been replaced in the free pronoun form. Possessive prefixes are phonologically reduced in Proto-Athapaskan and in Yeniseian.

7.2.5. Summary of linguistic evidence

As we have seen, there is substantial evidence of syntactic, morphological, phonological and lexical characteristics shared between ST, Yeniseian and ND. These are particularly strong in stable areas such as basic structural features including negation, prohibition, valency-increase and so on; also in SAP pronouns, lower numerals, basic kinship terms and so on.

7.3. Archaeological and Historical Correlates

It is generally agreed that the location of the Proto-ST speech community can be associated with the Neolithic Yangshao (仰韶) Culture (7000–5000 BP) in the upper Yellow River valley in north central China (Sagart et al. 2019; Zhang M. et al. 2019; Bradley 2022), or possibly somewhat earlier (Zhang H. 2020). Dates proposed in recent Bayesian lexical phylogenies for the separation of Proto-ST into its component subfamilies Sinitic and Tibeto-Burman are 5871 BP (Zhang M. et al. 2019), 7184 BP (Sagart et al. 2019) and 7983 BP (Zhang H. et al. 2020); note that many linguists do not accept the methodology behind such dates, two of which are substantially earlier than comparative linguistic and archaeological evidence suggest. The connection of Yangshao with PST is supported by the reconstruction of PST etyma for domestic animals including domestic DOG, PIG and COW and for crops including *SETARIA* MILLET, *PANICUM* MILLET, SOYBEAN and JOB'S TEARS (Bradley 1997, 2011, 2016, 2017, 2022), all of which are attested in the archaeological record in Yangshao sites. The earliest split within PST was most likely between Proto-TB, usually associated with Majiayao (马家窑) Culture (5300–4000 BP) to the southwest of the Yangshao Culture, and Proto-Sinitic, usually associated with the Longshan (龙山) Culture (5000–3900 BP) in the original Yangshao Culture area and spreading gradually to the east. Proto-Western TB or Proto-Tibetic can be associated with the middle Karuo (卡若) Culture (4700–4200 BP) to the southwest of Majiayao Culture and its later offshoots. This is summarised in Figures 7.2 and 7.3. It was only after the split into PTB and Proto-Sinitic that these groups started to grow rice, and so there is no shared PST etymon for RICE nor for other later crops such as BUCKWHEAT, BARLEY and WHEAT, nor for later domestic animals such as CHICKEN and HORSE (Bradley 2016, 2022).

As seen in Figure 7.1, the Yeniseian languages were spoken along the upper and middle Yenisei River in central Siberia. Due to extensive Russian contact from about 300 years ago, most of these languages are no longer spoken; the only one still spoken is Ket, which is critically endangered (Bradley and Bradley 2019: 141–145). Until fairly recently, these languages were also spoken much further south, extending into what is now northwest China; one of the groups within the multiethnic Xiongnu confederation (ca. 1700–1600 BP) spoke a Yeniseian language which is attested in names, vocabulary and a short text transmitted through Chinese sources (Vovin 2007). Until recently, modern Yeniseian societies were reindeer herders; this probably diffused from Uralic groups moving eastward. There was very close contact and extensive intermarriage between Yeniseian speakers and speakers of Uralic Samoyedic Selkup, so the modern genomes of Selkup speakers and Ket speakers are similar, and differ from the modern genomes of speakers of other Uralic languages further west and north (Jeong et al. 2019, Flegontov et al. 2019). Groups who speak or spoke Yeniseian languages do not cultivate any of the crops for which PST etyma are reconstructed, nor do they have domestic pigs or cows in their subarctic environment.

Table 7.8. SAP Possessives in Proto-ST, Yeniseian and Proto-Athapaskan.

	Proto-ST	Ket/Kott	Proto-Athapaskan
1	***ŋa +**	b-/ŋ-	***ʃi-/*tsi-/*sə-**
2	***naŋ +**	n-	***nə-**
1+2	***ni +**		***ne-**

The ND languages were spoken by the second last group to cross into North America from eastern Siberia; various dates circa 6000 BP are proposed in the literature for their arrival. Based on linguistic comparison within ND, the first group to separate from Proto-ND was the Tlingit of the coast at the southeastern tip of Alaska; the second was the Eyak, also formerly of southern Alaska. The residual

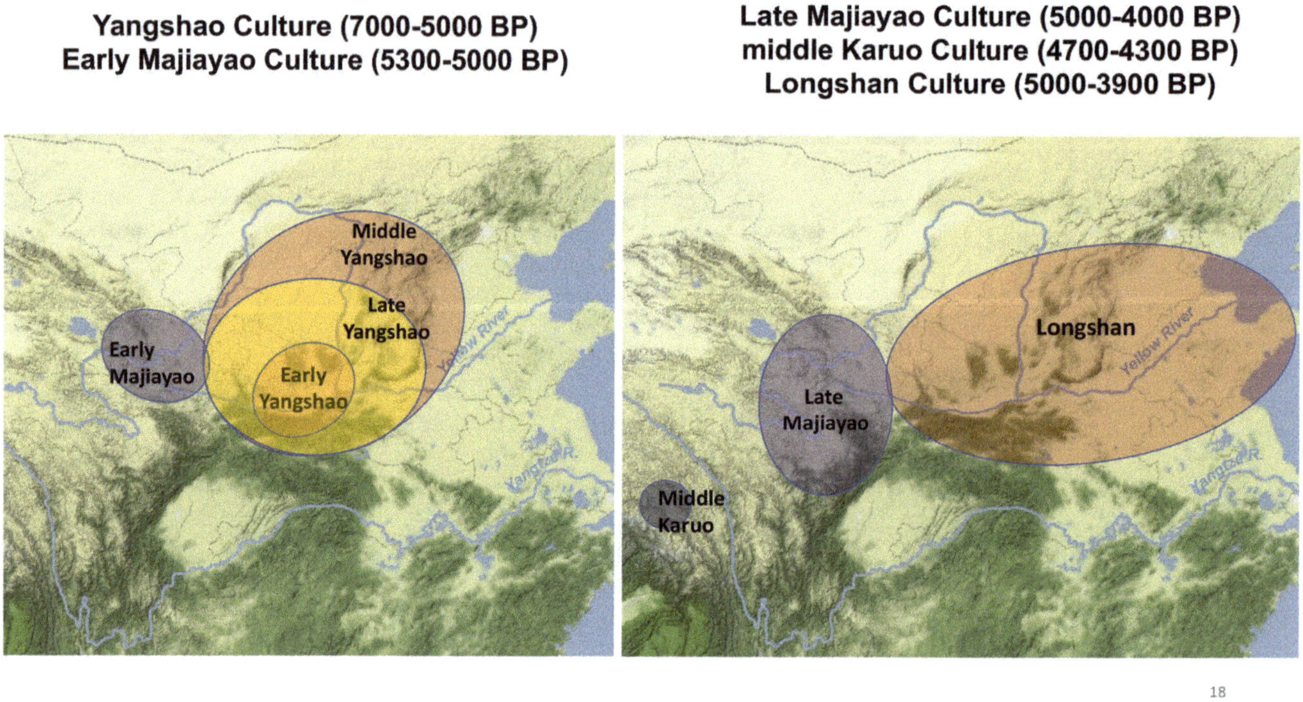

Figure 7.2. Distribution of archaeological cultures associated with Proto-ST and Proto-TB. Source: the author.

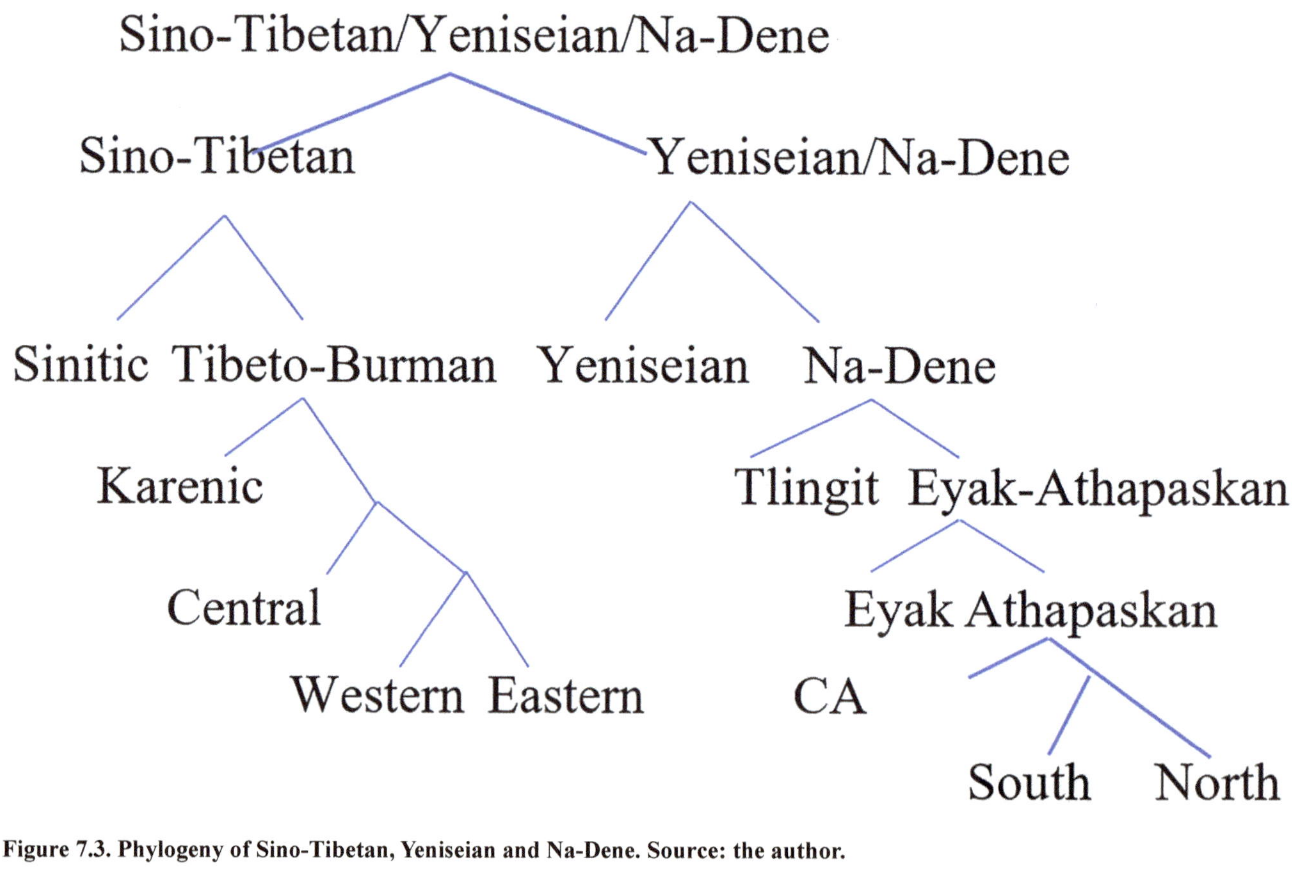

Figure 7.3. Phylogeny of Sino-Tibetan, Yeniseian and Na-Dene. Source: the author.

subgroup was Athapaskan, a cluster of languages which remained in contact for some time and then gradually spread across inland western Canada and later two branches in the western US, one mainly near the coast of northern California and another in the southwest, with the Navajo reaching their current location in the area where Colorado, New Mexico, Arizona and Utah meet not long before contact with European colonisers (Golla et al. 2007). ND groups did not grow any crops for which Proto-ST etyma are reconstructed.

The only domestic animal shared by groups speaking ST, Yeniseian and ND languages is DOG. Here there are two sets of cognates: one is a possible etymon shared between Yeniseian (Ket /ti:b/) and the first syllable of the ND form ***tleɲ**, though the sound correspondences are not regular. The other DOG etymon ***kwe(n)** is shared by all three families, including the ST form, the second syllable of the ND form, and with a meaning change in Ket to 'fox' as seen in Table 7.3 above.

7.4. Genetics

The genetic evidence for a connection between the modern groups who speak Yeniseian and ND languages is clear. Modern genomes derived from the Ancient North Eurasian genome are now widespread among modern Ket, Selkup and groups indigenous to the Americas (Jeong et al. 2019; Flegontov et al. 2019); this of course also supports the standard view that all the precolonial groups now indigenous to the Americas came originally from northeastern Eurasia. The Y haplogroup Q predominates among the Ket (Karafet et al. 2002), and the Y haplogroups Q and related C predominate among the speakers of ND languages who have an indigenous patriline (Malhi et al. 2008). For mtDNA, Ket mainly have haplogroup U4, U5 or F (Flegontov et al. 2016), while among speakers of ND languages with an indigenous matriline, haplogroup A is most frequent (Malhi et al. 2003).

The genomes of the speakers of the numerous and widespread ST languages are internally diverse, reflecting millennia of contact and assimilation in East Asia over at least the last six millennia, as well as longstanding migration into South and Southeast Asia. Overall, the East Asian genome predominates (Ning et al. 2020), but with a great deal of mixture with the genomes of speakers of Transeurasian languages to the north, Tai-Kadai-Austronesian and Miao-Yao languages to the southeast and Austroasiatic languages to the southwest (Bradley et al. forthcoming). According to van Driem (2021), the Y haplogroup O and the mtDNA haplogroups M and its derivatives are the most widespread among speakers of TB languages. The speakers of Sinitic languages are a more mixed population, but the ancient genomes from the Yangshao Culture are the precursor of the genomes which predominate among modern speakers of most TB languages and form a large proportion among speakers of Sinitic languages, particularly those in in the original core area of the Yellow River valley.

7.5. Social Structure

The dominant family pattern among most groups speaking ST and Yeniseian languages is named patrilineal clans, often supported by extended genealogies and strong sociopolitical and ritual connections with members of the same clan; conversely, most groups speaking ND languages are matrilineal. Many modern groups have preferential cross cousin marriage: father's sisters' sons marry mother's brothers' daughters, creating ongoing links between three patrilineal clans: one clan who provide your men with wives, your own clan, and another clan who provide your women with husbands (Benedict 1941). The most widespread pattern of terms for cousins in ST and ND languages follows the so-called Iroquois pattern (Murdock 1949), in which parallel cousins (children of one's father's brothers or mother's sisters) are terminologically distinct from cross cousins (children of one's father's sisters or mother's brothers). Modern Ket does not follow this pattern; the stem /-amdu-/ of the Ket forms /amdun/ 'male cousin' and /amdus/ 'female cousin' is cognate with the PST form ***mdu** 'parallel nephew/ niece', that is, the nephew or niece who is the child of a man's brother or a woman's sister and cannot marry one's child. The relevant ND cognate is ***u:d-e:** '(marriageable) cross cousin'. Note the meaning differences: cross cousin in ND, shifted down one generation and parallel nephew/ niece in ST, and all cousins in Yeniseian; also an apparent metathesis in Proto-ND. As seen in 2.4.1 above, cognates for various other etyma for kin terms such as parents and grandparents are shared between ST, Yeniseian and ND.

Among all three groups, after marriage, wives typically move to their husband's family; this virilocal residence also means that most political and ritual leaders are male elders. Burial customs also show similarities: burials are supine (lying on the back) and include personal artefacts such as pottery and tools, and graves are marked, maintained and respected. Of course, none of these cultural practices is limited to these three groups.

7.6. Conclusion

While some of the linguistic and cultural similarities between the ST, Yeniseian and ND language families may be accidental, they are very substantial. Given their locations, it is extremely unlikely that there has been recent contact between the three families. Thus, it appears that the ST, Yeniseian and ND language families are distantly related. This linkage probably predates the Neolithic in the sense of agriculture and animal husbandry, as there is no evidence of shared cognate vocabulary for any crops or for domestic animals other than one of two etyma for DOG. This suggests the phylogeny and approximate chronology shown in Figure 7.3.

In previous and forthcoming publications, I have traced the chronology of crops and domestic animals among speakers of the Sino-Tibetan languages through cognate and borrowed lexicon in these languages and related this to ST and TB migration history and phylogeny (Bradley 1997, 2011, 2016, 2017, 2018, 2022); the connections with archaeology and genetics will be presented in more detail in Bradley et al. (forthcoming).

Acknowledgements

My thanks to Andrey Kibrik, Ning Chao, Willem de Reuse and Edward Vajda for references and comments.

References

Baxter, William H. and Laurent Sagart (2014) Baxter-Sagart Old Chinese reconstruction, version 1.1 (20 September 2014). ocbaxtersagart.lsait.lsa.umich.edu, accessed 15 December 2021.

Bengtson, John D. (1991) Notes on Sino-Caucasian, Some Sino-Caucasian etymologies, Macro-Caucasian phonology, Macro-Caucasian: a historical linguistics, Hypothesis (abstracts), On Dene-Caucasian substratum in Europe, Some Macro-Caucasian etymologies. In Vitaly Shevoroshkin (ed.), *Dene-Sino-Caucasian Languages*, pp. 67–172. Bochum: Brockmeyer.

Bengtson, John D. (1994) Edward Sapir and the 'Sino-Dene Hypothesis'. *Anthropological Science* 102(3): 207–230.

Benedict, Paul K. (1941) Kinship in Southeastern Asia. PhD thesis, Harvard University.

Benedict, Paul K. (1942) Thai, Kadai and Indonesian: A new alignment in south-eastern Asia. *American Anthropologist* 44: 576–601.

Benedict, Paul K. (1972) *Sino-Tibetan: A Conspectus.* Cambridge University Press.

Benedict, Paul K. (1975) *Austro-Thai Language and Culture, With a Glossary of Roots.* New Haven CT: HRAF Press.

Benedict, Paul K. (1990) *Japanese/Austro-Tai.* Ann Arbor MI: Karoma.

Bradley, David (1979) *Proto-Loloish.* London: Curzon Press.

Bradley, David (1991) *彝语支源流*. (成都) 四川民族出版社. [Proto-Yi Branch Languages. Chengdu: Sichuan Nationalities Press].

Bradley, David (1997) What did they eat? Grain crops of the Burmic groups. *Mon-Khmer Studies* 27: 161–170.

Bradley, David (2011) Proto-Tibeto-Burman grain crops. *Rice* 4(3–4): 134–141.

Bradley, David (2016) 《山海经》与汉藏语系中的十二生肖, Chinese calendar animals in *Shanhaijing* and in Sino-Tibetan languages. In David Bradley and Pei Likun (eds), *《山海经》世界地理与中国远古文明, Shanhaijing World Geography and Ancient Chinese Civilisation*, pp. 7–17, 93–105. Beijing: Foreign Languages Press.

Bradley, David (2017) Ancient history of Sino-Tibetan in China. Paper presented at International Conference on Evolutionary Linguistics 9, Kunming 26–27 August. <academia.edu/38038361/ANCIENT_HISTORY_OF_SINO_TIBETAN_IN_CHINA>.

Bradley, David (2018) Subgrouping of the Sino-Tibetan languages. Paper presented at International Conference on Evolutionary Linguistics 10, Nanjing 27–28 October. <academia.edu/38038379/Subgrouping_of_the_Sino_Tibetan_languages>.

Bradley, David (2022) Phylogeny of Tibeto-Burman from plants and animals. In Mark Post, Stephen Morey and Toni Hüber (eds.), *Eastern Himalayan Prehistory.* Leiden: Brill.

Bradley, David and Maya Bradley (2019) *Language Endangerment.* Cambridge University Press.

Bradley, David, Liu Li, Rita dal Martello and Ning Chao (Forthcoming) Sino-Tibetan. In Martine Robbeets and Mark Hudson (eds.), *Oxford Handbook of Archaeology and Language.* Oxford University Press.

Campbell, Lyle and William J. Poser (2008) *Language Classification.* Cambridge University Press.

Čašule, Ilja (2016) *Evidence for the Indo-European and Balkan Origin of Burushaski.* München: Lincom Europa.

Čašule, Ilja (2017) *Burushaski Etymological Dictionary of the Inherited Indo-European Lexicon.* München: Lincom Europa.

Caveney, Geoffrey (2014) Sino-Tibetan ***ŋ-** and Na-Dene *kw/*gw/*xw: 1st person pronouns and lexical cognate sets. *Journal of Chinese Linguistics* 42/2: 461–487.

van Driem, George (2021) *Ethnolinguistic Prehistory: The Peopling of the World from the Perspective of Language, Genes and Material Culture.* Leiden: Brill.

Flegontov, Pavel, Piya Changmai, Anastassiya Zidkova, Maria D. Logacheva, N. Ezgi Altınışık, Olga Flegontova, Mikhail S. Gelfsand, Evgeny S. Gerasimov, Ekaterina E. Khrameeva, Olga P. Konovalova, Tatiana Neretina, Yuri V. Nikolsky, George Starostin, Vita V. Stepanova, Igor V. Travinsky, Martin Tříska, Petr Tříska and Tatiana V. Tatarinova (2016) Genomic study of the Ket: a Paleo-Eskimo related ethnic group with significant Ancient North Eurasian ancestry. *Nature Scientific Reports* 6: 20768.

Flegontov, Pavel, N. Ezgi Altınışık, Piya Changmai, Nadin Rohland, Swapan Mallick, Nicole Adamski, Deborah A. Bolnick, Nasreen Broomankhoshbacht, Francesca Candillo, Brendan J. Culleton, Olga Flegontova, T. Max Friessen, Choongwon Jeong, Thomas K. Harper, Denise Keating, Douglas J. Kennett, Alexander M. Kim, Thiseas C. Lamnidis, Ann Marie Lawson, Iñigo Olade, Jonas Oppenheimer, Ben A. Potter, Jennifer Raff, Robert A. Sattler, Pontus Skoglund, Kritsin Stewardson, Edward J. Vajda, Sergey Vasilyev,

Eizaveta Vezelovskaya, M. Geoffrey Hayes, Dennis H. O'Rourke, Johannes Krause, Ron Pinhasi, David Reich and Stephan Schiffels (2019) Paleo-Eskimo genetic ancestry and the peopling of Chukotka and North America. *Nature* 570: 236–240

Georg, Stefan (2007) *A Descriptive Grammar of Ket.* Leiden: Global Oriental.

Golla, Victor, Ives Goddard, Lyle Campbell, Marianne Mithun and Mauricio Mixco (2007) North America. In R.E. Asher and Christopher Moseley (eds.), *Atlas of the World's Languages*, 2nd edition, pp. 7–44. Routledge.

Greenberg, Joseph H., ed. by William Croft (2005) *Genetic Linguistics: Essays on Theory and Method.* Oxford University Press.

Holst, Jan Henrik (2017) *Die Herkunft der Buruschaski.* Aachen: Shaker.

Jeong Choongwon, Oleg Balanovsky, Elena Lukianova, Nurzhibek Kahbatkyzy, Pavel Flegontov, Valery Zaporozhchenko, Alexander Immel, Wang Chuanchao, Olzhas Ixan, Elmira Khussainova, Bakhytzhan Bekmanov, Victor Zaibert, Maria Lavryashina, Elvira Pocheshkhova, Yuldash Yusupov, Anastasia Agdzhouan, Sergey Koshel, Andrei Bukin, Pagbajabyn Nymadawa, Shahlo Turdikulova, Dlibar Dalimova, Mikhail Churnosov, Rosa Skhalyakho, Denis Daragan, Yuri Bogunov, Anna Bogunova, Alexandr Shtrunov, Nadezhda Dubova, Maxat Zhabagin, Levon Yepiskoposyan, Vladimir Churakov, Nikolay Pislegin, Larissa Damba, Ludmila Saroyants, Khadizhat Dibirova, Lubov Atramentova, Olga Utevska, Eldar Idrisov, Evgeniya Kamenshchikova, Irina Eveseeva, Mait Metspalu, Alan K. Outram, Martine Robbeets, Leyla Djansugurova, Elena Balanovska, Stephan Schiffels, Wolfgang Haak, David Reich and Johannes Krause (2019) The genetic history of admixture across inner Eurasia. *Nature Ecology and Evolution* 3: 966–976.

Karafet, Tatiana M., Ludmila P. Osipova, Marina A. Gubina, Olga L. Posukh, Stephen L. Zegura and Michael F. Hammer (2002) High levels of Y-chromosome differentiation among native Siberian populations and the genetic signature of a boreal hunter-gather way of life. *Human Biology* 74(6): 761–789.

Kotorova, Elisveta G. and Andrey Nefedov (2015) *Comprehensive Dictionary of Ket.* 2v. München: Lincom.

Leer, Jeff (1989) Directional systems in Athapaskan and Na-Dene. In Eung-Do Cook and Keren D. Rice (eds), *Athapaskan Linguistics: Current Perspectives on a Language Family*, 574–622. Berlin: Mouton de Gruyter.

Leer, Jeff (2011) Comparative Athapaskan Lexicon. Ms.

Malhi, Ripan Singh, Holly M. Mortensen, Jason A. Eshleman, Brian M. Kemp, Joseph G. Lorenz, Frederika A. Kaestle, John R. Johnson, Clara Gorodezky and David Glenn Smith (2003) Native American mtDNA prehistory in the American southwest. *American Journal of Physical Anthropology* 120: 108–124.

Malhi, Ripan Singh, Angelica Gonzalez-Oliver, Karl Britt Schroeder, Brian M. Kemp, Jonathan A. Greenberg, Solomon Z. Dobrowski, David Glenn Smith, Andres Resendez, Tatiana Karafet, Michael Hammer, Stephen Zegura and Tatiana Brovko (2008) Distribution of Y chromosomes among Native Americans: A study of Athapaskan population history. *American Journal of Physical Anthropology* 137: 412–424.

Murdock, George P. (1949) *Social Structure.* New York: Macmillan.

Nikolaev, Sergei (1991) Sino-Caucasian languages in America. In Vitaly Shevoroshkin (ed.), *Dene-Sino-Caucasian Languages*, pp. 42–66. Bochum: Brockmeyer.

Ning Chao, Li Tianjiao, Wang Ke, Zhang Fan, Li Tao, Wu Xiyan, Gao Shizu, Zhang Quanchao, Zhang Hai, Mark J. Hudson, Dong Guanghui, Wu Shiyao, Fang Yanming, Liu Chen, Feng Chunyan, Li Wei, Han Tao, Li Ruo, Wei Jian, Zhu Yonggang, Zhou Yawei, Wang Chuanchao, Fan Shengying, Xiong Zenglong, Sun Zhouyong, Ye Mailin, Sun Lei, Wu Xiaohong, Liang Fawei, Cao Yanpeng, Wei Xingtao, Zhu Hong, Zhou Hui, Johannes Krause, Martine Robbeets, Jeong Choongwon and Cui Yinqiu (2020) Ancient genomes from northern China suggest links between subsistence changes and human migration. *Nature Communications* 11: e2700.

Pritzker, Barrry M. (1998) *Native Americans: An Encyclopedia of History, Culture and Peoples.* 2v.

Ratliff, Martha (2010) *Hmong-Mien Language History.* [Pacific Linguistics 613]. Canberra: Research School of Pacific and Asian Studies, Australian National University.

Ruhlen, Merritt (1994) *On the Origin of Languages.* Stanford University Press.

Sagart, Laurent (1993) Chinese and Austronesian: evidence for a genetic relationship. *Journal of Chinese Linguistics* 21(1): 1–62.

Sagart, Laurent (2005) Sino-Tibetan-Austronesian: an updated and improved argument. In Laurent Sagart, Roger Blench and Alice Sanchez-Macas (eds.), *The Peopling of East Asia: Putting Together Archaeology, Genetics and Linguistics*, pp. 161–176. London: RoutledgeCurzon.

Sagart, Laurent (2017) A candidate for a Tibeto-Burman innovation. *Cahiers de Linguistique Asie Orientale* 46: 101–117.

Sagart, Laurent, Guillaume Jacques, Lai Yunfan, Robyn J. Ryder, Valentin Thouzeau, Simon J. Greenhill and Johann-Mattis List (2019) Dated language phylogenies shed light on the ancestry of Sino-Tibetan. *Proceedings of the National Academy of Sciences USA* 116(21): 10317–10322.

Scarpari, Maurizio (2017) Classical Chinese. In Rint Sybesma, Wolfgang Behr, Yuegou Gu, Zev Handel, C.T. James Huang and James Myers (eds), *Encyclopedia of Chinese Language and Linguistics,* v. I, pp. 608–620. Leiden: Brill.

Shafer, Robert (1952) Athapascan and Sino-Tibetan. *International Journal of American Linguistics* 18: 12–19.

Shafer, Robert (1957) Note on Athapascan and Sino-Tibetan. *International Journal of American Linguistics* 23: 116–117.

Shafer, Robert (1969) A few more Athapascan and Sino-Tibetan comparisons. *International Journal of American Linguistics* 35: 67.

Shevoroshkin, Vitaly and Alexis Manaster Ramer (1991) Some recent work on the remote relations of languages. In Sydney M. Lamb and E. Douglas Mitchell (eds.), *Sprung from Some Common Source: Investigations into the Prehistory of Languages*, pp. 176–199. Stanford University Press.

Starostin, George (2012) Dene-Yeniseian: A critical assessment. *Journal of Language Relationship* 8: 117–138.

Starostin, Sergei, translated by William Baxter III (1991) On the hypothesis of a genetic connection between the Sino-Tibetan languages and the Yeniseian and North Caucasian languages. In Vitaly Shevoroshkin (ed.), *Dene-Sino-Caucasian Languages,* 12–41. Bochum: Brockmeyer. Translation of 1984 publication in Russian.

Swadesh, Morris (1952) Review article of Athapascan and Sino-Tibetan, by R. Shafer. *International Journal of American Linguistics* 18: 178–181.

Swadesh, Morris (1971) *The Origin and Diversification of Language.* London: Routledge & Kegan Paul.

Toporov, Vladimir N. (1971) Burushaski and Yeniseian languages: some parallels. *Travaux Linguistiques de Prague* 4: 107–125.

Trombetti, Alfredo (1922/1923) *Elementi di Glottologia.* 2v. Bologna: Zanichelli.

Vajda, Edward J. (2004) *Ket.* [Languages of the World: Materials 204]. München: Lincom.

Vajda, Edward J. (2010) A Siberian link with Na-Dene languages. *Anthropological Papers of the University of Alaska* n.s. 5: 33–99.

Vajda, Edward J. (2012) A reply to G. Starostin. *Journal of Language Relationship* 8: 138–152.

Vajda, Edward J. (2013) Vestigial possessive morphology in Na-Dene and Yeniseian. *Alaska Native Language Center Working Papers* 12: 79–92.

Vajda, Edward J. (2018) Dene-Yeniseian: Progress and unanswered questions. *Diachronica* 35(2): 277–295.

Vovin, Alexander (2007) Did the Xiongnu speak a Yeniseian language? *Central Asian Journal* 44(1): 87–104.

Werner, Heinrich (2004) Yeniseic counting systems. In Edward J. Vajda (ed.), *Languages and Prehistory of Central Siberia*, pp. 121–127. Amsterdam: John Benjamins.

Wolfenden, Stuart N. (1929) *Outlines of Tibeto-Burman Linguistic Morphology.* London: Royal Asiatic Society.

Zhang Hanzi, Ji Ting, Mark Pagel and Ruth Mace (2020) Dated phylogeny suggests early Neolithic origin of Sino-Tibetan languages. *Nature Scientific Reports* 10: e20792.

Zhang Menghan, Yan Shi, Pan Wuyun and Jin Li (2019) Phylogenetic evidence for Sino-Tibetan origin in northern China in the Late Neolithic. *Nature* 569: 112–115.

Part II

Eurasian East-West Exchange in the Bronze and Middle Ages

8

By Steppe Highway or Mountain Corridors? Exploring the Archaeolinguistic Arguments for the Provenance of Western Eurasian Crops and Livestock in Central and East Asia

Rasmus G. Bjørn

Abstract: Wheat, barley, cattle, and horses are ancient domesticated crops and animals integral for the understanding of developments across Neolithic and Bronze Age Eurasia. It is well-established that they entered East Asia from Central Asia, but so far only vague ideas of their immediate provenance and cultural context have been presented, often in relation to the Tocharian and Iranic branches of Indo-European. In this chapter, I scrutinise the earliest comparative linguistic evidence for all four domesticates in relation to emerging chronologies from archaeology and genetics. It is concluded that neither wheat nor barley are of Indo-European descent, while the horse can be succinctly associated with the Iranic expansion with the Andronovo horizon. Only the East Asian words for cattle may plausibly be connected with an early stage of the Tocharian language carried to the region with Afanasievo.

Keywords: wheat, barley, cattle, horse, Indo-European, Central Asia, loanwords, exchange, Bronze Age, East Asia

8.1. Introduction

The Eurasian supercontinent is home to a number of important crops and livestock that have come to lay the foundation for past and present empires as well as the everyday cuisines of people across the globe. While the story begins at the intersection between humans and nature in the act of domestication, a subsequent string of intercultural exchanges brought the new crops and livestock into the hands and lives of people at opposite ends of Eurasia. In this chapter, the journey begins in the west and ends in the east, following the linguistic imprints of some of the most impactful Western Eurasian domesticates, namely the two Middle Eastern founder crops, wheat and barley, and cattle and the horse, to try to answer the question: by which route did these crops and animals travel – the Steppe Highway or the Inner Asian Mountain Corridor?

Language contact is one of the most revealing phenomena of prehistoric cultural exchange. Loanwords transfer through oral communication between speakers of different languages. There is a rising awareness among archaeobotanists of the prehistoric evidence enshrined in comparative linguistics (Fuller 2002; Stevens et al. 2016; Mikič 2019), and a recent paper employs loanwords from Indo-European languages to provide temporal and areal moorings to prehistoric language communities in East Asia (Robbeets et al. 2021). Comparative historical linguistics, however, is in a phase of rapid development, anchoring its relative chronologies to increasingly well-understood prehistoric cultures and migrations in an ongoing interdisciplinary exploration of prehistory (Haak et al. 2015; Peyrot 2019; Sagart et al. 2019; Robbeets et al. 2021). Further analysis of the comparanda (words that are being compared) and their real-world referents reveals some of the earliest Neolithic interactions in greater detail. It is important for archaeologists to know that the linguistic evidence is available and reflects these spreads; and for linguists to engage these difficult questions and frame the discussion in order for historical linguistics to claim its rightful seat at the table as human history is written. This chapter presents an up-to-date analysis of the linguistic material from the perspective of the Central Asian region that carried the words for barley, wheat, horse, and cattle into East Asia.

Common to all four of these domesticates is that scholars have, to some extent, argued for a western provenance for some or all of their words in the Transeurasian (Turkic, Mongolic, Tungusic, Koreanic and Japonic) and Sino-Tibetan (Sinitic, including Mandarin and Cantonese, and Tibeto-Burman) languages, often with an assumed Tocharian proto-form in mind. This has, in turn, played a part in the piece-by-piece reassembly of Eurasian prehistory, but with major advances and new papers central to the issue recently published, the ground has been fertilised for an updated discussion of the linguistic arguments paired with new perspectives from archaeology and genetics. It is the intent of this chapter to provide that synthesis. I will concern myself primarily with languages and cultures included in or bordering on Central Asia, broadly defined as the area between the Persian and Chinese empires as expertly treated already by Laufer (1919), but importantly including the steppe zone.

The primary goal, then, is to establish whether the words in the eastern languages can be considered borrowed from an Indo-European language to substantiate a Steppe origin or if they belong to cultural spheres of the linguistically less well-understood Inner Asian Mountain Corridor (IAMC).

8.1.1. Linguistic stratigraphy

Comparative linguistics revels in relative chronologies established on the basis of sequences of sound laws as well as established parent languages (Campbell 2013). The Bronze Age Indo-European migrations into Central Asia provide particularly salient pieces of evidence, as they both betray clear linguistic, genetic, and archaeological descent from Proto-Indo-European and the Yamnaya culture of the South Russian steppe zone (Anthony 2007; Peyrot 2019; Kuzmina 2007). The comparative method and internal reconstruction take us there by identifying regular descent from a common ancestor (e.g., Proto-Germanic **sibun* > Modern English *seven*), while loanwords attest to language contact (e.g., English *computer* → German *computer*). Contact hypotheses most often arise from similarity in form (sound sequence) and function (meaning), but will have to be substantiated to rule out chance resemblance or common descent (Anttila 1989: 25; Mair 1990: 43).

While linguists have made use of archaeobotanical and archaeozoological data to inform linguistic spread (Peyrot 2018; Mallory 2018; Witzel 2005; Bradley 2016; Stachowski 2008), research in the other direction has not been as prevalent, which is understandable considering the relative nature of comparative linguistic evidence. The revolution and proliferation of ancient DNA analyses have, however, helped link linguistic spreads more firmly in concrete chronologies, which consequently kicks the door wide open to the wealth of relative linguistic information. Stevens et al. (2017) do a great job in including linguistic evidence for the Central Asian crop dispersal, but they had yet to benefit from the composite analysis available only when considering all language families known from the region. This analysis, in turn, is only possible with the above-mentioned analyses of the individual language families. Based on the increasingly obvious assumptions on archaeolinguistic spreads, the analysis in the present chapter aims to do exactly that by bringing together the lines of linguistic evidence deduced from these articles into a stratigraphy for the region.

The present analysis follows a methodological protocol departing from previously proposed connections between different languages and between languages and archaeological cultures. Crucially, scholars connecting a word in language A with a similar word in language B often do not consider that other scholars may connect the word in language B with a third in language C and language D and so on. Similarly, new arguments may have been forwarded that connect language A to a datable and materially analysable archaeological culture that in turn allows for a more comprehensive analysis of the linguistic material. While chance resemblance may arise between any two words, similarity in form and function across several distinct but contiguous language communities significantly reduces chance phenomena and is taken as evidence in favour of contact phenomena. The older the linguistic data, the more difficult it usually becomes to provide clear answers, and most sets of comparanda have not received unanimous acclaim, despite methodologically sound attempts to connect them. It is thus all the more crucial to understand the data, its potential interpretation and its limitations.

8.1.2. Assumptions

I depart from the assumption of a Pontic-Caspian Steppe origin for the Indo-European branches of Central Asia, i.e. Tocharian and Indo-Iranic, importantly divorced from Southwest Asia. The hypothesis of a Proto-Indo-European homeland in the Western Steppe Zone enjoys the greatest support among Indo-Europeanists and manages to predict the genetic evidence by means of archaeological and linguistic evidence alone, more specifically the migration from the Steppes to both Europe and, more central to the questions at hand, Central Asia (Anthony 2007; Haak et al. 2015; Narasimhan et al. 2019). While the identification of the earliest speakers of the Anatolian branch remains unsettled, the Steppe provenance of both Tocharian and, later, Indo-Iranic continually receive supporting evidence from archaeology, genetics and linguistics, yielding the deeper provenance less important for Central Asia. Linguistically, the identification of a significant Uralic substrate in Tocharian (Peyrot 2019; Warries 2019) paired with recent genetic evidence of a rather late entry to the Tarim Basin (Zhang et al. 2021) help establish this pivotal corridor as a veritable terra incognita much longer, allowing a (string of) Central Asian substrate language(s) to have mediated the Neolithic impulses to China, in principle until the identification of Tocharian in the north and the Saka languages in the west (Kuzmina 2007). Other major breakthroughs include the identification of Yamnaya ancestry in the Afanasievo far into the eastern steppe zone of Mongolia (Honeychurch et al. 2021), suggesting that an early dialect of Indo-European, indubitably related to Tocharian, although not necessarily sharing its salient traits, was spoken far into the eastern steppe zone already in the middle of the third millennium BC. While the Eurasian steppes constitute the prototypical spread zone, the centrally located Pamir and Hindu Kush mountains provide contrasting residual zones (Nichols 1997) along with the comparative evidence of languages outside of the Central Asian zone. Nonetheless, proponents of a Transcaucasian or Anatolian homeland for Proto-Indo-European rely on the same lexical data, and the linguistic material obviously does not change with different assumptions, although the results of the analysis may.

Three Transeurasian language families emanated westward, and consequently into the purview of the Central Asian region, from the West Liao River Basin after 6000 BC: Turkic, Mongolic, and Tungusic (Robbeets et al. 2021). Despite a late uncontroversial proto-language towards the end of the first millennium BC, there is evidence that an early stage of Proto-Turkic was present in the Altai-Sayan region of Central Asia no later than 2000 BC (Golden 1998: 16-18, Bjørn 2022). This constitutes the earliest western reach of the Transeurasian languages. To the east of Turkic, Proto-Mongolic is much more recent, only safely associated with the period before the ascendency of Chinggis Khan and the Mongol Empire, but several loanwords shared with early Turkic identify protracted interactions in the eastern steppe zone (Robbeets et al. 2020: 760). Proto-Tungusic is found still further east, associated with the Iron Age in the Primorye and close to the Transeurasian homeland (Robbeets et al. 2020: 761–764).

As language isolates, Burushaski and Yeniseian do not lend themselves to systematic comparison with other languages beyond loanwords, and little can be inferred about their prehistory. It is consequently not known if the Burushaski language in the Hindu Kush continues a pre-agricultural community, the initial Neolithic migration, or if it only arrived with a secondary migration; or if Yeniseian once was spoken as part of the earliest steppe confederations (see Bradley this volume for possible deeper relations with Sino-Tibetan, that nonetheless do not interfere with the views presented here). The regional stratification of the earliest agricultural terminology in the region may help illuminate this difficult situation.

8.2. Data and Analysis

8.2.1. Crops

Contrasting with the millets and rice of a colder northern and a warmer southern East Asia, respectively, the Western Eurasian crops all originate in the same general ecozone of the Fertile Crescent in Southwest Asia. All the crops eventually came to cover the supercontinent from east to west, but it is not clear if the Fertile Crescent domesticates used the Steppe Zone as a staging ground for their eastern spread, or if they took a more direct route through the Inner Asian Mountain Corridor (IAMC).

8.2.1.1. Barley

Barley (*Hordeum vulgare*) was among the first Southwest Asian grain crops to make headway into Central Asia, owing to its greater resilience to the shifting conditions of the region (Spengler 2019: 115-139). Although an informative linguistic signal can be demonstrated, a layered string of subsequent developments has helped fertilise the ground for greater variation among the words for barley than those for wheat. Clearly discernible events include the development and spread of less labour-intensive naked barley (*H. vulgare var. nudum*) and the altitude adapted highland variety, as well as the still later preference for wheat for human consumption with the consequence that barley usage changed to animal fodder and brewing, triggering a shift back to the husked variant that produced a higher yield. Comparative linguistics allows the following dating and directionality for barley word traditions.

Proto-Indo-Iranic **yáwa-* 'grain, barley' can be dated to Sintashta ca. 2100 BC around the Southern Ural mountains (Kuzmina 2007; Anthony 2007). The word derives from Proto-Indo-European **yewós* 'grain (barley?)' (Kümmel 2017: 280; Mallory and Adams 1997: 236) and yielded Vedic Sanskrit *yáva-* 'grain, barley' as well as Common Iranic **yawa* 'grain, barley' with attested forms in Avestan *yauua-* 'barley' (Central Asia, ca. 1000 BC); Shughni *ǰav* 'barley' (Pamir modern); Munji *you* 'grain' (Pamir modern); Bactrian /yaw/ 'barley' (Bactria ca. 500 AD); Persian *ǰau* 'barley' (Iran 1000 AD); Ossetic *yæw* 'millet' (Caucasus modern). An early Iranic variant of the word was borrowed into Uralic (Holopainen 2019: 103–105) and Tocharian B *yap-* (Peyrot 2018:248–250, pace Adams 2013: 519–520). While the Indo-Iranic word may be Proto-Indo-European in provenance, its linguistic spread into Central Asia with Indo-Iranic (Sintashta and Andronovo) postdates the introduction of barley to Central Asia by at least a millennium for the earliest presence in the Pamir, and several millennia for the presence at Mehrgarh at 6500 BC.

Within Common Iranic (between 2100-1500 BC), another word for barley is found, **kas(a)-ka-* (> Khotanese *chaska*, Persian *kašk*, Munji *kosk*, Yazghulami *kusk/kåsk*, Shughni *čūšǰ*). It is likely internally derived and also borrowed into dialects of Uralic (Kümmel 2017: 281fn18), but it postdates **yawa-* and is similarly secondary to the search for the earliest spread of barley in Central Asia.

Proto-Turkic **arBa* (**B* designating either **b* or **p*) is the source of the Mongolic, Tungusic and Yeniseian words for barley. It may be more promising as it finds a difficult set of comparanda in a handful of Eastern Iranic words (Kümmel 2017: 281): **arpucyā* 'barley' (> Pashto *orb'əša* 'barley'; Ishkashimi *urvьs* 'barley') and **rucā-* 'barley' (> Khotanese *rrusā-* 'highland barley'; Khwarezmian *rsy* 'barley(corn)'). Since the two Iranic groups are mutually incompatible with a common proto-form, a borrowing from related substrate sources best explain the situation.

If ultimately related, the form most true to the source is the more morphologically complex **arpucyā*, which renders the simpler Proto-Turkic an unlikely candidate. Dating is difficult due to the late split of Proto-Turkic roughly 2000 years ago, while the Iranic forms demonstrably point to individual borrowings with a *terminus ante quem* set around the same time to account for the two classical languages of Central Asia, Khotanese and Khwarezmian. More suggestive is the region where the word is found: the Altai mountains for Proto-Turkic (Golden 1998: 16–18; Robbeets et al. 2020), Afghanistan (Pashto), the Pamir (Ishkashimi), and the western Tarim Basin (Khotanese), all within the Inner Asian Mountain Corridor, while Khwarezmian was spoken just south of the Aral Sea. It is thus conceivable that the word was borrowed from a language present in the IAMC prior to the arrival of Iranic speakers (Stachowski 2008: 13), and either similarly prior to Turkic presence in the Altai or making its way into the Altai region with the unknown language when Turkic was already there. The earliest barley in the Altai is dated to 3200 BC (Zhou et al. 2020). Although the word in most languages denotes the broad meaning, the early attestation in Khotanese 'highland barley' contrasts the inherited *chaska* '(regular) barley'. It is thus conceivable that the word originally referred to the highland variety more specifically, likely prevalent in the mountain region and naturally new and useful to incoming Iranic speakers hailing from the lowland steppe zone. For the Turkic speakers, however, this was the first kind of barley they encountered. Note that Tocharian B *tsänkana* 'naked barley' is a late borrowing from Middle Chinese *tcheng* 'blue/green (grain)' (Ching 2016: 384; Peyrot 2018: 254, pace Adams 2013: 803).

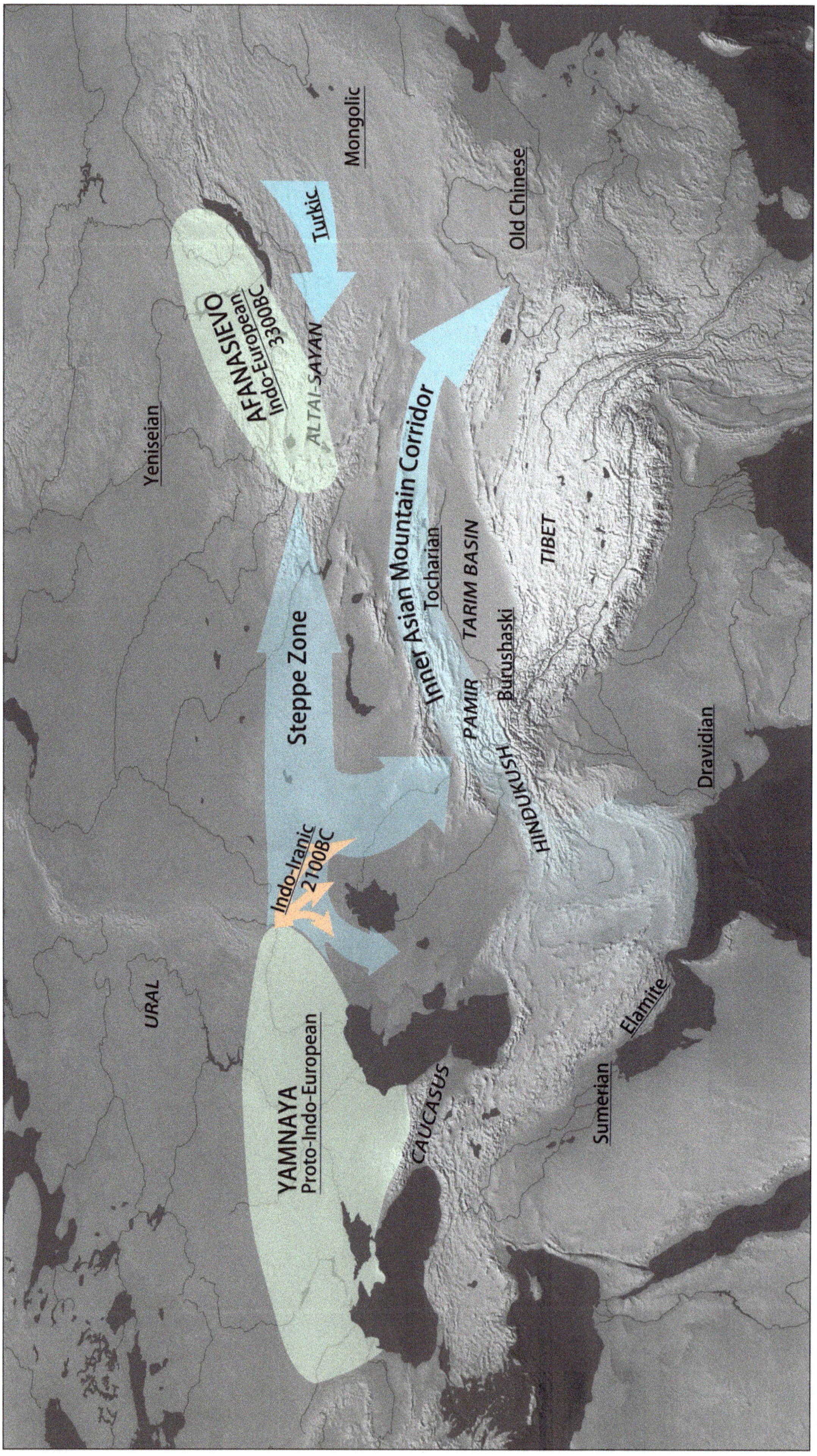

Figure 8.1. Central Asia and the languages instrumental in determining the linguistic spread of domesticated crops and animals. Source: map by J. Uchiyama.

Table 8.1. Sound comparisons for 'barley' in Turkic and Iranic.

Turkic	**a*	*r*	*B*	-	*a*
Iranic (A)	**a*	*r*	*p*	*uc*	*ā*
Iranic (B)	**-*	*r*	-	*uc*	*ā*

Another candidate for the oldest barley word in Central Asia is Burushaski *harí*, unfortunately without external comparanda in the wider region. A testament to some primacy is the possible internal association with very basic agricultural terminology, cf. Burushaski *harṣ* 'plow', *harkí* 'plowing and sowing, agriculture' (Berger 1998: 193–194).

Old Chinese **mə.rˤə* 'grain, wheat', in Mandarin used to denote both *dà-mài* 'barley' (literally, 'greater grain') and *xiao-mài* 'wheat' (literally, 'lesser grain'), constitutes a final and perhaps more hopeful candidate to attest to the original spread of barley into Central Asia (see also Deng, this volume). The Chinese form has been compared to Proto-Tungusic **murgi* 'barley' (Robbeets 2015: 29) and a small cluster of Indo-Iranic languages centred in the Hindu Kush in Nuristani **brā(k)-* 'flour' (> Waigali *brai/brē* 'flour', Kati *bŕē* 'flour', Ashkun *wŕēi/wlēi* 'flour') and Khowar *blan* 'barley', which has been derived from Proto-Indo-European, based on similar looking words in the Anatolian and Celtic branches of Indo-European (Blažek 2016), but the words are formally incompatible (Kümmel 2017: 281, fn.20) and at best attest to a wider Wanderwort phenomenon. A hypothetical common prototype for the source may take the form **mr̥(n)k* (*r̥* being a syllabic resonant as in Czech *krk* 'neck' or the freezing sound *brr*). Lastly, Proto-Turkic **bugday* 'wheat' has been drawn into the comparison (Savelyev 2017: 134–135; Stachowski 2008: 95), but the phonological match is by no means trivial and requires a development **m- > *b-*, which although known from Proto-Turkic is generally triggered by a following nasal. Even more problematic seems the internal consonant cluster with Proto-Turkic **-gd-* reflecting **-rK-* elsewhere (**K* representing a velar stop). On the other hand, the shift from barley to wheat may find some motivation in the semantic layering, if **arBa* 'barley' indeed was borrowed as 'highland barley', or a form of **mr̥(n)k-* borrowed directly from Old Chinese with the meaning. Old Chinese has a separate word **móu* 'barley', attested already in the *Shijing*, one of the earliest sources (Pulleyblank 1996: 13; see also Deng, this volume), but is under-studied in the literature, and should be treated by scholars in Sinitic to establish its earliest bearings before a proper composite analysis can be conducted. It is nonetheless tempting to consider it alongside **mə.rˤə* 'grain, wheat', that replaced **móu*, and the same word may then have been borrowed twice, first as the original 'barley grain', and later, from another source, as 'wheat' that gradually broadened its meaning with the waxing importance. None of the extant forms appear to account for the ultimate origin (Savelyev 2017: 134–135), which points to a substrate phenomenon of a word already present in the region by 3000 BC (Table 8.2).

Ultimately, three words aspire to the title as the earliest word for barley in Central Asia (Table 8.3).

8.2.1.2. Wheat

Wheat (*Triticum aestivum*) is the second early grain crop entering Central Asia from the west. Despite a throng of different variants interchangeably gaining prominence, there is a surprisingly clear linguistic signal for this cereal throughout Central Asia.

Table 8.2. Sound correspondences for 'barley'/ 'wheat'.

Indo-Iranic	**m*	*r̥*	*(n)*	*k*	-	-	*'barley'*
Tungusic	**m*	*ur*	-	*g*	-	*i*	*'barley'*
Turkic	**b*	*u*	*(n)*	*g*	*d*	*ay*	*'wheat'*
Old Chinese (A)	**m*	*ər*	-	ˤ	-	*ə*	*'wheat, grain'*
Old Chinese (B)	**m*	*óu*	-	-	-	-	*'barley'*

Table 8.3. Earliest words for barley in Central Asia.

Proto-form	Dating	Location	Attested in	Note
**arpuca*	> 500 BC	IAMC	Turkic, Iranic, Mongolic, Tungusic, Yeniseian	highland variant?
harí	unknown	Hindu Kush	Burushaski	native formation?
**mr̥(n)k-*	> 2500 BC	IAMC	Nuristani, Chinese, Tungusic, Turkic	

Table 8.4. Sound correspondences for 'wheat'.

Tocharian B	*k*	*a*	*n*	*t*	*i*	-	-	'bread'
Iranic	**g*	*a*	*n*	*T*	*u*	*m*	*a*	'wheat'
Indic	**g*	*o*	-	*dh*	*ū*	*m*	*a*	'wheat'
Brahui	*x*	*ō*	-	*l*	*u*	*m*	-	'wheat'
Elamite	*š*	*u*	-	*l*	*u*	*m*	-	'standing grain'
Burushaski	*g*	*u*	-	*r*	-	-	-	'wheat'
Old Tibetan	*g*	-	-	*r*	*o*	-	-	'wheat'

The earliest potentially datable testimony for 'wheat' in the Indo-Iranic languages with similarities in form and identity in function does not allow reconstruction to Proto-Indo-Iranic. A hypothetical, but irregular Common Iranic form can be extrapolated as **gan(T)u(H)ma-* (Kümmel 2017: 281-282), while the Indic languages continue the Sanskrit word **godhūma*, presumably reinterpreted as 'cow/earth smoke' through folk etymology (Wojtilla 1999: 229), a phenomenon especially common with loanwords (Anttila 1989: 92-93). Tocharian B *kanti* 'bread' has a similar form, but necessitates a semantic drift (Peyrot 2018). Similarities also with Hittite *kant-* 'wheat' and Ancient Greek *khóndros* 'groats of spelt and wheat' (Adams 2013: 146) make a Proto-Indo-European reconstruction tantalizing, but the irregularities remain. The most parsimonious explanation for the observed irregularities within such closely related languages is the early adoption of a word present throughout the region with local variation, here presumably the Central Asian heartland. It allows dating the presence in the region prior to the spread of the Indo-Iranic languages from Sintashta ca. 2100 BC. Outside of Indo-European, these words are often connected with Burushaski *gur* (Witzel 2009), which Berger (1998: 161) further tentatively associates with Old Tibetan *gro* 'wheat', but no further chronology is available from Burushaski alone. In Brahui, a North Dravidian language spoken in Balochistan, *xōlum* 'wheat' has been suggested as the form originally adopted by the Indic branch, potentially gaining pedigree through association with *šulum* 'standing grain' in Elamite, an ancient language of the Iranian Plateau potentially related to Dravidian (Southworth 2011). Needless to say, the words cannot not be reduced to a single proto-form, and the linguistic diversity in the region must have been significant (Palmér 2019: 55–57). Both Brahui and Burushaski may in their own right attest to older layers of the spread, but the exact connections are unclear. A hypothetical reconstruction to the language of BMAC has been suggested (Witzel 2009), but this would still not explain the Burushaski and Dravidian forms and a potential reach across Southwest Asia into Greece.

'Wheat' is in Tocharian B *ysāre*, literally the 'summer grain', but has been semantically broadened in Tocharian A *wsār* 'grain' (Peyrot 2018: 251–53). With no Proto-Indo-European pedigree or regional comparanda, it does not provide further evidence. The words in East Asia have already been treated above under barley.

8.2.2. Livestock

The second leg of the Neolithic revolution is livestock. The domestication of cattle, sheep, goats, and horses enabled pastoralism, which in the steppe zone and other places inhospitable to a sedentary lifestyle could be exploited without the crop package. Livestock also importantly introduces significant movable wealth and new concepts of ownership, hierarchies, and social prestige. Devoid of an academic tradition of comparanda between the language families of Central Asia, sheep and goats are left out of the current analysis.

8.2.2.1. Horse

The horse is native to the Eurasian steppe zone, and consequently presents a different perspective from the three other items treated here, all hailing from Southwest Asia. With riding, the horse eventually became a force of mobility that enabled both long-distance migration and pasture of greater herds of sheep, goats, and cattle, but its life as a domesticated animal started corralled, milked and slaughtered like cattle (Anthony 2007). Vehicular traction may well have enabled the Afanasievo migration ca. 3300 BC, but we know now that this cross-steppe movement of people and cattle happened a millennium before the Sintashta and later Andronovo spread a lineage of horses so useful that they came to supplant all other horse breeds in human possession (Librado et al. 2021). The nature of early horse use as well as the source of the East Asian words for the animal machine have been topics of justified debate (Honeychurch et al. 2021; Taylor 2021; Raulwing 2000; Janhunen 1998), touching upon the very dynamics that shaped Central and East Asia in the Bronze and Iron Ages.

Old Chinese **mˤraʔ* 'horse' has been compared to a string of East Asian languages, thus Proto-Tungusic **murin*, which was borrowed from Proto-Mongolic **morï-n*, and further Proto-Japonic **(m)umai*, Old Koguryŏ **meru*, and Proto-Amuric **muring* (Janhunen 1998). I will refrain from commenting on the finer details of East Asian language dynamics and instead focus on the source of the word in Old Chinese, where a hypothetical Proto-Tocharian **markos* has been discussed (Pulleyblank 1996). This solution has rightly been criticised; although the horse evidently ultimately comes from the west (Shaughnessy 1988;

Shelach-Lavi 2015; Rawson et al. 2020), the connection to **markos* is difficult as the word is a European regionalism, found only in the contiguous Celtic and Germanic branches, quite literally at the opposite end of the continent from the East Asian comparanda. It is notably not found in Iranic or Tocharian, that both, like all major branches, continue Proto-Indo-European **(h$_1$)eḱu̯os* 'horse' in Proto-Tocharian **yäkwe* and Proto-Indo-Iranic **áćwa-*. Attested continuants east of the Pamirs include Toch. B *yakwe*, Toch. A *yuk*, Khotanese *aśśa*, and a putative Old Steppe Iranian **ætswæ* (borrowed into Toch. B as *etswe* 'mule', Peyrot 2018). Instead, I propose that the term originates in the internally formed Common Iranic **bāra-ka-* 'beast of burden' that was used for 'horse' across the Steppe Zone, yielding Ossetic *bajrag* 'foal' in the Caucasus and Yazghulami *varag* 'horse' and Shughni *vōrǰ* 'horse' in the Pamir (Witczak and Novák 2016: 57; Morgenstierne 1974: 85-86; Abaev 1958: 232). The substitution of initial **b-* for **m-* in front of a liquid (*r* or *l*) is commonly observed in early loanwords into Old Chinese (Schuessler 2007: 66–67; Matisoff 2003: 133). The Iranic equestrian expertise spread with Andronovo to Central and East Asia at the right time for the appearance of horses (and chariots) in China and thus constitutes a much preferable source also of the horse word than a putative Tocharian form.

Although the source of the domesticated horse appears to have been identified, local words for the wild horse may be responsible for some of the variation seen in the Central Asian region. The primary branches in Uralic thus borrowed their terms independently, e.g. Proto-Samoyedic **juntз* from Common Turkic **junt* 'horse, mare' (if not both from a common substrate, Janhunen 1998: 416), and Proto-Ugric **low/luw* 'horse' from either Proto-Turkic 'transport animal', Proto-Tocharian 'animal' (Napolskikh 1996), or even the language of Botai (Parpola 2012: 295–296), while the Proto-Turkic form **(h)at* appears unique (if not from Old Iranic **ætswæ*).

8.2.2.2. Cattle

Cattle is the largest of the Southwest Asian domesticated animals. It was an integral part of the Proto-Indo-European subsistence strategy as reflected both in the Yamnaya culture of the western steppe zone (Anthony 2007) and its successor cultures in Central Asia, Andronovo (Kuzmina 2007) and Afanasievo (Honeychurch et al. 2021). Domesticated cattle were introduced to ancient China in the Yellow River Valley no later than 2300 BC (Chong 2020).

Proto-Indo-European **gʷṓu–* 'cow' belongs to a set of formally similar words from the Caucasus, Southwest Asia and Egypt, including Egyptian *gw* 'bull', Kabardian *gwaw* 'bull', Chechen-Ingush *govr* 'horse', Sumerian *gu*$_4$ ~ *gud* 'bull') (Mallory and Adams 1997: 135; Nichols 1997: 143; Bjørn 2017: 61–62). The Indo-European word is carried into Central Asia by both Proto-Indo-Iranic **gā́w-* and Proto-Tocharian **keu*. It has not escaped notice that the structure of Old Chinese **ŋʷu* resembles that of Proto-Indo-European (Mallory and Adams 1997: 135; Pulleyblank 1995 apud Sagart 1999: 194; Blažek 2013: 47–48), but a clear connection is lacking. As with the horse above, nasalisation of the initial obstruent in the transfer would suffice to formally account for the Old Chinese reconstruction, but unlike 'horse', however, there is no succeeding liquid to trigger the change, and the transfer is too early to be from Iranic and must be sought elsewhere. A native onomatopoeic formation remains a possibility (Sagart 1999: 192–194), but it has little intrinsic probative value as it hardly explains the widespread similarities and recent and less controversial introductions, including Proto-Mongolic *geün* 'bull' and further Manchu *geo* 'mare; cow' (Gamkrelidze and Ivanov 1995: 491–492). Both the Chinese and the Indo-European forms have also been compared to the similar Yeniseian **kuʔs* 'horse, cow' (Sedláček 2008: 238; Witzel 2003: 12–13) that also shows semantic broadening. Burushaski similarly shares formal similarity with an initial labial obstruent and a rounded back vowel *buja* (Berger 1998: 60), but here, like elsewhere, the lack of solid external comparanda makes further analysis difficult.

8.3. Discussion

Having thus explored the earliest layers of linguistic data for the spread of barley, wheat, horse and cattle to Central and East Asia, the possible interpretations can be addressed.

8.3.1. Crops

Unsurprisingly, the linguistic evidence supports the spread of barley through the Inner Asian Mountain Corridor, but with a number of different innovations also reflected in the varied vocabulary. Nonetheless, two forms shared between different language groups that cannot be explained in

Table 8.5. Comparanda for 'cow'.

Proto-Tocharian	**k*	*e*	*u*	-	-	'cow'
Proto-Mongolic	**g*	*e*	*ü*	*n*	-	'cow'
Manchu	*g*	*e*	*o*	-	-	'mare, cow'
Old Chinese	**ŋʷ*	-	*u*	-	-	'bull'
Burushaski	*b*	-	*u*	*j*	*a*	'cattle'
Proto-Yeniseian	**k*	-	*u*	*ʔ*	*s*	'horse, cow'

either of the language families point to an earlier Neolithic substrate that was borrowed into the known immigrant languages. The datable introduction of words for barley with the Indo-Iranic languages did not erase the previous layer(s) of agricultural terminology. While strictly hypothetical, it is possible that **arpuca* and **mr̥(n)k-* belong to the same Neolithic substrate. The form **mr̥(n)k-* is thus potentially the oldest word for barley carried from Southwest Asia into Central Asia. In the Central Asian mountains, however, the shift to highland barley introduced a new term **arpuca* that may have triggered a broadening of the original word to cover 'wheat', as seen in Turkic and, perhaps, separate borrowings into Chinese. Curiously, the Central Asian word for 'barley' may thus have become the East Asian word for 'wheat', likely through borrowing and broadening through different, but as yet unidentified languages of ancient Central Asia. Broadening is visible in the history of Chinese, while a different type of barley may have caused the shift in Turkic. This situation rendered the otherwise widespread Central Asian word for wheat obsolete in East Asia where no trace of **GanTum* is to be found.

8.3.2. Livestock

On a background of different local words for the wild horses, the concomitant introduction of the modern domesticated horse and a specialised word for it by Iranic speaking Andronovo can be precisely dated in East Asia, although the exact route from the West remains to be sorted out. This situation also provides a valuable perspective of the intricate prehistoric contact situation between Pre-Proto-Turkic and Pre-Proto-Mongolic: since Turkic appears to have been indifferent to both Tocharian and Iranic horses, the Mongolic borrowing suggests that the two communities did not adopt horses at the same time or from the same source.

An Indo-European source for cattle in East Asia is also possible, but it has to be found before the Andronovo expansion, and must consequently be associated with Afanasievo and the later Tocharian languages, which satisfies form and function as well as time and space. It is curious to note here as well that cattle words in East Asia also became associated with horses. Similar form and function in Burushaski lack a clear context and await further analysis of the archaeolinguistic situation in the Central Asian heartland; since similar forms are known from Ancient Southwest Asia, it cannot be excluded that Burushaski continues a tradition carried directly through the Inner Asian Mountain Corridor.

8.3.3. The importance of Tocharian

The identification of the borrowed word for barley, highland barley as well as a transparent and presumably recent native formation for wheat suggests that the Tocharian speech community adopted its agricultural vocabulary regionally, and from the datable sources no earlier than 2000 BC. The case is different for livestock, where both cattle and horse are inherited from Proto-Indo-European. Despite suggestions to the contrary, a secondary word for horse was probably not current in Proto-Tocharian to account for the East Asian words. Librado et al. (2021) now support earlier indications that domesticated horses only truly spread with Andronovo. This brings further clarity to the recent discussion by Honeychurch et al. (2021) and Taylor (2021) regarding the use of horses in Afanasievo; it is safe to say that they were not regularly ridden, and perhaps only corralled and pastured along with cattle, sheep and goats. This in turn might help explain the semantic shift from 'cow' to 'horse' in Tungusic and Yeniseian. It is not clear if they took turns with oxen harnessed to pull a wagon, as the shared, albeit limited, vehicle vocabulary suggests, viz. Proto-Indo-European 'wheel' > Tocharian 'wagon, cart, chariot', and 'convey' > 'way' (Anthony 2007), a feat perhaps already suggested by the crossing of the steppe zone. It thus appears that the Tocharian languages and their immediate ancestor Proto-Tocharian have been given too much credit for the western words in East and Central Asia; the cases for both 'barley' and 'horse' are thus unimpressive. This is perhaps not surprising given the amount of time the Tocharian speech community spent in the Central Asian cultural and linguistic centrifuge before emerging in the northern Tarim Basin in the first millennium AD. Instead, and perhaps more surprisingly, the linguistically identifiable cultural changes the Afanasievo migration introduced to Central and East Asia around 3000 BC were immediate and tied to archaeologically less salient concepts, namely 'seven', 'name-fame', 'sister-in-law', 'honey', 'metal' (Bjørn 2022), and, perhaps, 'cattle'.

8.4. Conclusions

There can be little doubt that the Inner Asian Mountain Corridor served as both the linguistic and physical trajectory for the early important food crops wheat and barley, but livestock appear to have entered East Asia more easily with Indo-European migrations across the Steppe Zone. The Central Asian mountains act as a semantic prism that broadens otherwise stable meanings to the west to provide cover terms for the newly introduced species. There is no evidence for barley or wheat being transmitted to East Asia with speakers of Indo-European languages, and a steppe route can thus be abandoned altogether in favour of earlier and little-known substrate languages across the Inner Asian Mountain Corridor. The horse was likely used across the steppe zone with little exchange between the different language groups before speakers of Indo-Iranic spread the modern domesticated horse into China and beyond. Cattle may owe many of their East Asian word forms to early Indo-European contacts, presumably with Afanasievo, but a direct trajectory from Southwest Asia through the Inner Asian Mountain Corridor cannot be excluded either. The linguistic data hold great promise for future insights grounded in the perspectives of the people who migrated and exchanged across Eurasia, the true value of which only is attainable through determined interdisciplinary analysis.

References

Abaev, V. I. (1958) *Istoriko-etimologićeskij slovar' osetinskogo jazyka* [Historical-etymological dictionary of the Ossetic language] *I, A-K'*. St. Petersburg: Nauka.

Adams, Douglas Q. (2013) *A Dictionary of Tocharian B - Revised and Greatly Enlarged.* New York & Amsterdam: Rodopi.

Anthony, David W. (2007) *The Horse, the Wheel and Language.* Princeton University Press.

Anttila, Raimo (1989) *Historical and Comparative Linguistics*, second revised edition. Amsterdam: John Benjamins.

Berger, Hermann (1998) *Die Burushaski-Sprache von Hunza und Nager.* Teil III Wörterbuch. Wiesbaden: Harrassowitz.

Bjørn, Rasmus G. (2017) *Foreign Elements in the Proto-Indo-European Vocabulary* (MA thesis/prize paper). University of Copenhagen (see also www.loanwords.prehistoricmap.com).

Bjørn, Rasmus G. (2022) Indo-European loanwords in Bronze Age Central and East Asia: Six new perspectives on prehistoric exchange in the Eastern Steppe Zone. *Evolutionary Human Sciences* 4: e23.

Blažek, Václav (2013) Indo-European zoonyms in Afroasiatic perspective. *Journal of Language Relationship* 9: 39–54.

Blažek Václav (2016) On Indo-European 'barley'. In Sandgaard Hansen *et al.* (eds), *Etymology and the European Lexicon. Proceedings of the 14th Fachtagung der Indogermanischen Gesellschaft, 17-22 September 2021, Copenhagen, 53–67.* Wiesbaden: Reichert.

Bradley, D. (2016) Chinese calendar animals in Shanhaijing and in Sino-Tibetan languages. In D. Bradley and P. Likun (eds.) *Shanhaijing World Geography and Ancient Chinese Civilization,* pp. 1–9. Beijing: Wai yu jiao xue yu yan jiu chu ban she.

Campbell, Lyle (2013) *Historical Linguistics: An Introduction.* Third edition. Edinburgh University Press.

Ching, Chao-jung (2016) On the names of cereals in Tocharian B. *Tocharian and Indo-European Studies* 17: 29–64.

Chong, Yu (2020) *The Origin of Cattle in China from the Neolithic to the Early Bronze Age.* Oxford: BAR.

Fuller, Dorian (2002) An agricultural perspective on Dravidian historical linguistics: archaeological crop packages, livestock and Dravidian crop vocabulary. In Peter Bellwood and Colin Renfrew (eds.), *Examining the Farming/Language Dispersal Hypothesis,* pp. 191–213. Cambridge: McDonald Institute for Archaeological Research.

Gamkrelidze, Thomas V. and Vjaceslav V. Ivanov (1995) *Indo-European and the Indo-Europeans* [English version by Johanna Nichols]. Berlin: Mouton de Gruyter.

Golden, Peter B. (1998) The Turkic peoples: a historical sketch. In L. Johanson and É. Csató (eds), *The Turkic Languages.* Routledge.

Haak, Wolfgang *et al.* (2015) Massive migration from the steppe was a source for Indo-European languages in Europe. *Nature* 522: 207–211.

Holopainen, Sampsa (2019) *Indo-Iranian Borrowings in Uralic: Critical overview of the sound substitutions and distribution criterion.* Doctoral dissertation, University of Helsinki.

Honeychurch, William *et al.* (2021) The earliest herders of East Asia: examining Afanasievo entry to Central Mongolia. *Archaeological Research in Asia* 26: 1–19.

Janhunen, Juha (1998) The horse in East Asia: reviewing the linguistic evidence. In Victor H. Mair (ed), *The Bronze Age and Early Iron Age Peoples of Eastern Central Asia,* pp. 415–430. Washington, DC: The Institute for the Study of Man.

Kümmel, Martin J. (2017) Agricultural terms in Indo-Iranian. In M. Robbeets and A. Savelyev (eds) *Language Dispersal Beyond Farming,* pp. 275–290. Amsterdam & Phildelphia: John Benjamins.

Kuzmina, Anna (2007) *The Origin of the Indo-Iranians.* Leiden: Brill.

Laufer, Berthold (1919) *Sino-Iranica: Chinese Contributions to the History of Civilization in Ancient Iran.* Chicago Field Museum of Natural History.

Mair, Victor H. (1990) Old Sinitic '*mʸag*', Old Persian '*maguš*', and English *'magician'. Early China* 15: 27–47.

Mallory, James P. (2018) The Indo-Europeans and agriculture. In G. Kroonen, J. Mallory and B. Comrie (eds), *Talking Neolithic,* pp. 196–241. Washington, DC: Institute for the Study of Man.

Mallory, James P. & Adams, Douglas Q. (1997) *Encyclopedia of Indo-European Culture.* Chicago and London: Fitzroy Dearborn.

Matisoff, James A. (2003) *Handbook of Proto-Tibeto-Burman: System and Philosophy of Sino-Tibetan Reconstruction.* University of California Press

Mikič, Aleksandar (2019) *Lexicon of Pulse Crops.* Boca Raton, London & New York: CRC.

Morgenstierne, Georg (1974) *Etymological Vocabulary of the Shughni Group.* Wiesbaden: Dr. Ludwig Reichert.

Napolskikh, Vladimir (1996) Proisckhozhdenie ugorskogo nazvanija loshadi. *Linguistica Uralica* XXXII(2): 116–118.

Narasimhan, Vagheesh M., *et al* (2019) The formation of human populations in South and Central Asia. *Science* 365: eaat7487.

Nichols, Johanna (1997) The epicentre of the Indo-European linguistic spread. In R. Blench and M. Spriggs (eds), *Archaeology and Language I: Theoretical and Methodological Orientations*, pp. 122–148. Routledge.

Palmér, Axel (2019) *Traces of 'Pre-Indo-Iranian': Chronological Layers and Structural Characteristics of Early Indo-Iranian Loanwords* [MA thesis]. Leiden: Universiteit Leiden.

Parpola, Asko (2012) The problem of Samoyed origins in the light of archaeology: On the formation and dispersal of East Uralic (Proto-Ugro-Samoyed). *Per Urales ad Orientem. Iter polyphonicum multilingue.Festskrift tillägnad Juha Janhunen på hans sextioårsdag den 12 februari 2012* (Suomalais-Ugrilaisen Seuran Toimituksia = Mémoires de la Société Finno-Ougrienne 264), pp. 287–298. Helsinki: Suomalais-Ugrilais Seura.

Peyrot, Michaël (2018) Tocharian agricultural terminology: Between inheritance and language contact. In G. Kroonen, J. Mallory and B. Comrie (eds), *Talking Neolithic: Proceedings of the workshop on Indo-European origins held at the Max Planck Institute for Evolutionary Anthropology, Leipzig, December 2–3,* 2013, pp. 242–277. Washington DC: Institute for the Study of Man.

Peyrot, Michaël (2019) The deviant typological profile of the Tocharian branch of Indo-European may be due to Uralic substrate influence. *Indo-European Linguistics* 7: 72–121.

Pulleyblank, Edwin G. (1995) The historical and prehistorical relationships of Chinese. In William S.-Y. Wang (ed), *The Ancestry of the Chinese Language,* pp. 145-194 (Journal of Chinese Linguistics Monograph Series No.8). Berkeley.

Pulleyblank, Edwin G. (1996) The earliest contacts between Indo-Europeans and Chinese. *International Review of Chinese Linguistics* 1(1): 1–25.

Raulwing, Peter (2000) *Horses, Chariots and Indo-Europeans.* Budapest: Archaeolingua.

Rawson, Jessica, Konstantin Chugunov, Yegor Grebnev and Limin Huan (2020) Chariotry and prone burials: reassessing Late Shang China's relationship with its northern neighbours. *Journal of World Prehistory* 33: 135–168.

Robbeets, Martine (2015) Proto-Transeurasian: where and when? *Man in India* 95: 921–946.

Robbeets, Martine; Juha Janhunen, Alexander Savelyev and Evgeniya Korovina (2020) The homelands of the individual Transeurasian proto-languages. In M. Robbeets and A. Savelyev (eds), *The Oxford Guide to the Transeurasian Languages,* pp. 753–771. Oxford University Press.

Robbeets, M., Bouckaert, R., Conte, M. *et al.* (2021) Triangulation supports agricultural spread of the Transeurasian languages. *Nature* 599: 616–621.

Sagart, Laurent (1999) *The Roots of Old Chinese.* Amsterdam: John Benjamins.

Sagart, Laurent *et al.* (2019) Dated language phylogenies shed light on the ancestry of Sino-Tibetan. *PNAS* 116(21): 10317–10322.

Savelyev, Alexander (2017) Farming-related terms in Proto-Turkic and Proto-Altaic. In M. Robbeets and A. Savelyev (eds), *Language Dispersal Beyond Farming,* pp. 123–154. Amsterdam: John Benjamins.

Schuessler, Axel (2007) *ABC Etymological Dictionary of Old Chinese.* University of Hawai'i Press.

Sedláček, Kamil (2008) The Yeniseian languages of the 18th century and Ket and Sino-Tibetan word comparisons. *Central Asiatic Journal* 52: 219–305.

Shaughnessy, E.L. (1988) Historical perspectives on the introduction of the chariot into China. *Harvard Journal of Asiatic Studies* 48: 189–237.

Shelach-Lavi, Gideon (2015) Steppe interactions and their effects on Chinese culture during the second and early first millennia BCE. In Michal Biran and Reuven Amitai (eds), *Nomads as Agents of Cultural Change: The Mongols and Their Eurasian Predecessors,* pp. 10–31. University of Hawai'i Press.

Southworth, Franklin (2011) Rice in Dravidian. *Rice* 4: 142–148.

Spengler III, Robert N. (2019) *Fruit from the Sands. The Silk Road Origins of the Foods We Eat.* University of California Press.

Stachowski, Kamil (2008) *Names of Cereals in the Turkic Language.* Kraków: Institute of Oriental Philology.

Stevens, Chris J. *et al.* (2016) Between China and South Asia: A Middle Asian corridor of crop dispersal and agricultural innovation in the Bronze Age. *The Holocene* 26(10): 1541–1555.

Taylor, William T. T. (2021) Building a scientific model for East Asian pastoral origins: A reply to Honeychurch *et al. Archaeological Research in Asia* 26: e100286.

Warries, Abel (2019) *Uralic Influence on the Tocharian Agglutinative Case System.* MA Thesis, Leiden University.

Witczak, K.T. and L. Novák (2016) A Pamir cereal name in Mediaeval Greek sources. *Studia Iranica* 45: 53-64.

Witzel, Michael (2003) Linguistic Evidence for Cultural Exchange in Prehistoric Western Central Asia. *Sino-Platonic Papers* 129.

Witzel, Michael (2005) Central Asian roots and acculturation in South Asia: linguistic and archaeological evidence from western Central Asia, the Hindukush and northwestern South Asia for early Indo-Aryan language and religion. In T. Osada (ed), *Linguistics, Archaeology and the Human Past*, pp. 87–211. Kyoto: Research Institute for Humanity and Nature.

Witzel, Michael (2009) The linguistic history of some Indian domestic plants. *Journal of Bioscience* 34(6): 829–833.

Wojtilla, Gyula (1999) The Sanskrit Godhūma apropos of a Short Excursion in Indo-European and Indo-Aryan prehistory. *Acta Orientalia Academiae Scientiarum Hungaricae* 52(3/4): 223–234.

Zhang, Fan, Chao Ning, Ashley Scott *et al.* (2021) The genomic origins of the Bronze Age Tarim Basin mummies. *Nature* 599: 256–261.

Zhou, Yong, Zhongxu Chen *et al.* (2017) Uncovering the dispersion history, adaptive evolution and selection of wheat in China. *Plant Biotechnology Journal* 16: 280–291.

Zhou, X., J. Yu, R.N. Spengler *et al.* (2020) 5,200-year-old cereal grains from the eastern Altai Mountains redate the trans-Eurasian crop exchange. *Nature Plants* 6: 78–87.

9

Central-Western Asian and European Cultural Influences on Gold and Silver Wares Discovered in China on the Eurasian Steppe

Jingming Zhang

Abstract: The Eurasian steppe straddles Europe and Asia, and is the largest steppe belt in the world. The northern steppes of China are an integral part of the Eurasian steppe and their interaction with each other constituted a medium of communication between East and West. This chapter will use gold and silver wares as a focus, taking the period of China's historical development as a vertical time line, and using the shapes, patterns, production techniques and period features in the gold and silver wares as points of observation to analyse the influence of western cultures on the character and style of gold and silver wares. At the same time, this chapter further explores the cultural exchanges between the ancient northern steppes of China in the Eurasian steppe and western countries and regions, reflecting the historical context of civilisational exchange and cultural integration between ancient China and other regions.

Keywords: Eurasian steppe, northern steppes of China, gold and silver ware, plastic arts, cultural exchange

The Eurasian steppe refers to the geographic region from the Great Khingan Mountains in the east, the Altai Mountains in the west, the Great Wall of China in the south, and Mongolia in the north. The grasslands north of the Great Wall lie in the eastern part of the Eurasian steppe. Within the area of modern China, the steppe includes Inner Mongolia, Xinjiang and the western regions of the three northeastern provinces of China, as well as the northern regions of Hebei, Shanxi, Shaanxi and Ningxia. Many sites and artefacts have been found on the Chinese steppes through archaeological research, and among them some gold and silver wares have Central and Western Asian and European cultural influences. Archaeological data show that the ancient gold and silver objects unearthed in the area of modern China are mainly located in the grasslands north of the Great Wall, the Central Plains and the southern regions. Among them, one of the earliest specimens was found on the grasslands north of the Great Wall, with a date equivalent to the late Xia dynasty, i.e., from the 17th to 16th centuries BC.[1] The gold and silver wares of this time did not yet show the characteristics of Central/Western Asian or European culture, but the bronze wares already reflected cultural exchanges between China and the West.

After the 7th century BC, the number of gold and silver wares found in the grasslands north of the Great Wall increased. In particular, the use of animal shapes, decorative patterns and production techniques gradually became consistent with those in other parts of the Eurasian steppes. This reflects the early cultural exchanges between China and the west across the steppes. After the 3th century AD, the historical stage of the Eastern and Western Jin and Sixteen Kingdoms to Song, Liao, Jin and Yuan dynasties was also a period of prosperity for Sino-Western mercantile and cultural exchange in the Eurasian steppe passage area. Observation of the shapes, patterns and production techniques of gold and silver wares discovered by archaeological excavation, in addition to the inheritance and integration of the cultures of various ethnic groups in China, shows that these wares incorporate Central and West Asian and European cultural elements. This reflects the historical situation of exchange and mutual learning between ancient civilisations and national cultural exchanges between China and other countries in the world.

Based on his earlier publications in Chinese (Zhang 2005, 2011), in this chapter the author takes gold and silver wares as the main subject, uses archaeological and historical research methods, and takes the development period of Chinese history as the vertical time line. The chapter examines plastic arts, patterns, craftsmanship, chronological characteristics and other factors as observation points in order to analyse the Western cultural characteristics and styles of gold and silver artefacts found in northern China. The author also discusses the cultural exchange between China and Western countries and regions in the ancient Eurasian steppe as reflected in archaeological discoveries of gold and silver objects from the 15th century BC to the early 12th century AD.

[1] Editors' note: this chapter uses the Great Wall as a convenient geographical marker without implying that an actual wall existed from the Bronze Age.

9.1. Cultural exchange between China and the West reflected in archaeological gold and silver wares from the 15th to 6th centuries BC

The earliest gold and silver artefacts discovered by archaeology in China can be traced back to the late Xia Dynasty, that is, the 17th to 16th centuries BC. The main gold and silver specimens are gold earrings unearthed from the tombs of Dadianzi in Aohanqi, Inner Mongolia (Institute of Archaeology, Chinese Academy of Social Sciences et al. 1996) and gold and silver earrings from Huoshaogou in Yumen, Gansu Province (Gansu Provincial Museum et al. 1981). The Dadianzi tombs belong to the Lower Xiajiadian culture and are representative of the early bronze culture of northern China distributed in the eastern part of the Eurasian steppe. The metal products unearthed from the tombs are mostly small ornaments, such as bronze earrings, finger rings, decorations at the top of sticks and gold earrings.

The gold earring unearthed from tomb No. 516 in Dadianzi is surrounded by gold wire in an oval loop; one end is flat and the other end has a round and blunt tip. Copper earrings with similar styles were found in sites and tombs of the Lower Xiajiadian culture. There is also a different shape of copper earring with a flat, trumpet-shaped mouth. This is a popular type in the grasslands north of the Great Wall. This type had a direct influence on similar objects found in the area south of the Great Wall in the Shang Dynasty. For example, the flat trumpet-shaped gold earrings unearthed from the Shang dynasty tombs in Liujiahe, Pinggu District, Beijing (Beijing Cultural Relics Management Office et al. 1977), directly inherit the shape of those of the Lower Xiajiadian culture and provide clues for studying the relationship between the Lower Xiajiadian and Shang cultures (Figure 9.1).

Trumpet-shaped earrings made of metal were also found in two tombs in the Almaty region of Kazakhstan. They are both flat and circular trumpet-shaped earrings. The flat trumpet-shaped earrings found in the grasslands north of the Great Wall continued in the eastern part of the steppes until the early Western Zhou dynasty. They have a completely different shape from the round trumpet-shaped earrings of the Andronovo culture found in Kazakhstan.

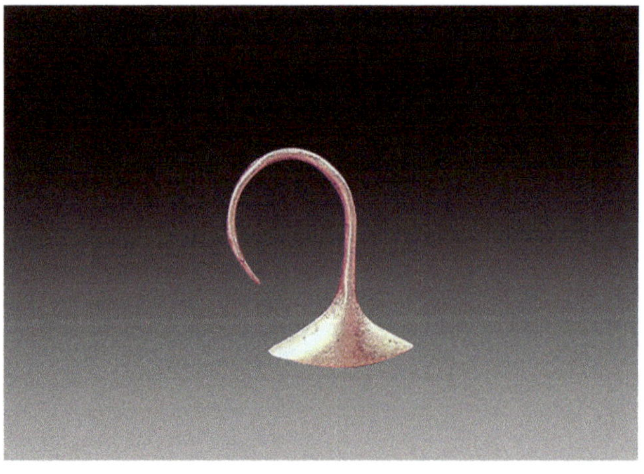

Figure 9.1. Flat trumpet-shaped gold earring from Liujiahe. Source: Photograph by the author.

Generally speaking, the flat trumpet-shaped earrings found in the steppes north of the Great Wall have similarities with the flat trumpet-shaped earrings in Kazakhstan, and the relationship between the two is very clear. Lin Yun (1992) has written that, 'This kind of exchange occurring between settled groups so far apart' means that 'there must be highly mobile people as the medium [of exchange]'. At that time, the Lower Xiajiadian culture mainly consisted of agricultural groups. There must have been nomadic non-East Asian people of the Eurasian steppe who spread this culture in the region of modern China. The precious metal culture spread from Kazakhstan through Xinjiang to the eastern steppe, a process which had a certain influence on the local bronzes.

From the 8th to the 6th centuries BC, equivalent to the late Western Zhou to the middle Spring and Autumn Period, the number of gold objects unearthed from the steppes north of the Great Wall gradually increased, with examples mainly found in tombs of the Upper Xiajiadian culture in southeast Inner Mongolia. The main sites where these objects were discovered were the Xiaoheishigou tomb (Xiang and Li 1995; Inner Mongolia Autonomous Region Institute of Cultural Relics and Archaeology et al. 2009), Nasitai tomb, and Nanshangen tomb (Zhaowudameng Cultural Relics Workstation, Liaoning Province et al. 1973), all in Inner Mongolia. The main category is single decorations.

The gold wares unearthed from Xiaoheishigou tomb include round gold artefacts with duck pattern ornaments, gold wire loop, gold ware with nail-shaped ornaments, gold clasps, a gold loop with leopard pattern ornament, a bronze knife with animal pattern and a wrapped gold handle (Figure 9.2), a bronze dagger with a gold loop handle and so on. A bent-leg horse-shaped gold ornament was unearthed from the tomb of Nasitai (Figure 9.3), and a spring-shaped gold earring was unearthed from the tomb of Nanshangen.

Among these types of gold ware, the styles of the gold loop with leopard pattern ornament and the bent-leg horse shaped gold ornament are in contrast with the Tagar culture distributed in the Minusinsk Basin in the middle reaches of the Yenisei River. In the 5th century BC, new patterns appeared in the Tagar culture, represented by deer with bent legs, a beast curled in a loop, a standing beast with an animal in its mouth, and fictional animals. Judging from the animal-patterned objects unearthed in the grasslands north of the Great Wall, the age of the bent-leg animals and loop-shaped beasts predates the Tagar culture. Examples here include the bent-leg horse shaped gold ornament from Nasitai in Ningcheng and the gold loop with leopard pattern ornament unearthed in Xiaoheishigou. The age of these objects is earlier than the Tagar culture. It is hard to say that this kind of animal pattern found on the steppes is influenced by Scythian culture from the Siberian 'wild beast pattern' and likely has its own origin. It is inferred that some 'beast patterns' of the Tagar culture are influenced by animal patterns from the eastern steppe.

Figure 9.2. Bronze knife with animal pattern and wrapped gold handle from Xiaoheishigou. Source: Photograph by the author.

Figure 9.3. Bent-leg horse shaped gold ornament from Nasitai. Source: Photograph by the author.

9.2. Characteristics of western culture in archaeological gold and silver wares from the 3rd century BC to the early 3rd century AD

From the 3rd century BC to the beginning of the 1st century AD, equivalent to the late Warring States and Western Han Dynasty, the Huns and other ethnic groups lived mainly on the steppes north of the Great Wall. Many archaeological sites and tombs have a large number of objects decorated with animals, a style which became a main feature of the Hun culture. The animal style of gold and silver ware is closely related to the 'beast pattern' of Scythian culture.

Among the objects of Scythian culture found in Rostov Oblast and the Autonomous Region of Kalmykia in southern Russia (Ancient Orient Museum and Asahi Shimbun 1991), gold and silver wares from the 5th to the 4th century BC are both similar to and different from late Warring States Hun gold and silver wares unearthed in the eastern steppes.

A gold necklace with a tiger's head, a gold ornament with a tiger biting a wild boar (Figure 9.4) and an iron sword with an inlaid gold handle were unearthed from Scythian sites. Other examples are a silver necklace with a tiger biting a sheep's head in its mouth unearthed from Yulongtai tomb in Zhungeerqi, Inner Mongolia (lnner Mongolia Museum et al. 1997), a silver deer ornament and sword handle with inlayed gold and silver unearthed from Nalingaotu tomb in Shenmu, Shaanxi Province (Dai and Sun 1983), a gold buckle with leopard biting boar pattern inlaid with gems (Figure 9.5), and a gold buckle with tiger biting cattle pattern (Figure 9.6) from Aluchaideng tomb in Hangjinqi, Inner Mongolia (Tian and Guo 1980). These artefacts are very similar to each other.

The necklaces are decorated with animal patterns on the ends, but the shapes are different. The round gold necklace with a tiger's head has only one circle, and there is a gap in the middle of the front. Meanwhile, two sides are decorated with opposite tiger patterns. The silver necklace with a tiger biting a sheep's head in its mouth is twisted into two circles, and one end is decorated with the pattern of a biting sheep's head. The gold ornament with the tiger biting wild boar motif and the gold buckle with leopard biting boar motif inlaid with gems are the same in theme, but the production process and the arrangement of the animals are different.

A gold ornament with a deer pattern (Figure 9.7) was found from tomb No. 2 at Kitokov in Rostov Oblast and a gold scabbard ornament with a lying deer pattern (Figure 9.8) was excavated from tomb No. 2 of Xigoupan in Zhungeerqi, Inner Mongolia (Yikezhaomeng Cultural Relics Workstation et al. 1980). Even if the craftsmanship or shape is almost the same, the posture is different. The gold bead necklace from the No. 8 tomb of the Five Brothers in Rostov Oblast has the same type of decoration as the gold bead necklace (Figure 9.9) unearthed from the tomb of Aluchaideng in Hangjinqi, but the shape is different. The former is connected with gold beads in a knotted gold wire, and the latter is formed by directly connecting the gold beads in a series. The human-shaped gold ornament unearthed from tomb No. 2 at Busochino in Rostov Oblast is similar to the antelope-shaped gold ornament unearthed from the Aluchaideng tomb in Hangjinqi in terms of craftsmanship and sacrificial functions, but again their shapes are different.

Figure 9.4. Gold ornament with a tiger biting a wild boar. Source: Re-drawn by the author from Ancient Orient Museum and Asahi Shimbun (1991).

Figure 9.5. Gold buckle with leopard biting boar pattern inlaid with gems from Aluchaideng. Source: Photograph from Zhang (2005).

Figure 9.6. Gold buckle with tiger biting cattle pattern from Aluchaideng. Source: Photograph from Zhang (2005).

There is no prototype for the monster-shaped pattern in China. Du Zhengsheng (1993) argues that, 'The origin of the monster motif of the cultural objects in northern China should be between the northern shore of the Black Sea and the Altai Mountains. It is difficult to trace its roots in China, where its appearance is later than the Black Sea to Altai, mostly after the late Warring States period.' Objects with this motif are found in the grasslands north of the Great Wall, the Central Plains, and the southern regions of modern China.

Figure 9.7. The gold bead necklace from tomb No. 2 at Kitokov. Source: Re-drawn by the author from Ancient Orient Museum and Asahi Shimbun (1991).

Figure 9.8. Gold scabbard ornament with lying deer pattern from tomb No. 2 at Xigoupan. Source: Photograph from Zhang (2005).

Figure 9.9. Gold scabbard ornament with lying deer pattern from grave No. 8 of the Tomb of the Five Brothers. Source: Photograph from Zhang (2005).

Figure 9.10. Gold ornament with deer-shaped monster from the tomb of Nalingaotu. Source: Photograph by the author.

A gold ornament with a deer-shaped monster has been excavated from the tomb of Nalingaotu in Shenmu (Figure 9.10). A gold ornament with a standing monster pattern, a gold scabbard ornament with a reclining monster, and a gold scabbard ornament with a monster pattern have been unearthed from tomb No. 2 in Xigoupan. The monster images often have an eagle beak, an animal body and a head with long horns. They belong to the category of illusory animals, and the source cannot be found in the animal patterns from the grasslands north of the Great Wall. But these images are widely found in Scythian and Altai art and they date earlier than those from the steppes north of the Great Wall.

Figure 9.11. Gold ornament with a monster pecking a tiger pattern from Taidi Tomb No. 1. Source: Photograph by the author.

In tomb No. 1 of Taidi, ancient Jiaohe city, Turpan, Xinjiang, a gold ornament with a monster pecking a tiger pattern was found (Figure 9.11) (Xinjiang Institute of Cultural Relics and Archaeology et al. 1999). The monster also had an eagle beak, an animal body and a long horn. It can be seen that the monster pattern was first popular in Scythian and Altai art and then spread to the grasslands north of the Great Wall. The monster pattern of Hunnic gold and silver wares was the result of this influence.

The impact of this culture spread to the Central Plains and southern regions, with examples including a gold belt ornament with an eagle beak on the animal body, a gold ornament with eagle beak on a sheep's head and a gold ornament with eagle beak on an animal head unearthed from tomb No. 30, Xinzhuangtou, Yixian (Hebei Province) (Hebei Provincial Institute of Cultural Relics 1996); a gold belt ornament with an eagle beak on the body of an animal from the Western Han tomb at Shizishan Xuzhou (Jiangsu Province) (Zou and Wei 1998); and the copper belt ornament with an eagle beak on the body of an animal from tomb No. 94 at Yinshanling, Pingle (Guangxi Province) (Guangxi Zhuang Autonomous Region Cultural Relics Task Force 1978). Therefore, some Hun animal shapes on gold and silver wares were developed through cultural exchanges between China and the west.

From the early 1st to the early 3rd centuries AD, during the Eastern Han dynasty, the Tuoba Xianbei and Wuhuan lived mainly in the steppes north of the Great Wall. The former used the most gold and silver wares. From the perspective of craftsmanship, many new techniques were introduced from the west.

An iron buckle covered by gold leaf with a divine beast pattern with inlaid gems was unearthed from the tomb of Taoheqi, Inner Mongolia (Figure 9.12) (Yi and Lu 1984). The main ornamentation was hammered into convex texture, and the detailed part uses the engraving method. This craft was introduced from Persia around the beginning of the 3rd century. The divine beast pattern is more standard than the monster pattern of the Huns. It has an eagle beak, a leopard body, a sheep horn and two wings, which show an inheritance from western culture or from the Tuoba Xianbei's worship of the gods who are meritorious to them. Therefore, winged divine beasts are widely popular in the art of various parts of Europe and Asia, and there are many types. Among them, the 'griffin' plastic art has a lot to do with the Chinese divine beasts. Xianbei gold and silver wares with divine beast patterns are similar to the popular 'griffin' in the grasslands of West Asia and Eastern Europe. They have an eagle's head and a beast's head, and also have long wings. In terms of artistic modelling, the Xianbei divine beast pattern does not completely absorb Western shapes, such as the tips of the wings facing backwards rather than forwards.

The inlay technology of Xianbei gold and silver wares also has characteristics of the times. Xia Nai (1983) wrote, 'The technique of inlaying was produced in the pre-Qin dynasties, but the inlaying of gemstones and beading flourished in the Jin Dynasty. There are also some objects which have set diamonds. These objects entered China and Southeast Asia from Greece and Rome to the east.'

The gold buckle with wild boar pattern (Figure 9.13) unearthed from the tomb of Lingpiyao in Helingeer, Inner Mongolia (Inner Mongolia Museum et al. 1984) has a surface inlaid with gemstones and turquoise, so that the wild boar pattern and inlay play the role of the main decoration at the same time. This is different from the traditional mosaic technique in the grasslands north of the Great Wall and undoubtedly has a certain connection with the cultural exchanges between China and the west. In addition, another object unearthed in the Lingpiyao tomb is a gold bowl with an eight-petal floral mouth, the style of which comes from the Iranian plateau region. A gold ornament with the pattern of a Hu 'western barbarian' fighting a beast (Figure 9.14) excavated from North Halatuda in Inner Mongolia is similar to the gold ornament with a pattern of a divine man riding a dragon excavated from the tomb in Sibrkand, Afghanistan (Figure 9.15) (Sarianidi 1979; Mei 1991). These examples all show that western culture had an influence on the gold and silver wares of the Xianbei during the Eastern Han Dynasty.

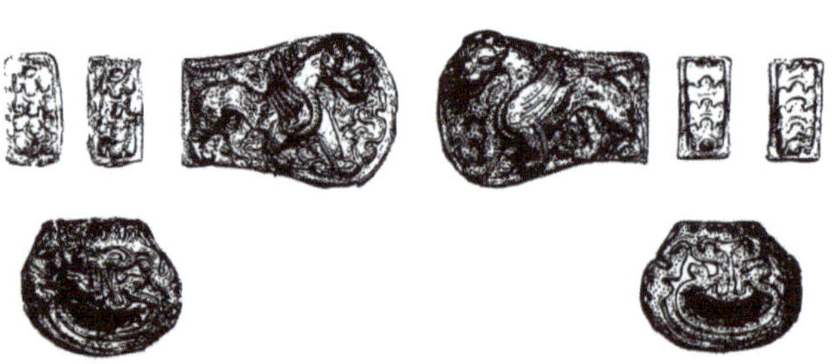

Figure 9.12. Iron buckle covered by gold leaf with divine beast pattern which inlaid gems from the tomb of Taoheqi. Source: Re-drawn by the author from a photograph.

Figure 9.13. Gold buckle with wild boar pattern. Source: photograph from Zhang (2005).

Figure 9.14. Gold ornament with a Hu 'western barbarian' fighting a beast. Source: Photograph from Zhang (2005).

Figure 9.15. Gold ornament with a pattern of a divine figure riding a dragon. Source: Re-drawn by the author from Sarianidi (1979).

9.3. Western cultural characteristics of archaeological gold and silver artefacts discovered from the late 3rd to late 5th centuries

In the latter part of the 3rd century AD, China began to enter the period of the Jin dynasty and the Sixteen Kingdoms. The Murong Xianbei and the Han people influenced by Xianbei culture in the northern steppe established the 'Three Yan' regimes in the western Liaoning area, namely Qianyan, Houyan and Beiyan. The Longcheng capital (present Chaoyang, Liaoning Province) became one of the main cultural centres in the eastern Eurasian steppes during this period. Many objects have been discovered in this area, and some gold and silver wares have characteristics of western culture.

In the tombs found in the Chaoyang area of Liaoning Province during the two Jin dynasties and the Sixteen Kingdoms, a kind of gold *buyao* crown ornament (Figure 9.16) is common. *Buyao* are a type of ancient jewellery worn by women, so named because they shake with each step. These ornaments have been found at Tiancaogou Jin tomb in Chaoyang (Chaoyang City Museum et al. 1997), Wangzifenshan Jin tomb (Liaoning Provincial Institute of Cultural Relics and Archaeology et al. 1997), Fangshen Village Jin tomb in Beipiao (Chen 1960), Qianyan tomb No. 1 in Shiertaizhuanchang (Liaoning Provincial Institute of Cultural Relics and Archaeology et al. 1997), Beiyan Feng Sufu tomb (Li 1973) and so on. These burials produced gold *buyao* crown ornaments with the shape of a flowering tree (Figure 9.17). This differs from those of the Eastern Han and the Southern and Northern Dynasties, and belongs to the Murong Xianbei, indicating ethnic and chronological consistency. These kinds of gold *buyao* crown ornaments are also found in the tomb of Sibrkand in Afghanistan, which has tree-shaped gold *buyao* crown ornaments and gold *buyao* crown ornaments using thin gold pieces. They have certain similarities in shape and craftsmanship, and especially those decorated with round gold leaves are even more similar.

Figure 9.16. Gold *buyao* crown ornament. Source: Photograph by the author.

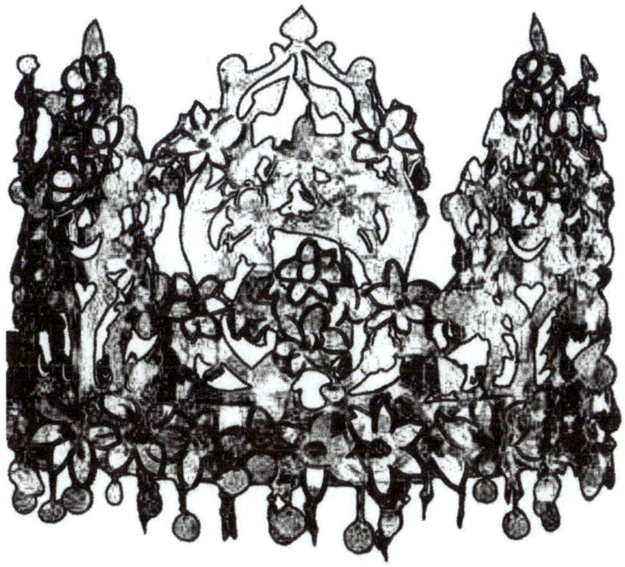

Figure 9.17. Gold *buyao* crown ornament in a flowering tree shape. Source: Re-drawn by the author from Sarianidi (1979).

Gold *buyao* crown ornaments were introduced to China during the Eastern Han. They were first popular in the Central Plains of China, becoming a major feature of Murong Xianbei gold jewellery later in time. A phoenix-shaped gold *buyao* crown ornament was unearthed from the Eastern Han Maolitu tomb, Inner Mongolia (Zhao 1999) and was used as a model for traditional Chinese phoenix birds. On the gold *buyao* crown ornaments of the Jin dynasty and the Sixteen Kingdoms, the decoration of birds, beasts and flowers is not seen, and only the branches are decorated with gold leaves. The gold *buyao* crown ornaments unearthed from the cellars of the Northern Wei Dynasty in Xihezi, Inner Mongolia (Lu and Chen 1984), are decorated with cattle and horse heads (Figure 9.18). This shows that the art of gold *buyao* crown ornaments changed over time. Therefore, it can be seen that after *buyao* were introduced from West Asia, they combined characteristics of Chinese culture with new innovations reflecting the exchange of Chinese and Western cultures. This art form was later introduced to the Korean Peninsula and Japan, and continued into the Liao and Jin periods in China.

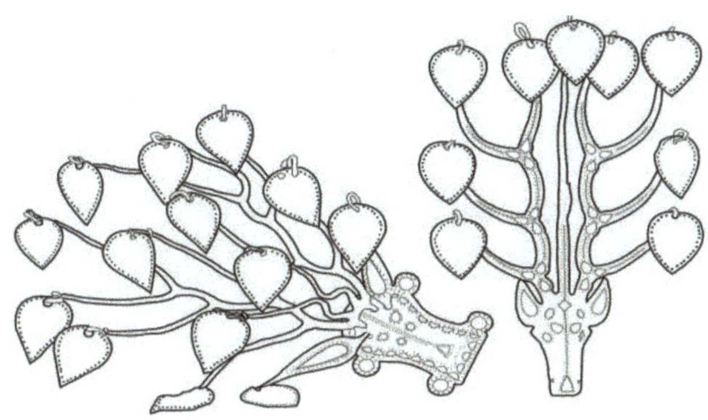

Figure 9.18. Gold *buyao* crown ornaments. Source: Re-drawn by the author from a photograph.

After the Tuoba Xianbei moved south from the Daxingan Mountains, they arrived in the Helingeer area south of the Daqing Mountains in present-day Inner Mongolia and established the Northern Wei Dynasty. Later, they moved to Pingcheng (now Datong, Shanxi Province). In this area, gold and silver wares with western cultural characteristics were found, for example, a silver bowl carved with gilded flowers or patterns, silver jars, a silver pedestalled cup with gilded flowers or patterns, and a silver bowl without patterns excavated from a tomb in the southern suburbs of Datong (Shanxi Provincial Institute of Archaeology et al. 1992). A wide silver cup with multi-lobed oval surfaces and a silver bowl carved with flowers or patterns were excavated from the Northern Wei city site in Datong (Datong City Museum et al. 1977). A silver plate with gilded flowers and hunting patterns, a silver ear cup with a linked-pearl motif and a silver pedestalled cup were excavated from Fenghetu tomb (Xia 1983) and Huagetatai tomb (Ma 1983). Among these finds is a silver bowl carved with gilded flowers or patterns excavated from a tomb in the southern suburbs of Datong (Figure 9.19). It has an open mouth, a slight carination under the mouth, a curved belly, and a rounded bottom. The mouth edge and upper belly are each decorated with a circle of linked-pearl motifs, and the belly is delineated by Acanthus leaf patterns into four zones, each with a circular ring, inside which there is the side profile of the head of a man wearing a circular hat, with deep eyes, a high nose and long hair draped over his shoulders.

These leaf pattern and circular ring decoration are common patterns in Sassanian Persian art, but the character of the circular hat belongs to the Central Asian style and is often found on Hephthalite coins, so its origin should be in Central Asia and may be related to the Hephthalites. The wide silver cup with multi-lobed oval surfaces excavated from the Northern Wei city site is divided into eight curves, with a flat mouth edge and two beasts decorated in the centre of the inner base (Figure 9.20). Its shape is similar to silver wares popular in Sassanian Persia. The silver plate with a hunting pattern excavated from the Fenghetu tomb (Figure 9.21) is in the same style as plates from Sassanian Persia.

Figure 9.19. Silver bowl carved with gilded flowers or patterns. Source: courtesy of Datong Museum of Shanxi Province.

Figure 9.20. Wide silver cup with multi-lobed oval surfaces. Source: photograph by the author.

Figure 9.21. Silver plate with hunting pattern. Source: photograph by the author.

Thus, at the beginning of the Northern Dynasties, some gold and silver wares in China already had stylistic characteristics from further west. These gold and silver wares were introduced to the Chinese belt of the Eurasian steppe through the steppe passage and had a certain influence on traditional Chinese gold and silver wares in terms of style and types of wares. It can be seen that the channel of cultural exchange between China and the west was very important for the development and evolution of Chinese gold and silver wares.

9.4. Chinese and Western cultural styles of archaeological gold and silver wares from the late 6th to late 9th centuries

During the Tang dynasty, cultural exchange between the Chinese part of the Eurasian steppe and the west occurred with societies including the Kushan empire, the Hephthalites, Sassanids, Byzantium, Sogdians and the Arab empire. These were countries and regions where the production of gold and silver wares were developed, and they had a great impact on gold and silver wares in China during this period.

The silver handled ewer, a silver plate with gilded lynx pattern, and an oval silver cup unearthed from the Lijiayingzi tomb in Inner Mongolia (Aohanqi Cultural Center 1978) are Sogdian silverware imported during the Tang. The mouth of the silver ewer (Figure 9.22) has a flow arc handle, and a bust of a Hu 'western barbarian' is welded where the upper end meets the edge of the mouth. This type of silver handled ewer can be seen in Central and West Asia, and is generally considered to be a Persian Sassanid artefact (Xia 1978). The upper end of the handle is placed on the mouth of the silver handled ewer, the neck is thick and short, the foot rim is short and fat, there is no knotty decoration, and the bottom rim is decorated with a circle of linked-pearl motifs around the foot, which is closer in style to Sogdian products.

The oval silver cup from Lijiayingzi lacks divided flaps or ears, and its shape is very special, differing from the oblong cups of Sassanid Persia and Central China. There are many figures holding such oblong cups for feasting in the Sogdian murals in Central Asia (Azarpay 1981). The shape of the Sassanian gold and silver oblong cups found in China is multi-lobed oval, as for instance the four-curved gold cups excavated in Xi'an (He et al. 1983) and the begonia-shaped gold cups excavated from a tomb in Yichuanyaling (Luoyang City's Second Cultural Relics Working Team et al. 1995). The shape is close to the Sassanian multi-lobed oval oblong cups: either they are imports or direct imitations. The oval silver cups excavated in Lijiayingzi are closer in shape to the Sogdian oblong cups.

The inner core of a silver plate with a gilded lynx pattern (Figure 9.23) is decorated with animal motifs, leaving a blank space around it. This practice was prevalent in the Sogdian region, and the silver plate may be a commercial product from the Sogdian region in China. All these objects are examples of the introduction of Sogdian silverware into the Chinese part of the Eurasian steppes during the Tang dynasty.

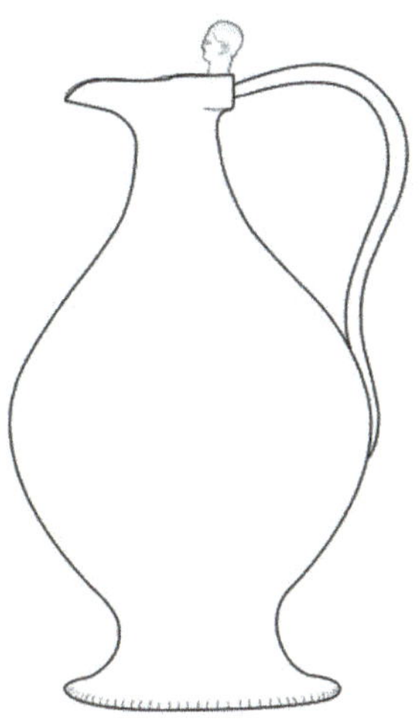

Figure 9.22. Silver handled ewer from Lijiayingzi. Source: Re-drawn by the author from a photograph.

Figure 9.23. Silver plate with a gilded lynx pattern. Source: Photograph from Zhang (2005).

Among the gold and silver wares of the Tang dynasty, many retain a distinctly Persian Sassanid style. In terms of shape, Tang oblong cups completely imitate the multi-lobed oval features of Sassanian oblong cups, but also have characteristics that are different from Persian Sassanian silverware, such as a deep body, open mouth and pedestal. The begonia-shaped silver cup with gilded *makara* sea-creature pattern in the collection of the Inner Mongolia Museum is a four-curved cup with a deep belly and a rounded foot, the inner rim of which is a full circle of linked-pearl motif; the bottom centre of the cup is decorated with a *makara* pattern (Figure 9.24). This has a distinctly Persian Sassanian silver character, in which the *makara* is based on a figure from ancient Indian mythology and Buddhist art. However, the cup has some decorative patterns of rings with single dots, which were characteristic of the layout of gold and silver wares in the Tang dynasty. Therefore, this silver cup is probably an imitation of the Persian Sasanian style silverware of the Tang.

Figure 9.24. Begonia-shaped silver cup with gilded *makara* pattern. Source: Photograph from Zhang (2005).

Some Roman-style objects were unearthed from Shuimoguo in Hohhot (Inner Mongolia Cultural Relics Task Force and Inner Mongolia Museum et al. 1975). Among them are gold coins, gold crown ornaments, gold rings with embedded gemstones and silver pedestalled cups. A gold coin with a pattern and inscription was cast in the reign of the Eastern Roman Emperor Leo I (AD 457–474); it differs from the Arabic imitations found in other parts of China.

The crocodile pattern on the gold crown ornament from Shuimoguo has a polka dot pattern carved around it (Figure 9.25). This pattern is not seen on traditional Chinese gold and silver wares and is probably imported from the west. The gold ring has an embedded gemstone engraved with a person with long hair portrayed as walking. This kind of ring was popular in ancient Rome and Persia.

The silver pedestalled cup from Shuimoguo has a convex edge on its belly (Figure 9.26), which is similar to finds from the Black Sea region and ancient Roman-Byzantine vessels from the perspective of shape. A gold and silver goblet unearthed on the outskirts of the city of Poltava on the Black Sea coast of Ukraine has a convex ridge on the belly (Tokyo National Museum et al. 1985). An agate pedestalled cup in the Pitti Palace Silver Museum in Italy, is a Byzantine product from the 9th to 11th centuries AD, also with a convex ridge on the belly. All of these pedestalled cups are very similar to the ones excavated from the Shuimoguo burial site. This type of pedestalled cup is supposed to have originally been a Roman-style vessel form, which was later used by Byzantium. Since the Persian Sassanid dynasty controlled the passage from China to Byzantium, the influence on the goblets in China may have been direct from Persia and some scholars believe that the silver goblet excavated from Shuimoguo was a Sassanid import (Xia 1978).

Figure 9.25. Gold crown ornament from Shuimoguo. Source: Photograph from Zhang (2005).

Figure 9.26. Silver pedestalled cup from Shuimoguo. Source: Photograph by the author.

All of this shows that the Byzantine empire carried out close exchange and trade with China. From a general point of view, the Sui and Tang dynasties were a prosperous period for the development of Chinese culture, and the peoples of the eastern Eurasian steppes were culturally linked to the four civilisations of Byzantium, Iran, India and China, acting as the transmitter of material and spiritual elements between east and west. Integrating different civilisations into their own cultures, the plastic arts of gold and silver ware incorporated various cultural factors which had been formed gradually, creating a greater development space for the heyday of gold and silver ware in the Liao and Jin Dynasties and cultural exchange between China and the West.

9.5. Western cultural elements in archaeological gold and silver wares from the early 10th to early 12th centuries

During this period, the Eurasian steppe region of China was controlled by the Liao dynasty and many gold and silver artefacts were found there. At the same time, the craftsmanship of Chinese gold and silver wares reached its heyday. The rapid development of gold and silver wares benefited from foreign cultural factors that played an important role.

The Liao dynasty established a powerful kingdom based on the steppes during this period. This was an achievement made by the joint control of the Khitan, Han and other ethnic groups. However, the limited culture, technology and resources of the steppe peoples objectively encouraged them to carry out foreign trade, driving cultural exchange between the Liao, the Central Plains and neighbouring peoples, and countries to the west. The gold and silver wares of the Liao not only incorporated a large number of cultural characteristics of the Tang dynasty, but also absorbed many elements of Turkic and western culture, with the result that the objects display certain features of Turkic, Sogdian or Persian culture.

The Turkic style of folded shoulders is often found in the gold and silver wares of the Liao dynasty. The gilded silver pot (Figure 9.27) excavated from the tomb of Yelyu Yuzhi (Inner Mongolia Institute of Cultural Relics and Archaeology et al. 1996) and the silver jug with folded shoulders (Figure 9.28) excavated from tomb No. 1 in Erbadi (Xiang 1984) are very similar to the gold cups with folded shoulders excavated from Turkic tombs located in the western part of the Minusinsk Basin in Russia. However, the decoration and engraved characters are in Chinese style, even though the shape imitates the Turkic style. The decoration is a linked-pearl motif, showing this gold and silver vessel also has Persian influences.

Figure 9.27. Gilded silver pot. Source: Re-drawn by the author from a photograph.

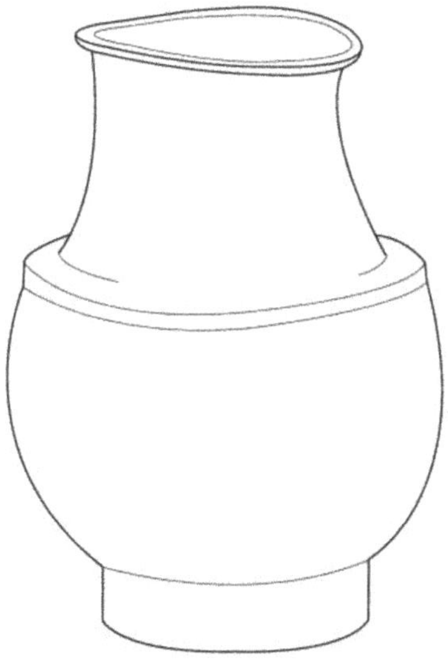

Figure 9.28. Silver jug with folded shoulders. Source: Re-drawn by the author from a photograph.

The cups and bowls of Sogdian silverware are mostly divided into curved or petal-shaped bodies, and most of the divided curved lines have wider grooves, some of them penetrating each other, with a large number of divided petals and rich variations. The surface of these objects is shaped into convex and concave undulations, giving them a strong three-dimensional effect. The curves of the gold and silver wares in the late Tang dynasty only made straight lines which were relatively shallow. Generally, the lines were not connected to each other, and some were even very short. The number of petal-shaped bodies is mainly four, five or six. The petal-shaped or multi-lobed oval gold and silver wares of the Liao dynasty mainly inherited the style of the late Tang, but some gold and silver wares obviously imitate Sogdian silverware. Many imitations of the Sogdian style appeared in the Liao dynasty (Figure 9.29).

Figure 9.29. Silver cup. Source: Re-drawn by the author from a photograph.

Persian-style gold and silver wares are mainly characterised by the linked-pearl motif. The linked-pearl motif is a decorative pattern from Persia, popular in the Sui and Tang dynasties after the 5th century. This kind of decoration is also found on pedestalled gold and silver cups wares in the early Liao dynasty, such as the silver cup with the dragon and pearl pattern unearthed from the Dayingzi Liao tomb in Chifeng (Former Rehe Provincial Museum Preparatory Group et al. 1956) (Figure 9.30). The bottom of the cup is decorated with the pattern of the linked-pearl motif. The decoration is the same as the Liao gold pedestalled cup with double-phoenix pattern (Figure 9.31). This cup appeared in the London antiquities market and was bought by collector Christian Deydier (Han 1994; Zhu 1998). It has a rounded mouth with linked-pearl motif. However, the shape of the cup is similar to the pedestalled cup in the hands of the Hephthalite figures in the 5th to 6th century AD frescoes discovered by Balalyk in Central Asia. Therefore, although the decorative patterns of these two pedestalled cups have an obvious Persian style, the shape of the vessels is still influenced by other nations.

The formation of the transportation routes of the steppe Silk Road connected China with many countries and regions in Central and West Asia. Therefore, along with the smooth development of the steppe Silk Road, the exchange of Chinese and Western cultures was promoted. In this process, the role of the Sogdians distributed in the Amu Darya and Syr Darya was extremely prominent and far-reaching. The Sogdian region is a hub of east-west traffic and a relay station between north and south. It is the region with the most direct and closest relations with China. According to historical records, from the 4th century AD, a large number of Sogdians immigrated to China as they gradually spread eastwards. Until the 8th century, the immigration east of Turpan linked not only the middle part of the Silk Road along the Hexi Corridor to Xi'an and Luoyang, but also the north part of the Silk Road via Turpan and Hami (Xinjiang), Juyan (Inner Mongolia) to Datong (Shanxi Province) and then to Chaoyang (Liaoning Province). In today's archaeological excavations, these areas all have historical remains of Sogdians.

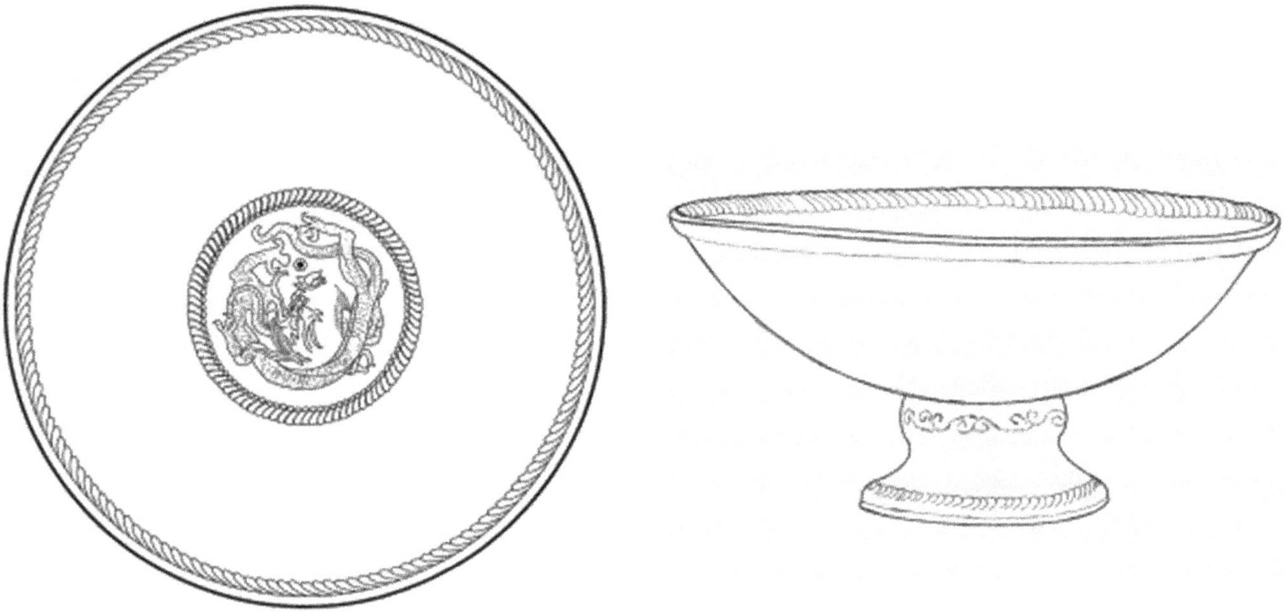

Figure 9.30. Silver pedestalled cup with dragon and pearl pattern. Source: Re-drawn by the author from a photograph.

Figure 9.31. Gold pedestalled cup with double-phoenix pattern. Source: Re-drawn by the author from a photograph.

Therefore, some Sogdian products and imitations appeared in the gold and silver wares of the Tang dynasty, or those wares were influenced by Sogdian silverware in shape, pattern and craftsmanship, the result of cultural exchange across the steppe Silk Road. Until after the eleventh century, this style of Sogdian silverware could still be seen in the gold and silver wares of the Liao and Jin dynasties.

9.6. Conclusions

The gold and silver wares unearthed in the eastern Eurasian steppes with western cultural influences had already appeared during the Xia and Shang dynasties. The author believes that the migration of ethnic groups led to this cultural dissemination. The introduction of flat trumpet-shaped earrings into China was likely caused by cultural transmission from the migration of nomads to the east.

From the late Western Zhou to the middle Spring and Autumn eras, the shapes and patterns of ornaments with bent-leg animals and with loop-shaped beasts found in southeast Inner Mongolia were similar in style to the Tagar culture distributed in the middle reaches of the Yenisei River. The production time of the ornaments should be earlier than the Tagar culture. This is the result of the westward spread of the eastern steppe culture in China.

From the late Warring States period to the Western Han dynasty, the animal shapes and decorations in the gold and silver wares of the Huns formed a trend of mutual exchange with the Scythian culture. In particular, the origin of the monster pattern could not be found in China, but it prevailed in Scythian and Altai art. Clearly the monster pattern spread from west to east to the northern steppes of China, and then into the Central Plains and southern regions of China.

In the Eastern Han, the animal decorations of Xianbei gold and silver wares inherited cultural elements from Central Asia, West Asia and Europe on the basis of the Huns. Examples include the divine beast pattern and the 'griffin' popular in West Asia and the Eastern European steppes. These wares have many similarities, but have some slight differences in shape, indicating the shapes were not completely absorbed and imitated. Meanwhile, Xianbei gold and silver wares also absorbed the engraving method and technology of inlaying with gems introduced from Persia and Greece or Rome.

During the period of the Jin dynasty and the Sixteen Kingdoms, gold *buyao* crown ornaments were popular in the southeastern part of the northern grassland area of China. These ornaments were related to the technique of the gold *buyao* crown and decorated crowns with gold leaves found in West Asia. During the Northern Wei dynasty, gold and silver wares with characteristics of Persian Sassanian silverware were influenced by Central and West Asian production processes or were directly imported from the west.

During the Sui and Tang dynasties, gold and silver wares with characteristics of Byzantium, Persia, the Sogdians and other countries or regions were discovered in the southeast and central parts of Inner Mongolia. This shows that during this period, commercial and cultural exchanges between China and the west were very close.

In the Liao dynasty and later, the unprecedented prosperity and development of gold and silver wares were inseparable from the nourishment of cultures such as the Turkic, Sogdian or Persian. The shape of folded shoulders, multi-petal shapes, and linked-pearl motifs are all marks of gold and silver ware production and decorative craftsmanship from those countries and regions.

It can be seen that the gold and silver wares found in archaeological excavations in areas of the Eurasian steppe connected with China have absorbed certain factors of west Eurasian culture in terms of shape, pattern and production techniques. This is witness to a long history of cultural exchange between China and the west. At the same time, the gold and silver wares are a concrete manifestation of the exchange and development between the northern peoples of ancient China and other regions, as well as a microcosm of the continuous development and innovation under the integration of world civilisations.

Acknowledgement

This research was supported by Fundamental Research Funds for the Central Universities from North Minzu University (2020KYQD12) for 'An empirical study on ethnic migration, cultural exchange and mutual learning and Chinese cultural identity on the Grassland Silk Road'.

References

Aohanqi Cultural Center (1978) Gold and silver wares unearthed from Lijiayingzi in Aohanqi. *Archaeology* 2: 117–118 (in Chinese).

Ancient Orient Museum and Asahi Shimbun Company (Eds.) (1991) *The Treasures of Nomadic Tribes in South Russia*. Tokyo: Asahi Shimbun (in Japanese and English).

Azarpay, Guitty (1981) *Sogdian Painting: The Pictorial Epic in Oriental Art*. University of California Press.

Beijing Cultural Relics Management Office et al. (1977) Shang Dynasty tombs discovered in Pinggu, Beijing. *Cultural Relics* 11: 1–8 (in Chinese).

Chaoyang City Museum *et al*. (1997) Tiancaogou Jin tomb in Liaoning Chaoyang. *Cultural Relics* 11: 33–41 (in Chinese).

Chen, Dawei (1960) Brief report on the excavation of Jin tombs in Fangshen village in Beipiao, Liaoning Province. *Archaeology* 1: 24–26 (in Chinese).

Dai, Yingxin and Jiaxiang Sun (1983) Cultural relics of the Huns unearthed in Shenmu, Shaanxi. *Cultural Relics* 12: 23–30 (in Chinese).

Datong City Museum et al. (1977) Brief report on the excavation of the Northern Wei city site in Datong, Shanxi. *Cultural Relics* 9: 11–14 (in Chinese).

Du, Zhengsheng (1993) The animal patterns of Eurasian Steppe and the investigation of ancient northern nationalities in China. *Collected Papers of the Institute of History and Language* (Volume 64). Nanjing: Jiangsu Ancient Books Publishing House (in Chinese).

Former Rehe Provincial Museum Preparatory Group et al. (1956) Excavation report of Dayingzi Liao tomb in Chifeng County. *Acta Archaeologica Sinica* 3: 1–36 (in Chinese).

Gansu Provincial Museum et al. (1981) Thirty years of cultural relics and archaeological work in Gansu Province. *Thirty Years of Cultural Relics and Archaeological Work*, pp. 139–153.

Guangxi Zhuang Autonomous Region Cultural Relics Task Force (1978) Yinshanling Han tomb in Pingle. *Acta Archaeologica Sinica* 467–496 (in Chinese).

Han, Wei (1994) The research on inscriptions of gold and silver wares in the Taiping period of the Liao Dynasty. *The National Palace Museum Monthly of Chinese Art* (Taiwan): 4–22.

He, Lin et al. (1983) The gold cup of Tang Dynasty discovered in Xi'an. *Cultural Relics* 9: 14 (in Chinese).

Hebei Provincial Institute of Cultural Relics (1996) *Yanxiadu*. Beijing: Cultural Relics Publishing House (in Chinese).

Inner Mongolia Autonomous Region Institute of Cultural Relics and Archaeology *et al.* (2009) *Xiaoheishigou-Excavation Report of Xiajiadian Upper Cultural Site*. Beijing: Science Press.

Inner Mongolia Cultural Relics Task Force and Inner Mongolia Museum et al. (1975) Foreign silver coins unearthed near Hohhot. *Archaeology* 3: 182–185 (in Chinese).

Inner Mongolia Cultural Relics Work Team *et al.* (1962) The tomb during the Northern Wei Dynasty in Meidai Village, Hohhot, Inner Mongolia. *Archaeology* 2: 86–87.

Inner Mongolia Institute of Cultural Relics and Archaeology et al. (1996) Report on the excavation of the Yelyu Yuz's tomb in the Liao Dynasty. *Cultural Relics* 1: 4–32 (in Chinese).

Inner Mongolia Museum et al. (1984) Goldware unearthed from the tomb of the Northern Wei Dynasty in Lingpiyao Village of Helingeer. *Steppe Cultural Relics* 3: 52–54 (in Chinese).

Inner Mongolia Museum *et al.* (1997) The tomb of the Huns in Yulongtai, Zhungeerqi, Inner Mongolia. *Archaeology* 2: 111–114 (in Chinese).

Institute of Archaeology, Chinese Academy of Social Sciences et al. (1996) *Report on the Excavation of the Dadianzi - Xiajiadian Lower Culture Site and Cemetery*. Beijing: Science Press.

Institute of Archaeology, Chinese Academy of Social Sciences et al. (1980) *Sui and Tang tombs in the Suburb of Chang'an in Tang Dynasty*. Beijing: Cultural Relics Publishing House (in Chinese).

Li, Yaobo (1973) The tomb of Feng Sufu during the Beiyan Dynasty in Xiguanyingzi, Beipiao county, Liaoning. *Cultural Relics* 3: 2–19 (in Chinese).

Liaoning Provincial Institute of Cultural Relics and Archaeology et al. (1997) Main achievements of archaeological excavations in 1987 and 1990 of Wangzifenshan tombs in Chaoyang. *Cultural Relics* 11: 4–18 (in Chinese).

Liaoning Provincial Institute of Cultural Relics and Archaeology et al. (1997) Brief report on the excavation of Shiertaizhuanchang 88M1 tomb in Chaoyang. *Cultural Relics* 11: 19–32 (in Chinese).

Liaoning Provincial Institute of Cultural Relics and Archaeology et al. (1986) Beiling Liao tomb in Kazuo, Liaoning Province. *Liaohai Cultural Relics Journal* 1: 32–51 (in Chinese).

Liaoning Provincial Institute of Cultural Relics and Archaeology et al. (2015) *Lamagou Liao tomb in Lingyuan*: 27–28. Beijing: Cultural Relics Publishing House 9 (in Chinese).

Lin, Yun (1992) Archaeological exploration of Donghu and Shanrong. *Papers of the Third International Symposium on Archaeology around the Bohai Sea.*

Lu, Sixian and Tandong Chen (1984) Golden ornaments of ancient northern nationalities unearthed in Damaoqi. *Cultural Relics* 1: 81–83 (in Chinese).

Luoyang City's Second Cultural Relics Working Team *et al.* (1995) Yichuanyaling tomb of the home grandmother of Qi Guo during the Tang Dynasty. *Cultural Relics* 11: 24–44 (in Chinese).

Ma, Yuji (1983) Brief report on the excavation of Huagetatai during Northern Wei Dynasty in the Xiaozhan Village, Datong. *Cultural Relics* 8: 1–4 (in Chinese).

Mei, Cun (1991) The discovery of Daxia golden treasures. *Cultural Relics World* 6: 44–48 (in Chinese).

Sarianidi, Viktor Ivanovič (1979) The golden nobles of Shibarghan. *Time Magazine* (2 July).

Shanxi Provincial Institute of Archaeology et al. (1992) A brief report on the excavation of the tombs of the Northern Wei Dynasty in the southern suburb of Datong. *Cultural Relics* 8: 1–13 (in Chinese).

Tian, Guangjin and Suxin Guo (1980) Relics of the Huns found in Aluchaideng, Inner Mongolia. *Archaeology* 4: 333–338 (in Chinese).

Tokyo National Museum, Osaka City Museum of Fine Arts and Nihon Keizai Shinbun (Eds.) (1985) *Culture Contacts between East and West in Antiquity and Middle Ages from USSR*. Tokyo: Nihon Keizai Shinbun (in Japanese).

Xia, Nai (1978) Sassanian cultural relics unearthed in China in recent years. *Cultural Relics* 2:111–116 (in Chinese).

Xia, Nai (1983) A study of Sassanian silver plate unearthed from the tomb of Fenghetu during the Northern Wei Dynasty. *Cultural Relics* 8: 5–7 (in Chinese).

Xiang, Chunsong (1984) Liao tombs No. 1 and No. 2 in the Erbadi, Keshiketengqi. *Steppe Cultural Relics* 3: 80–90 (in Chinese).

Xiang, Chunsong and Yi Li (1995) Investigation and clearance report of the stone coffin tomb in Xiaoheishigou, Ningcheng. *Cultural Relics* 5: 4–22 (in Chinese).

Xinjiang Institute of Cultural Relics and Archaeology et al. (1999) Report on the excavation of Goubei Taidi No.1 tomb in ancient Jiaohe city, Turpan. *Cultural Relics* 6: 18–25 (in Chinese).

Yi, Kejian and Sixian Lu (1984) Relics of the Northern Wei Dynasty unearthed from Tumotezuoqi. *Steppe Cultural Relics* 3: 51–55 (in Chinese).

Yikezhaomeng Cultural Relics Workstation *et al.* (1980) The Huns′ tomb in Xigoupan. *Cultural Relics* 7: 1–10 (in Chinese).

Zhang, Jingming (2005) *Ancient Gold and Silver Wares from the Grasslands of Northern China.* Beijing: Cultural Relics Publishing House (in Chinese).

Zhang, Jingming (2011) *Research on Gold and Silver Wares of the Liao Dynasty*. Beijing: Cultural Relics Publishing House (in Chinese).

Zhao, Yaxin (1999) The phoenix-bird-shaped crown ornament of the Xianbei was discovered from the Maolitu tomb in the Kezuohouqi. *Cultural Relics* 7: 39–40 (in Chinese).

Zhaowudameng Cultural Relics Workstation, Liaoning Province et al. (1973) Stone coffin tomb in Nanshangen, Ningcheng County. *Acta Archaeologica Sinica* 2: 27–40 (in Chinese).

Zhu, Tianshu (1998) The Gold and Silver wares of the Liao Dynasty. Beijing: Cultural Relics Publishing House (in Chinese).

Zou, Houben and Zheng Wei (1998) The gold belt with buckle of the Western Han Tomb in Shizishan, Xuzhou. *Cultural Relics* 8: 37–43 (in Chinese).

10

West Eurasian cultural impacts on the Liao dynasty: an archaeological analysis

Hui Wang

Abstract: Living in the trans-Eurasian steppe zone, the Khitan people had unique advantages in the process of cultural and material exchange due to the mobility of their activities and frequent foreign contacts. Meanwhile, the *Hu* people living in the territory of the Liao also contributed to the spread of culture from the west to some degree. Artefacts and murals found in archaeological sites provide important evidence to understand the cultural exchange between the Liao dynasty and the regions to the west. By analysing objects excavated from hundreds of tombs and settlement sites belonging to the Liao dynasty, this research summarises the influence of the culture from western regions on the design of murals and porcelain, the techniques of silver and gold wares, the introduction of plants and glassware, as well as people's religious beliefs during the period. The chapter also notes differences in social hierarchy in the diffusion of objects and ideas from western Eurasia into medieval China.

Keywords: Liao dynasty, Khitan culture, western border zone of China, tomb murals, funerary objects

The Liao dynasty (AD 907–1125) was established in the northern region of ancient China by the ethnic Khitan. The Khitan, Han and other ethnic groups lived in the area under the control of the Liao. The government of the Liao Dynasty ruled over a vast territory during its heyday; the westernmost part of its territory crossed the Altai Mountains and reached the Ertis River (Figure 10.1). The western regions of Eurasia discussed in this chapter include the vast area west of the westernmost territory of the Liao dynasty. As a nomadic people who lived along the trans-Eurasian land route, cultural and material exchanges were very important for the social and economic development of the Khitan due to their lack of natural resources on the grasslands, their focus on pastoralism, and the high mobility of Khitan populations. Analysis of excavated objects from tombs and settlement sites belonging to the Liao dynasty shows that the daily life of the people was influenced by the cultures of the western regions of Eurasia.

10.1. Historical background

As a northern nomadic people, clear records about the Khitan can first be found in the *Wei shu* (the 'Book of Wei', written by Wei Shou during the Northern Qi Dynasty). According to this work, the earliest history of Khitan dates back to the middle of the fourth century. The Khitans had a long and complex development before Yelü Abaoji 耶律阿保机 established the Liao dynasty in Chifeng (in present-day Inner Mongolia) in 907. The Khitans had already lost control over the management and organisation of the Silk Road after the An Lushan Rebellion of AD 755–763) (Wu and Cheng 2014). In order to resume the Khitan management of the Silk Road, the Liao government reopened the grassland through use of immigration, the establishment of management institutions and diplomatic activities, as well as effective maintenance, management, improvement and control. After that, interaction between the Khitans and people of the 'Western Regions' (西域 *Xiyu*) in cultural exchange and commercial trade became increasingly close and frequent. This situation was also more convenient for official communications between the Liao dynasty and the 'Western Regions'. Many records in the *Liao shi* (the 'History of the Liao', written by Tuo Tuo and others during the Yuan dynasty) reflect exchange and communication between the Liao and parts of western Eurasia at that time. For example, 'in August of AD 938, the emperor Taizong of Liao (辽太宗, a.k.a. 耶律德光 *Yelü Deguang*) gave an Uyghur envoy (回纥 *Huihe*) the special position of *Kuoli Yuyue* (阔里于越) in the Liao government and also gave him a bow and arrow, clothes, horses and flags (特授回鹘使阔里于越，并赐旌旗、弓剑、衣马、馀赐有差) (Tuo Tuo et al. 2017: 53). The objects and murals excavated in archaeological sites of the time also bear witness to the cultural influence of the 'Western Regions' on Liao society.

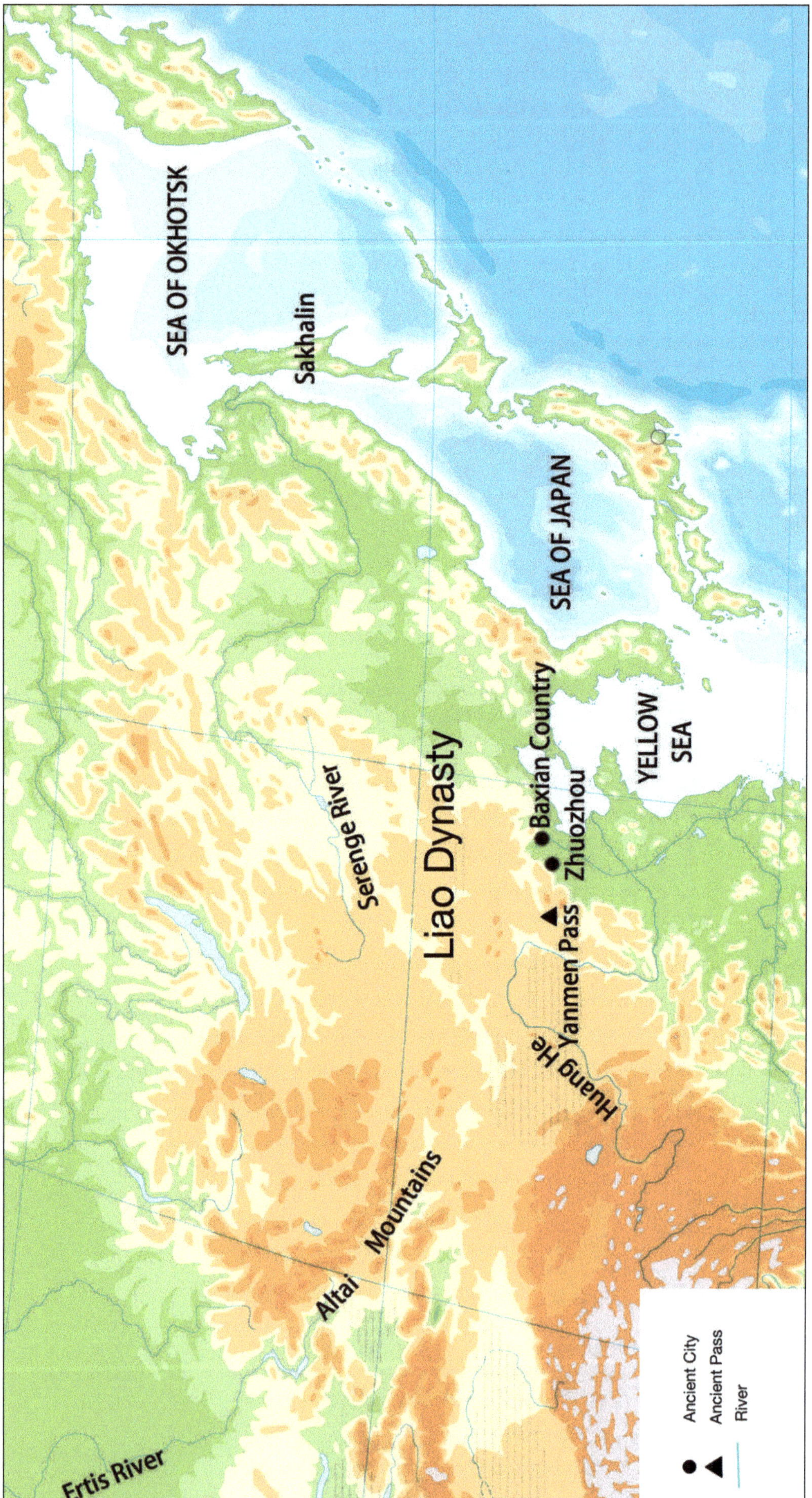

Figure 10.1. The location of the Liao Dynasty. Source: map by the author.

10.2. Archaeological research on the cultural influence of the 'Western Regions' during the Liao dynasty

Artefacts and murals found in archaeological sites provide important evidence to understand cultural exchange between the Liao and regions to the west. Among these artefacts, some are exotic products from the west, while some are imitation products made by local craftsmen. For example, glassware is regarded as direct evidence of communication between the Liao dynasty and the countries and regions of the Islamic world or the Mediterranean. Gold and silver wares imitating western styles have been discovered in many tombs of high-level Liao nobles. Meanwhile, the aesthetic tastes of the Liao people at that time were greatly influenced by many trends from the 'Western Regions', leading to new market demands and technological changes in handicraft industries. Furthermore, some cultures from western Eurasia had already been introduced to ancient China before the Liao dynasty and had a long-lasting influence on Liao society. For instance, Buddhism had a major influence on the Khitans, both nobles and commoners. Nestorian crosses were discovered in several tombs of ordinary Khitans. Furthermore, several foreign plants were introduced into the Liao dynasty by the Khitans, such as watermelon, enriching the people's dietary life at that time. Moreover, some detailed information not found in written sources is recorded by the tomb murals due to the custom of painting murals in tombs to describe the real lives of deceased people and their social situation when they were alive. These materials demonstrate that there was a close interaction between the Liao dynasty and the 'Western Regions'.

Relying on analysis of the objects, patterns, styles and shapes that did not originate in ancient China, archaeologists study their origins and the timing and routes by which they reached the Liao dynasty in order to understand their influence on the Liao people. On one hand, archaeologists pay attention to research into the methods, transportation routes, and timing of exchanges and communications. For the objects discovered in urban and living sites, archaeologists summarise the characteristics of foreign cultures and sort them according to their time of arrival. On the other hand, archaeologists also focus on studying the origin of objects through technical methods and analysing cultural elements. Research on glass, gold and silver and other artefacts discovered in tombs and living sites belonging to the Liao dynasty has become more and more thematic and detailed. Here, the author selects the six most typical archaeological discoveries to demonstrate the cultural traits of the 'Western Regions' during the Liao dynasty.

10.3. The *Hu* (胡) people

The meaning of *Hu* is 'foreign' or 'not local'. However, the meaning of the Hu people during the Liao dynasty is quite different from that of other periods due to the special historical background. Chinese scholars have conducted further research and discussions on the concept of the Hu people during the Han and Tang dynasties, but paid little attention to research on these people during the Liao. The Hu in this chapter include all the people from the vast area west of the westernmost territory of the Liao Dynasty. Rong (2020) proposed that the Hu include the nomadic tribes from the north and other tribes from the 'Western Regions' during the Han and Tang. Yang (2011) also discussed the meaning of the Hu during the Song dynasty at the same time as the Liao and mentioned that the Khitans were also regarded as Hu by the Song. However, according to the records in the *Liao shi*, the Khitans not only claimed themselves to be descendants of Emperor Yan and Emperor Huang, but also referred to the Liao as *zhong guo* or 'Middle Kingdom' (Zhao 2010). Thus, it is clear that the concept of the Hu is relative and changed in different cultural and historical environments. The Hu during the Liao dynasty discussed in this chapter included people from the vast area west of the Liao territory.

There are many images of people found on Liao murals and objects. However, it is difficult to identify images of the Hu people. Most scholars judge images of the Hu based on facial features, clothing and hairstyles. However, the method of identifying Hu only by facial features with deep eye sockets and high noses, an approach widely used in existing research, is not objective enough (Yang 2008). For example, the images of Hu playing musical instruments and dancing (Figure 10.2) on the wooden coffin discovered in Yelü Yuzhi's tomb (Inner Mongolia) were recognised as people from Bohai in former studies, due to features of their hairstyles and costumes compared with people of the Tang, Five Dynasties and Liao (Qi Xiaoguang et al. 1996; Ba 2006: 66). Later, Shao Guotian and Feng Enxue discussed and refuted this view from two different angles. Based on the Dala drum in the hands of the musician, Shao (2001: 309–318) proposed that the musician in the image probably should be a Hu person from the 'Western Regions'. Feng Enxue took the musician's moustache as the key clue and conducted a comparative study with the images of Hu gatekeepers (Figure 10.3) found in the tombs in Xuanhua (Hebei province). He pointed out that these people both have moustaches shaped like swallow's tails. Thus, Feng (2011) believed that the images of the musicians found in Yelü Yuzhi's tomb represent Hu people from the 'Western Regions' with Mongoloid characteristics. Moreover, Feng also puts forward some suggestions for judging and distinguishing images of the Hu during the Liao dynasty. Hu people characterised by deep eye sockets and high noses, beards and curly hair in the Bao Mountain tomb (Inner Mongolia) have Caucasoid features (Figure 10.4). Another kind of image of Hu people found in tombs in Xuanhua (Hebei province) and Yelü Yuzhi's tomb have Mongoloid features (Feng 2011).

Figure 10.2. ***Hu*** **musician playing a Dala drum (荅腊鼓) found in Yelü Yuzhi's tomb. Source: Re-drawn by the author from a photograph published in the site report (Qi et al. 1996).**

Figure 10.4. ***Hu*** **gatekeeper found in the tomb located in** ***Bao*** **mountain (宝山). Source: Re-drawn by the author from a photograph published in the site report (Qi et al. 1998).**

Figure 10.3. ***Hu*** **gatekeeper found in Zhang Kuangzheng's tomb in Xuanhua. Source: Re-drawn by the author from a photograph published in the site report (Hebei Provincial Institute of Cultural Relics 2001: 31).**

Images of Hu people were also discovered on some clothing accessories, such as a jade belt belonging to the early Liao dynasty, now collected in the Aohanqi Museum in Inner Mongolia (Figure 10.5). The jade belt is made up of nine jade plaques, each engraved with different images. These jade plaques represent a scene of playing Qiuci (Kucha) 龟兹 music, which includes showing a *Huteng* dance (胡腾舞), playing a *Paiban* (拍板), playing a flute, playing a *Sheng* (笙), drinking, beating a *Jilou* drum (鸡娄鼓), playing a Pipa (琵琶), and beating a *Maoyuan* drum (毛元鼓) as well as playing a *Bili* (荜篥). All of the people depicted have deep eye sockets and high noses, curly hair, and wear long boots. Thus, according to the view of Lü (2015), the jade belt was most likely made in the 'Western Regions' and then spread to the Liao instead of being made by local Liao craftsmen. Nevertheless, the jade belt is still evidence of cultural exchange and interaction between the west and the Liao. These images are direct evidence of the interaction between the Liao dynasty and the 'Western Regions'.

10.4. Gold and silver ware during the Liao dynasty

During the Tang, the design and style of gold and silver wares were largely inspired by the cultures of the 'Western Regions'. The gold and silver wares of the Liao dynasty also contain many cultural traits from the west since the

Figure 10.5. Images of *Hu* people playing *Qiuci* music. Source: Re-drawn by the author from photographs in Lü (2015).

Liao learned a great deal about technology, pattern, style and shape from Tang gold and silver working. As a result, it is no surprise that many characteristics of gold and silver wares produced in western Eurasia can be found in the gold and silver wares of the Liao. Meanwhile, it was common for Liao artisans to imitate the gold and silver wares of the 'Western Regions'. However, local craftsmen during the Liao did not completely aim at reproducing western-style gold and silver wares or those of the Tang, but added new elements to their products. Faced with the rich gold and silver artefacts found in Liao sites and tombs, Zhu Tianshu (1998), Zhang Jingming (2011) and Wang Chunyan (2020) have made extensive research on these materials, summarised their features in different periods, and analysed changes influenced by the cultures of the 'Western Regions' (see also Jingming Zhang, this volume).

The gold and silver cup found in Yelü Yuzhi's tomb (Inner Mongolia) is the product of Liao craftsmen adding innovative elements on the basis of imitating Sogdian gold and silverware. The mouth edge of the cup is designed as a heptagonal shape and the cup body can be naturally divided into seven planes to decorate the seven figures in the ancient Chinese story *Zhulin Qixian* ('The Seven Sages of the Bamboo Grove') (Figure 10.6), instead of imitating the traditional Sogdian cups typified by handles and octagonal mouth edges. Moreover, the cup greatly weakened the importance of the ring handle and only retained a very small pad to place fingers on. This contrasts sharply with the ring-shaped handles commonly found on Sogdian gold and silverware cups.

The two identical gold and silver handled-cups discovered in the tomb in Turki (吐尔基) mountain were not only typical products made by local craftsmen but also successful imitations of Sogdian gold and silverware styles. According to other objects discovered in the tomb, the tomb owner was most probably one of the highest noble women at that time (Tala and Zhang 2004). The objects retain a more traditional style of Sogdian gold and silverware, such as octagon mouth edges, ring handles, finger pads, and the Lianzhu pattern (联珠纹, meaning a series of beaded decorative patterns), as well as multi-curved plane segmentation. The images of the decoration of the cup body are divided into eight pictures. The costumes and hairstyles of the people are all in Han Chinese styles (Fig. 10.7). Furthermore, some gold and silverware with many western cultural features were discovered in tombs belonging to the Liao dynasty, such as the tomb in Daiqintala (代钦塔拉), Youai (友爱) village and the tomb of Princess Chenguo (陈国公主: the princess of the Liao Dynasty whose fief was Chen). These artefacts reflect the change in people's aesthetic taste and the market reform of the handicraft industry during the Liao dynasty. Meanwhile, these discoveries also suggest that gold and silverware were only used by rich people or nobles.

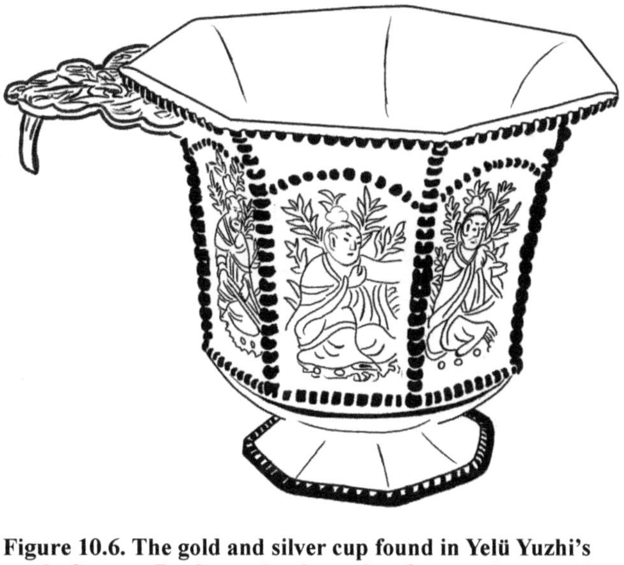

Figure 10.6. The gold and silver cup found in Yelü Yuzhi's tomb. Source: Re-drawn by the author from a photograph in Yang (2004: 173).

Figure 10.7. Images of the gold and silver cup body from the tomb in Turki Mountain. Source: Re-drawn by the author based on photographs published by Zhang (2017).

10.5. Phoenix head pots

The gold and silver wares of western Eurasia also had a strong influence on the design style and shape of ceramics in ancient China. One example is the pot with a phoenix head (凤首壶 *fengshou* pot), which was a kind of popular ceramic during the Tang Dynasty. The name comes from the craftsman's special design for the cap and the top of the neck of the pot. The top of the neck near the pot's mouth is shaped like a phoenix head, as for instance the Sancai three coloured porcelain pot belonging to the Tang dynasty, excavated in Gangu county, Gansu Province (Figure 10.8). Craftsmen also designed the cap of the pot into a phoenix head figurine, for example the Qing Ceramic (青瓷) pot of the Tang dynasty in the collection of the Palace Museum in Beijing (Figure 10.9). These pots have a similar shape to the Hu bottles of the 'Western Regions', *i.e.*, a thin and long neck which looks like a bamboo knot, a ring foot, bulging shoulders and a long bottle body. The silver Hu bottle cover with gold discovered in Li Xian's tomb from the Northern Zhou dynasty (Figure 10.10) was a product of the Persian Sassanid dynasty and was introduced to the Northern Zhou by commercial trade or official contact (Wu 1987). This object provides more detailed information about the comparative study of pots with phoenix heads and the Hu bottles with respect to shape, design style, patterns as well as decorative skills. Meanwhile, many images of pots with phoenix heads found in Liao tombs show that they followed the phoenix head pot style of the Tang and that these objects remained very popular ceramic products during the Liao.

Figure 10.8. Tang *sancai* pot with bird's head excavated in Gangu in the collection of the Museum of Gansu Province. Source: Re-drawn by the author from published photographs.

Figure 10.10. Silver *Hu* bottle excavated from the tomb of the Li Xian couple, Northern Zhou dynasty. Source: Re-drawn by the author from photographs in Ningxia Museum and the Guyuan Museum of Ningxia (1985).

Figure 10.9. Qing Ceramic pot with phoenix head from the Tang dynasty in the collection of the Palace Museum. Source: photographed by Gong Yiwen.

A large number of murals in Liao tombs present scenes of people using phoenix head pots in their daily life. For example, the tomb mural of Xiawanzi (Inner Mongolia) shows a pot with a phoenix head being heated on a fire (Figure 10.11). Liao tomb murals in Xuanhua (Hebei) contain more images of these pots. In the mural in tomb No. 6, a Khitan man holds a pot with a phoenix head in his arms and the cap of the pot is designed as a phoenix head figurine (Figure 10.12). Another image discovered in tomb No. 5 depicts a scene of making tea, with a Khitan man holding a pot with a phoenix head and pouring water into a cup (Figure 10.13). According to these images, the neck of the Liao pots was longer than those of the Tang and the cap was usually designed as a phoenix head figurine at that time. Peng Shanguo pointed out the difference between the phoenix head pots and images of phoenix head pots in murals. Peng (2006) proposed that the phoenix heads are designed under the mouth of the pot and the top of the neck in pots discovered in the Liao region. This is quite different from the style in the murals where the cap of the pot is designed as a phoenix head figurine.

10.6. The introduction and cultivation of the watermelon during the Liao

Food remains and tomb murals provide further clues for research into west Eurasian influences on the culture, diet and life of people during the Liao dynasty. However, food residues are rarely reported from Liao archaeological sites. Thus, tomb murals provide important evidence for understanding the dietary customs at that time. Images of foods originating in western Eurasia are usually found in tombs decorated with murals, such as the images of watermelons discovered in tomb No. 1 at Bao Mountain, Inner Mongolia (Figure 10.14) and the mural tomb in Zhaitang, Beijing (Figure 10.15). According to written sources in the chapter *Xianbei Ji* (陷北记, 'Record of Life in the Liao Dynasty'), written by Hu Qiao during the Liao, and published in the *Wudai shi* ('History of the Five Dynasties), written by Ouyang Xiu during the Northern Song Dynasty, watermelon is not indigenous to ancient China and was introduced and planted near Liaoshangjing city (辽上京) during the early Liao. Hu Qiao (2015) recounts his experience when passing through Liaoshangjing and describes the introduction and planting of watermelon during the Liao:

> Going east from Liaoshangjing about 20 km I arrived in Zhenzhu zhai (真珠寨), and the people living here just had the custom of eating vegetables. The following day, I went further to the east and found the terrain slopes gently and more trees and plants in the environment. The people living in this area can get an opportunity to eat watermelon. I heard that the seeds of watermelon were acquired when Emperor Taizu of Liao (辽太祖, a.k.a. *Abaoji*) vanquished the Uighurs (回鹘, a.k.a. Uyghurs) but the people of Liao developed a special method called *Niupeng Fugai* (牛棚覆盖, meaning to cover the small pit where watermelon seeds are planted with cow dung in order to improve the soil temperature) to grow the watermelons. As a result, they can get the sweet fruit of watermelons as large as the winter melons of the Central Plains region (Ouyang Xiu, Vol.73 III: 906).
>
> (自上京东去四十里，至真珠寨，始食菜。明日东行，遂入平川，多草木，始食西瓜。云契丹破回纥得此种，以牛粪覆棚而种，大如中国冬瓜而味甘。)

Figure 10.11. Mural of phoenix head bottle discovered in the No. 5 Xiawanzi tomb. Source: Re-drawn by the author from photographs in Aohanqi Banner Museum et al. (1999).

Figure 10.12. Mural of phoenix head bottle discovered in tomb No. 6 in Xuanhua. Source: Re-drawn by the author from photograph in Hebei Provincial Institute of Cultural Relics (2001: 48).

Figure 10.13. Mural of phoenix head bottle discovered in tomb No. 5 in Xuanhua. Source: Redrawn by the author from a photograph in Hebei Provincial Institute of Cultural Relics (2001: 74).

Figure 10.14. Image of watermelons in the mural of tomb No.1 of Yang Mountain. Source: Re-drawn by the author from a photograph published in Shao (1999).

Figure 10.15. The mural of Zhaitang tomb in Beijing. Source: Re-drawn by the author from a photograph in Lu and Zhao (1980).

The watermelon seeds discovered in the archaeological sites of Liaoshangjing (Inner Mongolia) support the authenticity of this record. However, scholars have different opinions about the place where the Emperor Taizu of Liao obtained the watermelon seeds. Combined with the records in the *Liao Shi,* some scholars believe that the watermelon seeds were obtained by Emperor Taizu of Liao when he attacked Futu City (浮屠城) in Qocho (高昌回鹘, Gaochang Uighur) (Gu 1998; Huang 2005; Wu 2015; Yang Fuxue et al. 2021). However, Cheng Jie (2017) puts forward a different point of view. He believes that the watermelon seeds of Mobei were introduced by Manichaeism missionaries from Central Asia between AD 763 and 840, and that Emperor Taizu of Liao obtained the seeds when he came to the ancient capital (present-day Өвөрхангай аймаг in Mongolia) of Mobei (漠北) on the upper reaches of the Orkhon River.

10.7. The religious influence of the 'Western Regions' on the society of the Liao

Funerary objects involving Buddhism, Taoism, shamanism and Nestorianism were usually found in tombs during the Liao dynasty. These materials suggest the complex religious beliefs of the people at that time. In particular, neither Nestorianism nor Buddhism were indigenous to ancient China, but they still had a great influence on the people of the Liao. Before the establishment of the Liao dynasty, most Khitans were believers in shamanism. Following the influence of Han Chinese culture, more and more Khitan people accepted Buddhist culture after the establishment of the Liao. The rapid development of Buddhism during the Liao stemmed from the strong promotion by emperors and governments, aiming to control and repress the enormous power of shamanism. Relying on their great religious influence amongst believers, the leaders of the shamanism group led by Shen Sugu (神速姑) and Yulu Dugu (余庐睹姑) had previously supported the coup during the early period of the Liao dynasty. Thus, the Liao emperor chose to promote Buddhism because it was more suitable to control people and society. The influence of shamanism remained mainly amongst some Khitans in the lower levels of society. As a result, many tombs belonging to Khitan nobles and ordinary people were influenced by the culture of Buddhism. Funerary objects containing Buddhist features were discovered in the tombs of princess Chenguo and Yelü Yuzhi, as well as the tombs of the Zhang family.

The content of the epitaphs found in the tombs of the Zhang family (Hebei Province), shows that the family members included Khitan and Han Chinese, and that most of them were Buddhist believers, which is the reason why the funerary artefacts and design style were influenced by Buddhist culture. For example, several wooden coffins decorated with handwritten Buddhist scriptures in Chinese and Sanskrit on their surfaces were found in these tombs (Figure 10.16). Furthermore, most owners of these tombs accepted the Buddhist custom of cremation. A few murals depicting scenes of chanting Buddhist scriptures were also found in these tombs (Figure 10.17) and a string of wooden prayer beads was discovered in tomb No. 7. In the mural, some volumes of scriptures were placed on a long table and a maid lights a candle to illuminate the whole room. Cranes and auspicious clouds, regarded as patterns symbolising longevity in Taoism, were painted on the left side of the mural. The tomb occupant who will chant the Buddhist scriptures does not appear in the mural, but the whole image seems to depict the maid making preparations before the chanting. The mural suggests that the tomb owner had a special room dedicated to chanting Buddhist scriptures and that Buddhism played an important role in the daily life of the tomb owner. However, the cranes and auspicious clouds, as well as the chanting room furnishings appearing on the mural at the same time, suggest the amiable relationship between Buddhism and Taoism during the Liao dynasty. Another mural in this tomb cannot be ignored: it depicts a scene of three people playing Go together (Figure 10.18). Li Qingquan proposed that, based on the clothes and hairstyles of the three people, these are believers of Buddhism, Taoism and Confucianism. This mural can be regarded as key evidence to confirm the harmonious relationship among Buddhism, Taoism and Confucianism during the Liao. Considering that some members of the Zhang family were Khitan, we might expect to find some Khitan influence in these tombs but there was no evidence of the influence of shamanism. This shows that the Khitan family members were deeply influenced by Buddhism and Han Chinese culture.

Nestorianism was introduced into ancient China during the Tang in AD 635, but its development was seriously damaged due to Emperor Wuzong of Tang's (唐武宗) suppression of Buddhism, and Nestorianism gradually disappeared. However, the archaeological materials unearthed from some Liao tombs imply that Nestorianism still had a few believers during the Liao dynasty. For example, Nestorian sculptured stones and crosses in tombs belonging to the Liao dynasty were discovered in Anshan (鞍山) in Liaoning Province. The Japanese scholar Torii Ryūzō conducted many archaeological investigations in northeast China. In 1928, he discovered two sculptured stones depicting the story of the 'nativity' and the story of the 'tree of life' in Anshan (Torii 1942: 59–60, 62), as well as seven crosses in a tomb containing the bones of seven people from the period of Emperor Shengzong of Liao (辽圣宗, 983–1031). These discoveries are still the main materials for further research on Nestorianism during the Liao dynasty and also make up for the lack of detailed historical records regarding Nestorianism at that time. According to the account by Torii Ryūzō, the tomb containing the seven crosses was located near the east gate of 219 garden in present-day Anshan (Li and Zhai 2017). But the influence of Nestorianism was limited to the ordinary Khitan people and the number of believers was small during the Liao dynasty (Chen 2014). Meanwhile, we cannot ignore the fact that Nestorianism was also influenced by the local culture of the Liao and Han Chinese. For example, in the image of the sculptured stone depicting the story of the 'nativity', the baby was lying in a cradle full of Khitan features.

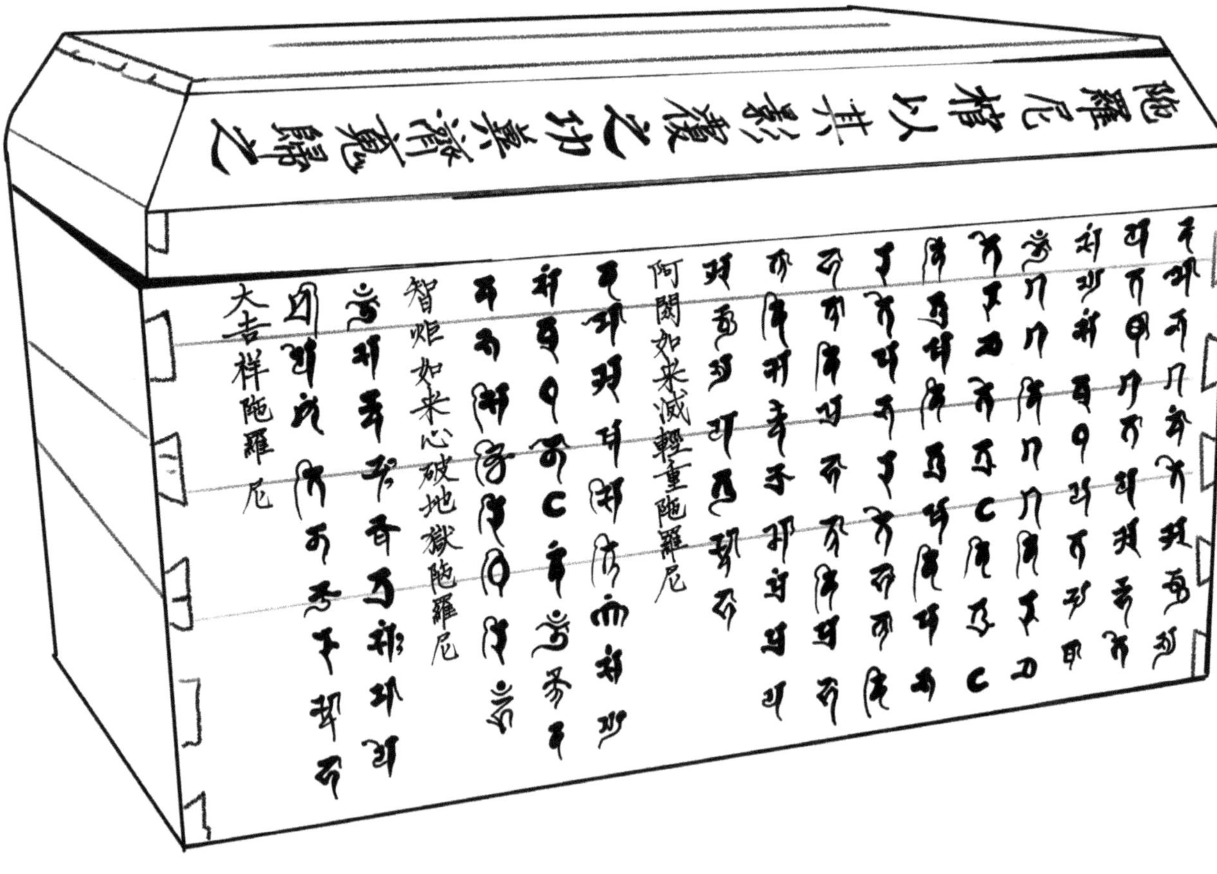

Figure 10.16. Wooden coffin with Buddhist scriptures in Chinese and Sanskrit from Tomb No. 7 at Xuanhua. Source: Re-drawn by the author from a photograph in Hebei Provincial Institute of Cultural Relics (2001: 143).

Figure 10.17. Tomb mural from Xuanhua. Source: Re-drawn by the author from a photograph in Hebei Provincial Institute of Cultural Relics (2001: 16).

Figure 10.18. Mural of three people playing Go together found in tomb No.7 in Xuanhua. Source: Re-drawn by the author from a photograph in Hebei Provincial Institute of Cultural Relics (2001: 150).

10.8. Glassware during the Liao dynasty

The history of native glass products in ancient China dates back to the Spring and Autumn period. These glass products have four characteristics: (1) most are lead glass; (2) most draw on the styles and shapes of jade artefacts; (3) most were used for amusement, watching or religious ceremonies; and (4) they are colourful but easily broken (An 2021). Many glass products are discovered in tombs, temples and pagodas of the Liao dynasty, such as the tomb of the princess Chenguo, princess Pingyuan (平原), Yelü Yuzhi, and the Geng Yanyi couple as well as the tombs in Turki Mountain, Yemaotai (叶茂台), Dayingzi (大营子) and Dule (独乐) Temple in Ji (蓟) county (Tianjin). According to their function and type, these Liao glass products are mainly containers and ornaments (Fu 2006). Based on analyses of the social identity of the owners of these glass products, it is clear that high-class nobles, important officials, and very rich people were able to possess exquisite glassware during the Liao dynasty.

Meanwhile, archaeologists propose that some glassware are exotics from 'Western Regions' instead of native Liao products because of their special style, patterns and manufacturing techniques (An 1990; Ma 1994; Sun 1999; Zheng 2010). Ma (1994) put forward the theory that there were probably three routes for glass products introduced into the Liao: (1) from Liaoshangjing (辽上京) or Liaozhongjing (辽中京) to Zhenzhou (镇州, present Mongolia) to Qocho (高昌, a.k.a. Gaochang); (2) from Liaoshangjing or Liaozhongjing to Juyan (居延) to Qocho; (3) and from Kaschgar (喀什噶尔) to Liaoshangjing. Ma believed that the ceramics made in Liao were also introduced to other regions through these same three routes. As regards the special square-shaped glass plate discovered in the tomb in Yemaotai, Ma (2002) proposed that it is a condiment tray and was introduced to the Liao from Iraq or Egypt since its shape and style are very similar to a condiment tray found in the British Museum. By comparing the style, shape and pattern with the glass collected in the Corning Museum of Glass and the Victoria and Albert Museum, Fu (2006) proposed that the glassware discovered in the Dule Temple is most probably early Islamic glass. Based on the manufacturing techniques and style, Zheng (2010) proposed that the stem cup discovered in the tomb in Turki Mountain is also Islamic glass. The glass discovered in the Princess Pingyuan's tomb is a further example of Islamic glass (Qi and Li 2018: 193). Using different research methods, An Jiayao made more detailed studies of the glass found in the tomb of the princess Chenguo based on component analysis and traditional archaeological methods. He found that two glass cups with rings are products from the Iranian plateau dating to about the ninth century; a glass bottle with a high neck and carved pattern is Iranian from the ninth to tenth centuries; a glass bottle with high neck and nipple pattern is most probably from Egypt or Syria; a glass plate with nipple pattern is probably from tenth or early eleventh century Byzantium; and another two glass bottles with high necks are most probably from Central Asia (An 1990; An 1993: 179-186). Due to the similar shape and style with the glass bottle discovered in the tomb of the princess Chenguo, An (1990) proposed that the glass bottle found in Dule (独乐) Temple is also Islamic. These glass products provide reliable evidence regarding cultural exchange and trade between the Liao dynasty and western Eurasia.

10.9. Conclusions

The archaeological sites and objects belonging to the Liao dynasty suggest that the culture of the 'Western Regions' had a great influence on Liao society, although some western cultural elements were introduced early in the dynasty while some were later. During the Five Dynasties, a large number of Hu people from the 'Western Regions' lived in the central plains of ancient China. After the fall of the Five Dynasties, the Liao administrated several original areas of the Five Dynasties and some Hu people moved to live under the Liao. As a result, it is not surprising that a large number of images of Hu people appear in Liao tombs. These people naturally brought their special cultures to the area of the Liao dynasty. This just reflects the complex social structure and situation at that time. It cannot be ignored that some cultures of the 'Western Regions' and some exotic objects do have obvious social hierarchical differences in the process of dissemination. This is clear in the dissemination of glassware and Nestorianism. Islamic glassware was regarded as a luxury during the Liao dynasty and is found almost only in tombs of high-level nobles. According to current archaeological materials, cultural influences of Nestorianism are only found in the tombs of Khitan commoners, not in the tombs of wealthy people and nobles. This shows that the believers of Nestorianism at that time had a very clear social class. Moreover, some cultures of the 'Western Regions' were gradually being integrated with the local cultures of the Liao, creating the development and innovation of local handicrafts. As a result, the designs and workmanship of gold and silverwares reached a peak during the Liao dynasty. More importantly, the cultures of the 'Western Regions' had a large influence on daily life and spiritual beliefs.

References

An, Jiayao (1990) On the early Islamic glassware unearthed in China in recent years 试探中国近年出土的伊斯兰早期玻璃器 *Kaogu* 12: 1116–1126.

An, Jiayao (2014) Dainty and exquisite, colourful in the East and West world: Domestic glass and imported glass found in Chinese archaeology 玲珑澄澈，缤纷东西——中国考古发现的国产玻璃和进口玻璃. *Wenming* 11: 26–39.

An, Jiayao (2021) The silk road and glassware 丝绸之路与玻璃器. *Wenwu Tiandi* 12: 87–92.

Aohanqi Museum *et al.* (1999) Briefing on the cleaning of Xiawanzi Liao tomb in Aohanqi 敖汉旗下湾子辽墓清理简报. *Caoyuan Wenwu* 1: 67–84.

Ba, Jingkan (2006) *Music and Dance During the Liao Dynasty* 辽代乐舞. Shenyang: Volumes Publishing Company.

Chen, Guang'en (2014) On the spread of Nestorianism in the Liao and Jin Dynasties 试论景教在辽金的流传. In Ma Mingda and Zong'an Ji (eds.), *Jinan Shixue* 暨南史学 (Vol.10), pp. 135–145. Guilin: Guangxi Normal University Press.

Cheng, Jie (2017) Study on the time, source and way of watermelon introduction into China 西瓜传入我国的时间、来源和途径考. *Journal of Nanjing Normal University (SOCIAL SCIENCES EDITION) (Nanjing Shida Xuebao Shehui Kexue Ban)* 4: 79–93.

Fu, Ning (2006) Islamic glassware of Liao Dynasty discovered in Inner Mongolia and the foreign trade and cultural exchange in Liao Dynasty内蒙古地区发现的辽代伊斯兰玻璃器——兼谈辽时期的对外贸易和文化交流. *Neimenggu Wenwu Kaogu* 2: 78–88.

Gu, Wenbi (1998) Questions about the origin and spread of cultivated watermelon in China (abstract) 中国种植西瓜的起源和传播问题(摘要). *Nongye Kaogu* 1: 378.

Hebei Provincial Institute of Cultural Relics (2001) *Excavation Report of the Liao Dynasty Frescoed tombs at Xuanhua: Report of Archaeological Excavationa from 1974–1993* 宣化辽墓:1974–1993 年考古发掘报告 II. Beijing: Cultural Relics Publishing House.

Hu, Qiao (2015) Liao dynasty: *Xianbeiji* 陷北记. In Xiu, Ouyang (ed) (2015) *History of the New Five Dynasties* 新五代史. Beijing: Zhonghua Book Company.

Huang, Shengzhang (2015) Introduction and development of watermelon in China 西瓜引种中国与发展考信录. *Nongye Kaogu* 1: 266–271.

Li, Gang and Sidan Zhai (2017) Exploration of the crosses made of pottery found in the tombs in Anshan 对鞍山墓葬出土的陶制十字的探究. In Liu Ning et al. (eds.), *Liao Jin History and Archaeology* (*Liaojin Lishi yu Kaogu*) 8: 65–73.

Lu, Qi, and Fusheng Zhao (1980) Brief report on the excavation of mural tomb in Zhaitang, Beijing 北京市斋堂辽壁画墓发掘简报. *Wenwu* 7: 23–27; 102–103.

Lü, Fuhua (2015) Discussion on the jade belt with the *Hu* people music-dance pattern during the Liao Dynasty and related problems 辽代胡人乐舞纹玉带及相关问题探讨. *Journal of Northeast Normal University (Philosophy and Social Sciences)* (*Dongbei Shida Xuedao Zhexue Shehui Kexue Ban*) 1: 177–181.

Ma, Wenkuan (1994) Islamic glass unearthed from Liao tomb and Liao tower and the relationship between Liao and Islamic world 辽墓辽塔出土的伊斯兰玻璃——兼谈辽与伊斯兰世界的关系. *Kaogu* 8: 736–743.

Ma, Wenkuan (2002) Islamic glass seasoning plate unearthed from the early Liao tomb in Yemaotai, Faku 法库叶茂台早期辽墓出土的伊斯兰玻璃调味方盘. *Zhongguo Lishi Wenwu* 3: 46–48; 96.

Ningxia Museum and The Guyuan Museum of Ningxia (1985) Brief report of Li Xian couple's tomb during the Northern Zhou Dynasty in Guyuan, Ningxia Province 宁夏固原北周李贤夫妇墓发掘简报. *Wenwu* 11: 1–20; 97–100.

Peng, Shanguo (2006) The type, origin, and direction of ceramic phoenix head pot in Tang Dynasty 唐代陶瓷凤首壶的类型、渊源与流向. *Zhongyuan Wenwu* 4: 56–59.

Qi, Dongfang and Yusheng Li (2018) *Ancient Chinese Material and Cultural History: Glass* 中国古代物质文化史・玻璃器. Beijing: Kaiming Press.

Qi, Xiaoguang, Zhiyong Gai and Yanshang Cong (1998) A brief report on the excavation of the Liao Mural tomb in Bao Mountain, Chifeng, Inner Mongolia 内蒙古赤峰宝山辽壁画墓发掘简报. *Wenwu* 1: 73–95.

Qi, Xiaoguang, Jianguo Wang & Yanshang Cong (1996) Brief report on the excavation of Yelyu Yuzhi's tomb during the Liao Dynasty 辽耶律羽之墓发掘简报. *Wenwu* 1: 4-32; 97–100.

Rong, Xinjiang (2020) Epitaph writing of *Hu* people in medieval China 中古入华胡人墓志的书写. *Wenxian* 3: 121–137.

Shao, Guotian (1999) A briefing report on the cleaning of Liao tombs No. 1 to No. 3 in Yang Mountain, Aohanqi Banner 敖汉旗羊山1—3号辽墓清理简报. *Caoyuan Wenwu* 1: 1–38.

Shao, Guotian (2001) The research on the musicians' images of the Xiaozhang murals in Yelü Yuzhi's tomb 辽耶律羽之墓小帐壁画'伎乐图考'. In The Chinese Society for the Ancient Capital Studies et al. (eds.), *Studies of Chinese Ancient Capitals* (*Zhongguo Dugu Yanjiu*) (Vol.18), pp. 324–335. Beijing: Chinese International Press.

Sun, Ji (1999) Discovery and study of Western artifacts in China since 1949 建国以来西方古器物在我国的发现与研究. *Wenwu* 10: 69–80.

Tala and Yaqiang Zhang (2004) Liao Dynasty tombs in Tuerji Mountain, Tongliao City, Inner Mongolia 内蒙古通辽市吐尔基山辽代墓葬. *Kaogu* 7: 2; 50–53; 106–108.

Yang, Boda (2004) *The Complete Collection of Chinese Gold and Silver Ware, Glass Ware and Enamel Ware* 中国金银玻璃珐琅器全集. Shijiazhuang: Hebei Fine Arts Publishing House.

Torii, Ryūzō (1942) *Sculptured Stone Tombs of the Liao Dynasty*. Peking: The Harvard-Yenching Institute (in English).

Tuo Tuo et al. (2017) Yuan Dynasty: *Taizu Ji* 太祖纪. In Liu Pujiang et al. (eds.), *Liao shi (History of Liao Dynasty)* 辽史 (Vol. 54): p. 53. Beijing: Zhonghua Book Company.

Wang, Chunyan (2020) *The Study of the Gold and Silver Ware during the Liao Dynasty* 辽代金银器研究. Beijing: Science Press.

Wang, Dafang (1998) 'Watermelon image' of tomb No.1 in Yangshan, Aohan Banner and the introduction of watermelon in Khitan and the unearthed ancient 'watermelon seeds' in China 敖汉旗羊山1号辽墓'西瓜图'—兼论契丹引种西瓜及我国出土古代'西瓜籽'等问题. *Neimenggu Wenwu Kaogu* 1: 40–44.

Wu, Di (2015) Introduction of watermelon into China and dissemination of its cultivation techniques西瓜引种中国及其栽培技术的传播. *Nongye Kaogu* 6: 249–252.

Wu, Yuhuan and Jiajing Cheng (2014) Liao Dynasty's control and management of the Silk Road 辽代对草原丝绸之路的控制与经营. *Qiusuo* 7: 158–162.

Wu, Zhuo (1987) A study on the gilt silver pot unearthed from Li Xian's tomb during the Northern Zhou Dynasty 北周李贤墓出土鎏金银壶考. *Wenwu* 5: 66–76.

Yang, Fuxue, Jiajing Cheng and Erdan Langna (2021) Textual research on the question about the route of watermelon from Qocho (高昌回鹘, Gaochang

Uighur) to Khitan 西瓜由高昌回鹘入契丹路径问题考辨. *Journal of the Silk Road Studies* (*Sichou Zhilu Yanjiu Jikan*) 2: 257–268.

Yang, Hong (2008) The *Hu* people and tomb figures of *Hu* people's image during the Tang dynasty 李唐王朝的胡人与胡俑. In Yingfeng Fan (ed.), *Qianling Wenhua Yanjiu,* pp. 52–53. Xi'an: San Qin Publishing House.

Yang, Rui (2011) The research on the *Hu* people during the Song dynasty 论宋代的胡人. *Zhonguo Bianjiang Shidi Yanjiu* 1: 87–94.

Zhang, Jingming (2011) *Study on Gold and Silver Wares of the Liao Dynasty* 辽代金银器研究. Beijing: Cultural Relics Publishing House.

Zhang, Tong (2017) A brilliant hymn on horseback: Appreciation of gold and silver wares found in the Liao Tomb in Turki Mountain 马背华器，流光溢彩——吐尔基山辽墓金银器鉴赏. *Wenwu Jianding yu Jianshang* 5: 4–12.

Zhao, Yongchun (2010) The research on the Liao people's views of *Zhong Guo* or 'Middle Kingdom' 试论辽人的"中国"观. *Wen Shi Zhe* 3: 78–90.

Zheng, Yanyan (2010) The glass stem cup during the Liao Dynasty 辽代玻璃高足杯. *Zhongguo Bowuguan* 3: 82–83.

Zhu, Tianshu (1998) *The Gold and Silver Ware During the Liao Dynasty* 辽代金银器. Beijing: Cultural Relics Publishing House.

11

Image analysis of drinking scenes in banquet murals in Liao Dynasty tombs

Jie Zhang

Abstract: The nobility of the Liao dynasty had a set of prescribed rituals for drinking tea, drinking alcohol and eating food. The usual order was drinking alcohol before drinking tea. Great feasts were held at significant sacrifices, festive celebrations and diplomatic activities. However, these large-scale ceremonies are confined to historical documents and there are no examples to verify their authenticity today. In Liao mural tombs, many images of drinking tea and drinking alcohol appear. They show more of the daily life of individuals and differ from the ceremonial norms of the Liao court, appearing more plebeian. This chapter uses the banquet scene images in mural tombs to analyse the phenomenon of drinking tea and drinking alcohol in the Liao dynasty. The chapter argues that the banquet scenes are a facet of the life in Liao society shared between the nobility and commoners.

Keywords: art history, Liao dynasty, mural tombs, banquet scene images, tea drinking, drinking alcohol, social life

The first discovery of wall murals in tombs of the Liao dynasty (AD 916–1125) was made as early as the 1920s. So far, more than 100 mural tombs have been scientifically excavated and restored. Most of these tombs are distributed in Chifeng and Tongliao (Inner Mongolia), Chaoyang (Liaoning Province), Zhangjiakou (Hebei Province), Datong (Shanxi Province) and Beijing. The highest concentration of tomb murals is in Chifeng, Zhangjiakou and Datong.

The murals are mostly discovered in the tomb passages, corridors, side walls and lintels of the tomb door, and on the four walls of the coffin chamber. The content of the murals includes very diverse scenes of animal grazing, hunting, banquets, drums, bustling music scenes, the armour of guards of honour, vehicle and horse travel, miscellaneous games, polo, doormen, attendants, immortals, door gods, astrological charts, landscapes, animals and flowers. Among these scenes, banquets are one of the most common themes.

In archaeological reports, there is no uniform name for food-related images in Liao tomb murals. Common images are cooking, the kitchen and chef, food and drink preparations, eating and drinking, banquets and tea ceremonies. These behaviours all represent preparation for feasts, and here I collectively term them 'banquet scenes'. Among them, most images are related to liquor or tea, reflecting the common customs of drinking alcohol and drinking tea at that time.

In the light of their historical development, the tombs of the Liao dynasty are divided into three phases: early, middle and late. This chapter uses these three phases to analyse the images of drinking tea and drinking alcohol in banquets in the tomb wall murals. Liao banquets were held for major sacrifices, festival celebrations, diplomatic etiquette and other major activities. From this period on, a set of rules and etiquette was developed for drinking both alcohol and tea. At present, scholars have not yet agreed on the types of alcohol drinking in the Liao dynasty, generally considered to include wine and liquor. Drinking tea usually meant green tea, but Liao people liked to drink alcohol and tea as part of the same banquet, usually consuming alcohol before tea.

The images of banquets in the Liao tomb murals show mostly scenes of personal daily life rather than major events in the palace. The images always show the consumption of alcohol and tea, and the preparation of the associated utensils. These scenes are closer to daily life, and differ from the etiquette of drinking alcohol at the Liao court. Therefore, from the perspective of art history, the images of drinking alcohol and tea in Liao tomb banquet murals reflect the everyday side of the dynasty's social life.

11.1. The content of the banquet scene images in the tombs of the Liao

The issue of the chronology and periodisation of tombs and wall murals in the Liao dynasty has been discussed in many academic circles. Basically, the chronological phases of mural tombs are classified according to the stages of Liao tombs. Wang Qiuhua (1994) holds the view that the tombs can be assigned to two periods, whereas Xu Pingfang (1986) and Li Yiyou (1991) propose three periods. Yang Xingyu and Zheng Chengyan (2010), however, suggest a division into five stages in three periods, while Dong Xinlin (2004) claims four periods. These classifications were based on two methods, namely the chronology of Liao tombs and the chronology of Liao tomb murals. Especially for the staging of the murals in the Liao tombs, the earliest murals should be the upper limit, and the latest murals should be the lower limit, and then be combined with historical clues to determine the stages. This means we should determine the specific ages of the Liao murals from archaeological reports, then compare with historical materials from the Liao dynasty in order to classify the chronology.

Summarising the views of the above scholars, the author divides the mural tombs into Early, Middle and Late stages by taking the historical period of the Liao dynasty as the vertical time line. The Early stage is from Emperors Taizu to Jingzong (907–983), the Middle stage from Emperors Shengzong to Xingzong (983–1055), and the Late stage from Emperors Daozong to Tianzuo (1055–1125). In this way, the integrity of the mural materials can be maintained, and they can also be compared with the historical culture and social customs of various stages in the Liao dynasty, providing a good reference for discussing the subjects, techniques, layout and cultural connotations of Liao tomb mural images.

The number of mural tombs found in the Early Liao is relatively small, and the painting skills are rather simple. The main mural tombs from this stage include Baoshan, Shuiquangou, Yelyu Yuzhi, Erbadi, Youai, Niangniangmiao, Yemaotai tomb No. 7, Zhaoduba, Zhao Dejun's tomb, and Xu Congyun couple's tomb. Among these, Baoshan No. 1 tomb has an exact date of AD 923 and can be said to be one of the earliest Liao tomb murals discovered so far. The early tomb murals reflect few banquet scenes, such as tables for banquets, cooking, food preparation, drinking preparation or attendants holding food. Images of drinking alcohol or tea were not observed.

The number of mural tombs discovered in the Middle Liao sharply increased, and the painters became more and more skilled. Representative mural tombs include Baiyinhanshan Han's family tombs, Qingdong Mausoleum, Dishuihu, princess Chen couple's tomb, Yangshan tomb No. 1, Kulun tomb No. 6, Nailingao, Xiao He's tomb, Geng Yanyi couple's tomb, and Qingyundian tomb No. 1. Images of banquets in the tomb murals include scenes of cooking, chefs, food preparation, drinking preparation, banqueting and eating. There are extensive scenes of drinking alcohol and tea.

The number of mural tombs discovered in the Late Liao further increased, with superb painting techniques and outstanding realistic styles. Representative mural tombs include Kulun Nos. 1, 2 and 7 tombs, Beisanjia, Yangshan No. 2 and No. 3 tombs, Qijia No. 3 and No. 5 tombs, Lamagou, Xiao Yi's tomb, Mutouchengzi, tomb No. 2 of Baiwanzhuang, the group of Wohuwan tombs, Pingquan, Zhang Kuangzheng's tomb, Zhang Wenzao's tomb, and Zhang Shiqing's tomb. The images of the tomb murals have a fixed 'copybook' finish, and there are many images of banquets on the tomb walls. The images of banquet scenes include cooking and food preparation, drinking and tea preparation, drinking tea, tea ceremonies, scenes of people drinking alcohol together, scenes of drinking and listening to music, and scenes of heating the liquor. There were many images of drinking alcohol and tea during this period.

Judging from the artistic techniques of the mural images, the Early Liao focused on freehand brushwork, with more representations of nomadic life and domestic scenes of the Khitan people; there were few scenes of drinking alcohol or tea at banquets. In the Middle period, the freehand brushwork became more realistic. The content of the Khitan tomb murals is different from that of tombs of the Han nationality. In the Khitan murals, images only showing tables for banquets no longer appeared; drinking alcohol and tea are more common in household life scenes. In the Late period, a realistic style dominated and the techniques used were quite skilled. Drinking alcohol and tea during banquets became common subjects. The images had a fixed format, indicating that artistic 'copybooks' were already popular among painters, with identical or similar themes appearing in tomb murals in different regions.

11.2. Examination of the drinking images of the banquet murals in Liao tombs

Drinking alcoholic beverages held an important role in the daily life, festival celebrations, religious rituals, major ceremonies, interpersonal communication and other activities in the northern China grasslands. Drinking has a long history in the region. Judging from Late Neolithic *zun* excavated from the Xiaoheyan culture Naitaidi site (Inner Mongolia) and *he* excavated from the Zhukaigou culture (also in Inner Mongolia)—both pottery vessels for drinking—we can imagine that liquor was already popular at that time. During the Xia, Shang and Zhou dynasties, the types and materials of wine vessels increased; *jue*, *gui* and *bei* show that the brewing and drinking of liquor were further expanded. In the subsequent artefacts of the northern nomads, a large number of wine vessels of various types were found, indicating the prevalence of brewing and drinking. There are many images of drinking in the tomb murals of the Liao dynasty, showing the preparation and serving of liquor, as well as drinking utensils, which reflect the drinking traditions of the Liao.

In the Early Liao tombs there were few drinking images in the banquet murals; only images of drinking vessels and pouring the liquor were shown. For example, in the mural of the hall on the north wall of tomb No. 1 in Baoshan (Qi et al. 1998) there is a rectangular table with plates, bowls, a pedestalled cup and chopsticks placed thereon. In this mural, the pedestalled cup can be considered as a drinking utensil.

A female attendant drawn on the woodblock serving as the door for a wooden tent in the Youai tomb (Miao 1996) holds a large ring-foot tray with both hands at chest level. On the tray is a small cup with a lotus petal-shaped base. This image can be understood as a gesture of presenting liquor for the tomb owner (Figure 11.1). The pedestalled cups and small cups with lotus petal-shaped bases shown in the tomb murals have petals and round mouths, which are not found in the excavated artefacts from the Early Liao. Most Liao dynasty drinking vessels were petal-shaped cups with ring feet and petal belly cups with round mouths. The excavated pedestalled cups have a shallow dish, a short and thin handle, and a small ring foot. By contrast, the pedestalled cups in the murals have large ring feet, which is probably an exaggerated creative expression. Although there is no scene of drinking behaviour in this tomb, the drinking vessels and female attendant show that the custom of drinking in this period was popular.

In the Middle Liao, images of drinking preparation and serving liquor were common in the murals of banquet in tombs, for example, the image of the kitchen on the southeast wall of Qijia No. 1 tomb (Aohanqi Museum et al. 1999a). In addition to the cooking scene, there are bowls, plates, cups, chopsticks and other utensils on the table, including small cups for liquor (Figure 11.2). There is an image of drinking preparation on the southwest wall of tomb No. 2 at Qijia (Aohanqi Museum et al. 1999a). On the right side of the table is a begonia-shaped plate with two small cups; on the left side is a matching lotus-shaped insulated bowl used to keep liquor warm and a handled ewer. Also, a rack on the ground on the left side of the square table has two rows of liquor jars with clay seals at their mouths.

Figure 11.1. Female attendant from the Youai tomb. Source: Re-drawn by the author from Miao (1996).

Figure 11.2. Kitchen scene mural from Qijia No. 1 tomb. Source: Re-drawn by the author from Aohanqi Museum et al. (1999a).

There is an image of drinking preparations on the eastern side of the tomb wall corridor of Tomb No. 2 in Baiyinhanshan Han's family grave (Tala et al. 2002). The second person holds a bottle with a small mouth and big belly with both hands; the third person holds the insulation bowl and a handled ewer. The image of drinking preparations on the southeast wall of tomb No. 5 in Xiawanzi (Aohanqi Museum et al. 1999b) shows, from right to left, the first person holding a large bowl in both hands, the second person holding a tray with two small cups in both hands, and the fourth person's left shoulder against the drinking jars. On the square table in front of the four persons, there are lotus-shaped insulation bowls and handled ewers, and plates are placed on both the left and right sides. There are small cups, watermelons, peaches and pomegranates on the table (Figure 11.3). The image of the female attendant in the Xisanjia tomb (Liaoning Provincial Institute of Cultural Relics and Archaeology et al. 2010) shows the attendant holding a plate with two small cups to serve her master food and drink.

Figure 11.3. Mural showing preparations for drinking alcohol from tomb No. 5 in Xiawanzi. Source: Photograph by the author.

The tomb murals of this period still lack drinking scenes; there are only images of drinking utensils, drinking preparations and serving the liquor, but the number of such images is higher and the content is more diverse than in the Early period. Drinking preparations are more commonly seen through drinking and storage utensils such as small cups, insulation bowls, handled ewers and liquor jars. The lotus-shaped insulation bowls and handled ewers in the images are generally used together. Boiling water is poured into the insulation bowl, the handled ewer is then placed in the bowl in order to warm the liquor. This method, suited to the cold winters of the north, is still used today. The liquor jars mostly have straight mouths, short necks, wide shoulders, bellies with curved, flat bottoms with clay seals on their mouths; the jars are inserted into a rack. Tableware, drinking utensils and various fruits are placed on the table. The use of fruits as a side dish for drinking is also a special custom in the Liao dynasty.

In addition to the murals of drinking preparations and serving liquor, there were also drinking scenes in banquets depicted in Late Liao tombs. An example is the mural of drinking preparation on the west wall in the Xiawanzi No. 1 tomb (Aohanqi Museum et al. 1999b). From left, the first person holds a begonia-shaped plate with a small cup in both hands. The second person holds a jar taken from the rack in both hands and appears to be smelling its mouth, perhaps in order to check its quality. The third person holds a plate in his left hand, and wipes the plate with a square dishcloth in his right hand. The rack is placed in front of the second person. There should have been two jars with sealing clay on them; one bottle is still inserted in the rack, while the other has been removed by the second person. There is also a square table in front of the third person, with a stack of inverted bowls and food boxes placed thereon, reflecting preparations for drinking (Figure 11.4). In the banquet scene in the eastern side of the tomb wall, the host sits on a wooden chair and looks ahead with a female attendant standing behind him. Another female attendant on the right side of the table holds an insulation bowl and handled ewer in both hands, serving the host with drink. On the table next to the host, there are two long plates with fruits, and a small cup and chopsticks are placed at the side of the host (Figure 11.5).

Figure 11.4. Mural showing preparations for drinking alcohol from the tomb of Xiawanzi No. 1. Source: Photograph by the author.

Figure 11.5. Banquet scene mural from the tomb of Xiawanzi No. 1. Source: Photograph by the author.

The images of food and drinking preparations in Xiao Yi's tomb (Wen 1989) show two attendants standing at a table, each holding a small cup with a handle in both hands. A jar is placed on the square high table, the mouth of the jar is exposed with a spoon handle, and two small cups with handles are placed next to the jar. In the image of the banquet scene in the Mutouchengzi tomb (Liaoning Provincial Institute of Cultural Relics and Archaeology et al. 1995), under a square table, one person is lying and two people are sitting on the ground, all of them apparently in a drunk state. The table is set with plates, cups and a handled ewer. The jar is in front of the table, and two people hold small cups and a handled ewer behind the table.

In the image of the banquet scene in the Jiuchang tomb in Zhuolu County (Liu and He 1987), the owner of the tomb talks with the person in front of him. Two male attendants stand behind; one attendant holds a bowl and plate in his hand, and the other holds a food box. Some musicians are playing music in order to enhance the entertainment of the tomb owner. There is a square table next to the tomb owner, where dishes, bowls, basins, dishes and other eating and drinking utensils are set. Especially, two jars stand under the table, showing that the owner of the tomb will enjoy abundant drinking with the people around him. There are murals of playing music and a banquet scene on the southwest wall of the back room of Han Shixun's tomb in Xuanhua (Liu and Yan 1992). A woman on the right is sitting on a round chair with a cup in both hands, and a table with many plates full of exquisite food stands in front of her. A male attendant is serving the woman with an empty long plate in his hands. A musician is playing a three-stringed instrument and the other two people face this musician. One is an old man, the other younger, but both smile and clap at the wonderful music. There are three jars on a table in front of the people. In this scene of drinking and listening to music, all of the people present create a lively atmosphere for the pleasure of the woman (Figure 11.6).

Figure 11.6. Banquet scene from Han Shixun's tomb in Xuanhua. Source: Re-drawn by the author from Liu and Yan (1992).

Judging from the banquet scenes in the tombs of the Late Liao, images of drinking preparation and serving liquor follow the style of the Middle phase. The contents of the images are similar, and the realistic features are very obvious, reflecting the lifestyle of the tomb owner before his or her death. During this period, a large number of banquet scenes and drinking images appeared, among which there are many scenes of drinking. Some show the owner of the tomb drinking alone, others the owner drinking with relatives and/or friends. The banquet scene in the tomb of Han Shixun is accompanied by listening to music while drinking, but the three-stringed musical instrument is not recorded in the *Liao History Yuezhi* (Tuotuo et al. Yuan Dynasty 1974d). It was probably a folk instrument introduced into the Liao by people from the Song dynasty. Fruit is placed on the table in the banquet scene of Xiawanzi Tomb 1, indicating that the custom of using fruit as a side dish while drinking was still popular.

The appearance of these scenes and customs about the prevalence of liquor is inseparable from the aristocracy of the Liao dynasty because drinking liquor was a ritual and sociable behaviour for them. The central government of the Liao set up a liquor-making office, an official organisation dedicated to the production of liquor and the management of related affairs. There were also many workshops specialised in distilling liquor among the common people. During the Lantern Festival, Liao Emperor Muzong is said to have watched lanterns in the market and drank with officials using liquor from the folk distilleries (Tuotuo et al. Yuan Dynasty 1974c). The presence of both official and private liquor-making institutions and distilleries promoted the prevalence of drinking customs in the Liao. This provided a realistic basis for the banquet scenes that appear in Liao tombs, and the appearance of liquor in the murals also explains the prosperity of the liquor industry at the time.

11.3. Analysis of tea drinking images in the banquet murals of Liao tombs

In the Liao tomb murals, there are many images of tea drinking and related customs, including tea preparation, pouring the tea, boiling tea, tea ceremonies and so on. Due to the cold and dry climate, the northern region where the Liao dynasty was located was not suitable for growing tea. Therefore, neither the local area nor the Yanyun area—today's Beijing, Tianjin, Shanxi and Hebei Provinces—produced tea, because the scarcity of rain and the short day-length means the region is not suitable for growing tea. As a result, all the tea used by the Liao needed to be imported. The tea came mainly from the Tang, the Five Dynasties, Song and other dynasties in the Central Plains of China by trade and exchange in markets, or was brought through gifts, by official envoys or even through private smuggling.

Although the Liao History and other documents recorded the tea drinking customs of the Liao and archaeological discoveries have also produced tools for milling, storing, boiling, steaming and drinking tea, we have so far not been able to understand the specific tea drinking methods and activities of the period. Therefore, the images of tea drinking in the Liao tomb murals provide important evidence for the study of tea culture at that time.

Judging from the murals of the Early Liao tombs, there are few tea drinking images, only depictions of tea utensils and serving tea. An example is the image of the banquet table on the north wall in the north corridor of Baoshan No. 1 tomb (Qi et al. 1998). There are eleven pieces of tableware on the square table, of which two petal-shaped bowls are tea utensils (Figure 11.7). So far this is the earliest tea utensil found in the murals of Liao tombs. There is also an image of attendants on the east wall of the tombstone room in Baoshan No. 2 tomb (Qi et al. 1998). The female attendant on the north side is holding a gold cup for tea drinking. This kind of tea utensil was also found in gold and silver in the Early Liao.

The male attendant on one side of the wooden curtain door of the Youai tomb (Miao 1996) holds a handled ewer with a tower shaped lid in his right hand and a bowl with four petals in his left hand. He seems to be in the process of serving tea for his host (Figure 11.8).

There are banquet images on the left and right walls in the front of the tomb of Zhaoduba (Cai and Li 2004). In the middle of the image on the left wall is a square dining table with a lotus-shaped phoenix head bottle, a three-legged cup with a tray and a flower mouth type cup. There is also a square table in the middle of the image on the right wall. On the table there is a lotus petal-shaped phoenix head bottle, a lotus petal-shaped handled ewer with a cover, a cup with a tray and a fancy mouth cup. These are all utensils for boiling water, boiling tea and drinking tea (Figure 11.9).

Images of tea preparation are common in the murals of Middle Liao tombs. In the image of service for the tomb owner on the east wall of the corridor in the Kangyingzi tomb (Xiang 1979), a sitting attendant holding a handled ewer passes it to another attendant who is sitting on the ground. In front of the two people, a square table has a handled ewer and a petal-shaped cup. The archaeological report concludes it is an image of preparations for drinking alcohol, but from the perspective of the type of utensils in the image, it is more likely to be a tea preparation scene.

In the image of tea preparation on the southwest wall of Yangshan No. 1 tomb (Aohanqi Museum et al. 1999c), five adults surround a square high table set with tea utensils and fruit. On the table are four sets of cups, one pot with a cover, one plate and one bowl containing fruit and dates. In front of the table, a child is sleeping on a box made of bamboo strips. Another child is boiling soup. In the foreground of the image is a three-legged brazier and a two-handled ewer with a long neck (Figure 11.10).

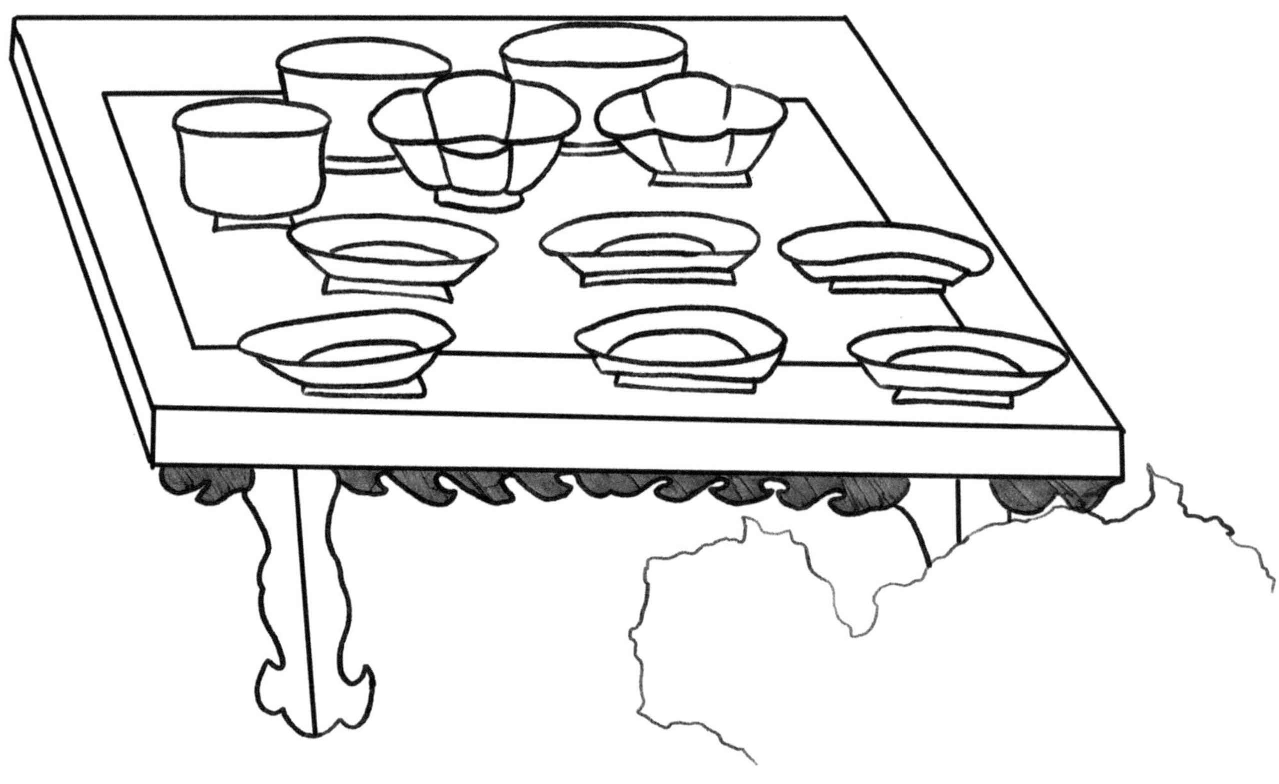

Figure 11.7. Mural of a banquet table from the Baoshan No. 1 tomb. Source: Re-drawn by the author.

Figure 11.8. The male attendant from the Youai tomb. Source: Re-drawn by the author from Miao (1996).

In the image of tea drinking preparations on the southeast wall of tomb No. 2 of Qijia (Aohanqi Museum et al. 1999a), one person holds a cup with a tray which he is ready to put on the table in front of him. There are bowls and plates on the square table. On one side of the table is a three-legged brazier with a handled ewer with a long neck. This scene shows the person is preparing tea.

In the Late Liao, scenes of tea preparation, tea drinking and tea ceremonies in the banquet murals in Liao tombs are more abundant. There are some large-scale scenes related to tea drinking. On the southwest wall of tomb No. 2 in Xiabali (Tao et al. 1990), there is a large-scale tea ceremony image. Bowls, plates and a cup with a tray are placed on a table. A female attendant behind the table stands with her hands crossed on her chest, and a male attendant lifts a tray with cups. Beside the table, there is a child who is holding a round fan and is bending over to fan the fire. There is a three-legged brazier next to the child and a handled ewer of boiling water (Figure 11.11). These people constitute a scene of a tea ceremony prepared for their host.

Although the number of tea utensils, tea preparation and serving scenes, and images of tea ceremonies in Early Liao tomb murals is relatively small, that does not necessarily mean that tea drinking was not popular at the time. Judging from historical texts, tea drinking among northern nomads become more and more popular from around the time of

Figure 11.9. The mural of banquet images from the tomb of Zhaoduba. Source: Re-drawn by the author from Cai and Li (2004).

Figure 11.10. Mural showing tea preparation from Yangshan No. 1 tomb. Source: photograph by the author.

the Northern Dynasties. According to the Record of Lu Yu in the *Xin Tang shu*, 'Since then [the Northern Dynasties], drinking tea has become a trend in society. At that time, when the Huihe [Uyghur] people entered the market, they always exchanged tea or other things by horse' (Xiu et al. 1975). This shows that drinking tea was popular among the Uyghur Huihe people in the Tang.

At the beginning of the founding of the Khitan, the emperor sent envoys to the Southern Tang for the exchange of tea, and the later Jin and Northern Han dynasties also sent envoys to the Liao state to offer tribute for tea. According to the Record of Khitan in the *Nan Tang shu*, in the first year of Huitong (938), Emperor Taizong of the Liao and the King of Dongdan each sent envoys to the Southern Tang to exchange sheep and horses for tea and medicines. In the third year of Huitong (940) in the summer, 'The Jin dynasty sent envoys to offer tribute for tea and medicine' (Tuotuo et al. Yuan Dynasty 1974a). In the fourth year of Yingli (954), the 'Northern Han dynasty sent envoys to offer tribute for tea and medicine' (Tuotuo et al. Yuan Dynasty 1974b). These records show that tea was a very popular and sought-after commodity among nomadic groups to the north of China.

Tea-related events were common in the major ceremonies of the Khitan. For example, tea, fruit and cake were used to pay tribute to the gods and the earth in the mountain worship ceremony, as well as in the emperor's birthday rituals, the empress's birthday celebration ceremonies, the sacrificial ceremony at the beginning of spring, the ceremony of giving a banquet in honour of the Song dynasty envoys, and so on. The emperor provided the tea and drinking tea became an important part of the

Figure 11.11. The mural of tea ceremony from the No. 2 tomb in Xiabali. Source: Re-drawn by the author from Tao et al. (1990).

ceremony. It can be seen that the custom of drinking tea had been formed in the Early Liao.

From the perspective of tea-drinking methods, the Early Liao inherited *sencha* from the preceding Tang and Five Dynasties. According to Lu Yu's *The Classic of Tea*, drinking *sencha* was divided into two processes. The first step was boiling the water, the second brewing the tea. People always liked drinking the soup and powder together after making tea. There are no images of *sencha* in the Early Liao tomb murals, but tea utensils were excavated in the tombs. A mixed silver-iron pot, dragon-patterned silver boxes, silver-handed pots, silver spoons, silver and gilt silver cup holders, silver bowls, silver statues with flower mouths, and celadon bowls were unearthed in the tomb of Panyanggou, Balingzouqi (Ma et al. 2016). Among these, some objects are utensils used for ordering tea, but there is still an absence of boiling utensils for tea-making. This phenomenon can be explained by the way of drinking tea at that time being dominated by infused *sencha*. However, judging from the lotus petal-shaped phoenix head bottle and the lotus petal-shaped handled ewer with a cover in the banquet images of the Zhaoduba tomb, boiling utensils used for making tea had already appeared by the Early Liao.

After the Middle Liao, boiling utensils used for making tea became popular due to the influence of the Song dynasty. The method was to boil the water first and then pour it into a utensil with tea powder to drink. This was a common tea drinking method for people of the Song. In the images of tea preparation, tea drinking and tea ceremonies in the tomb murals, they commonly used phoenix-head bottles and long-necked bottles placed on iron braziers or else placed the handled ewers on a table. These are all boiling utensils used for making tea. Zhu Yu of the late Northern Song recorded the custom of drinking tea in the Liao in Volume I of his *Pingzhouketan* (Zhu Song Dynasty 2012). The recorded method of Liao tea drinking must have been common after the Middle period, confirming the use of boiling utensils for making tea in the Middle and Late Liao.

Among the tomb murals of the Late Liao Dynasty, the tea ceremony images serve as the best proof to show tea drinking customs at the time. Almost all the Liao tombs found in Xiabali have tea ceremony images. The tea ceremony shows the method of boiling utensils used for making tea as well as the tea-making process in the Liao Dynasty (Figure 11.12).

According to historical records, cakes of tea were the main tea varieties imported into the Liao area in the Song dynasty; these included Tuan tea and milk tea. The cakes of tea shown in Liao murals are rectangular. There are also many utensils used in the tea ceremony in the murals, such as tea mills, sieves, brushes, braziers, tea stoves and so on. This culture of tea is also reflected in grave goods. The iron tea mill, iron brazier and iron stove unearthed from the Liao dynasty Daiqintala (Xinganmeng Cultural Relics Workstation et al. 1997) and Maiwanggou tombs (Inner Mongolia Institute of Cultural Relics and Archaeology et al. 1997) provide archaeological material in support of the murals. The objects unearthed from these tombs and the content of the murals together reproduce the whole picture of the tea ceremony of the Liao dynasty.

11.4. Conclusion

The reason why images of drinking alcohol and tea appeared in the tomb murals of the Liao was due to the growing evaluation of this kind of tradition by the upper class of the Liao dynasty. The related content of liquor and tea has been recorded many times in the literature.

When Song envoys celebrated the emperor's birthday or celebrated the New Year, the Song would send tribute to the Liao emperor, including drinking utensils, liquor and tea. The emperor would consequently hold a large banquet

Figure 11.12. Mural of a tea ceremony from the Xiabali tomb. Source: Photograph from Liu et al. (1995).

with drinking alcohol, eating and drinking tea conducted in sequence. On the one hand, this shows that liquor and tea were favoured by the elites of the Liao. On the other hand, it also shows that a complete set of liquor and tea customs had been established in the upper class of the Liao dynasty. Liquor and tea became an official bridge between the Liao and Song people.

The banquet images of drinking alcohol and tea in the Liao tombs reflect the common people's preference for liquor and tea. This is rarely reported in the historical literature, but it reflects the daily life of Khitan and Han commoners in the Liao.

Judging from the murals of the Early, Middle and Late Liao, the images of drinking alcohol and tea displayed changes in content from simple to complex. The scenes were gradually enriched. In the Late period, liquor and tea seemed to have been integrated into Liao society and people became used to the associated customs and traditions. Moreover, fruit and snacks were an important complement while drinking alcohol and tea and these foods also contributed to the social culture of the Liao dynasty.

References

(All references are in Chinese)

Aohanqi Museum et al. (1999a) Aohanqi Qijia Liao tomb. *Steppe Cultural Relics* 1: 46–66.

Aohanqi Museum et al. (1999b) Briefing on the cleaning of Xiawanzi Liao tomb in Aohanqi. *Steppe Cultural Relics* 1: 67–84.

Aohanqi Museum et al. (1999c) A briefing report on the cleaning of Liao tombs No. 1 to No. 3 of Yangshan in Aohanqi. *Steppe Cultural Relics* 1: 1–38.

Cai, Qiang and Daoxin Li (2004) Liao Dynasty mural tomb of Zhaoduba in Chaoyang, Liaoning Province. *Northern Cultural Relics* 2: 64–66.

Dong, Xinlin (2004) A brief discussion on the tomb formations and stages of Liao tombs. *Kaogu* 8: 62–75.

Inner Mongolia Institute of Cultural Relics and Archaeology *et al.* (1997) A brief report on the excavation of the Liao Dynasty cemetery at Maiwanggou in Ningcheng. In Wei, Jian (ed) (1997) *Collection of Inner Mongolia Cultural Relics and Archaeology* (Vol. 2): 609–630. Beijing: Encyclopedia of China Publishing House.

Li, Yiyou (1991) An introduction to the system about the Khitan people in the Liao Dynasty. *Collection of Research Articles on Archaeological Culture in Eastern Inner Mongolia*: 80–102. Beijing: China Ocean Press.

Liaoning Provincial Institute of Cultural Relics and Archaeology et al. (2010) A brief report on the excavation of the tombs during the Liao Dynasty in Xisanjia, Chaoyang. *Stories of Relics* 1: 35–40.

Liaoning Provincial Institute of Cultural Relics and Archaeology et al. (1995) Mural tomb of Mutouchengzi during the Liao Dynasty in Chaoyang, Liaoning Province. *Northern Cultural Relics* 2: 31–34.

Liu, Jianzhong and Yong He (1987) Brief report on excavation of Liao Dynasty mural tombs in Zhuolu, Hebei Province. *Kaogu* 3: 242–245.

Liu, Haiwen and Cheng Yan (1992) Han Shixun's tomb during the Liao Dynasty in Xiabali, Hebei Province. *Wenwu* 6: 1–11.

Liu, Haiwen, Jingzhai Li and Cheng Yan (1995) Liao Dynasty Mural tombs in Xuanhua, Hebei Province. *Wenwu* 2: 4–28.

Ma, Fenglei, Shixun Jiang, Weichun Cui et al. (2016) Tombs of Panyanggou during the Liao Dynasty in Balinzuoqi, Inner Mongolia. *Kaogu* 3: 30–44.

Miao, Runhua (1996) Youai Liao tomb in Balinyouqi, Inner Mongolia. *Wenwu* 11: 29–34.

Qi, Xiaoguang, Zhiyong Gai and Yanshang Cong (1998) A brief report on the excavation of the Liao mural tomb in Baoshan, Chifeng, Inner Mongolia. *Wenwu* 1: 73–95.

Tala, Ma Fenglei, Yaqiang Zhang and Qingyu Wang (2002) An excavation report on the cemetery of the Han family in Baiyinhanshan during the Liao Dynasty. *Steppe Cultural Relics* 2: 19–42.

Tao, Zongye, Zhongyu Liu, Zhao Xin et al. (1990) Xiabali Liao and Jin mural tombs in Xuanhua, Hebei Province. *Wenwu* 10: 1–19.

Tuotuo et al. (1974a) Yuan Dynasty: The second half of the biographic sketches of Emperor Taizong. In Feng, Jiasheng, Chenshu (eds.), *History of Liao Dynasty* (Vol. 4): 47. Beijing: Zhonghua Book Company.

Tuotuo et al. (1974b) Yuan Dynasty: The first half of the biographic sketches of Emperor Muzong. In Feng, Jiasheng, Chenshu (eds.), *History of Liao Dynasty* (Vol. 6): 72. Beijing: Zhonghua Book Company.

Tuotuo et al. (1974c) Yuan Dynasty: The second half of the biographic sketches of Emperor Muzong. In Feng, Jiasheng, Chenshu (eds.), *History of Liao Dynasty* (Vol. 7): 83–85. Beijing: Zhonghua Book Company.

Tuotuo et al. (1974d) Yuan Dynasty: Yuezhi. In Feng, Jiasheng, Chenshu (eds.), *History of Liao Dynasty* (Vol. 54): 881–898. Beijing: Zhonghua Book Company.

Wang, Qiuhua (1994) The stages of the decoration about the tombs of the Khitan nationality during the Liao Dynasty. *Northern Cultural Relics* 1: 43–47.

Wen, Lihe (1989) Xiaoyi's tomb during the Liao Dynasty in Yemaotai, Faku, Liaoning Province. *Kaogu* 4: 324–330.

Xiang, Chunsong (1979) Liao tombs painting material found in the Zhaowuda area of Liaoning Province. *Wenwu* 6: 22–32.

Xinganmeng Cultural Relics Workstation et al. (1997) Briefing on the cleaning of the Daiqintala Liao tomb in Keyouzhongqi. In Wei, Jian (ed.), *Collection of Inner Mongolia Cultural Relics and Archaeology* (Vol. 2), pp. 651–667. Beijing: Encyclopedia of China Publishing House.

Xiu, Ouyang, Song Qi et al. (1975) Song Dynasty: Lu Yu Biography. In Zhonghua Book Company (ed.), *New Tang Book* (Vol. 196), p. 5612. Beijing: Zhonghua Book Company.

Xu, Pingfang (1986) Tombs in the Liao Dynasty. *The Encyclopaedia of China* (Archaeology Volume). Beijing: Encyclopedia of China Publishing House.

Yang, Xingyu and Chengyan Zheng (2010) Taoism is natural and inclusive: concurrently discussing the features of murals in Liao Dynasty tombs in Inner Mongolia. *Chinese Museum* 3: 58–63.

Zhu,Yu (2012) Song Dynasty. In Li, Weiguo, Keqin Gao (eds.) (2012) *Pingzhouketan* (Vol. 2), p.10. Shanghai: Shanghai Classics Publishing House.

12

Rice: from medicine to food in Roman and medieval Italy

Claudio Pelloli

Abstract: Today, Italy is the largest rice producer in the European Union with an area of 229,547 ha and a net production of 1,468,790 t of unprocessed rice. However, for a very long time rice in Italy was used primarily as a medicinal plant and only marginally as food. At the same time, it was also a luxury food, used for extravagant dishes or as a minor ingredient on the tables of the rich. This chapter discusses why rice was used especially as a medicinal plant for many centuries and from where ancient people learnt this use. The essay also considers the changes in customs, behaviour and attitudes which might explain the shift from luxury food/ medicinal plant to staple food.

Keywords: Italy, Roman period, Middle Ages, rice, medicinal plant, staple food

12.1. Introduction

Rice has been connected to human consumption for thousands of years. Nevertheless, this connection has taken different forms in both time and space. Based on the current state of research, for example, in China, where rice was domesticated around the seventh millennium BC, and in Korea, where it was cultivated from 1500 BC, rice was used as a staple food (Jianh et al. 2007; Cohen 2011; Kim and Park 2020). In the Mediterranean basin, while rice has been known since the fourth to third centuries BC, its cultivation only spread many centuries after the birth of Christ and for two thousand years rice was used more as a medicine than as food (Ballarini 2014: 20; Gavinelli and Gilardi 2017: 423). However, it is unclear why in Italy (and generally in the Mediterranean basin) rice was first used above all as a medicine and only centuries later it became a staple food. Also unclear is from where the ancient Greeks and Romans learnt the medical use of rice. To try to explain these aspects the author will focus on a circumscribed territory, i.e., the Italian peninsula.

12.2. Rice as a Medicinal Plant

In the Mediterranean basin and more generally in Europe, rice had an important role as a medicinal plant for centuries. The medicinal role of rice in recent history was confirmed by, among other scholars, Christiaan Eijkman (1858–1930) and Kazimierz Funk (1884–1967). While studying beriberi, a disease caused by thiamine deficiency, Eijkman noticed a connection between this illness and bird feed in East Asia. Birds who lived off white rice had similar symptoms to humans affected by beriberi. He argued that there was nutritional value in the rice bran and that if people ate unpolished rice, the disease would disappear. Starting from this research, in 1912 Kazimierz Funk obtained from rice caryopses a new organic amino compound he called the 'soul of life', in Italian *anima della vita*, i.e., vitamins (Ghiardo 2020: 196–197).

Several ancient authors[1] that give us information about rice as a medicinal plant suggest to use it for different aches and conditions, from dysentery to dermatological uses, and also as a food for people afflicted with pulmonary diseases or sore throats. We could remember Pedanius Dioscorides, a Greek botanist and doctor who lived during the second half of the first century AD; his *De Materia Medica* gives us some information about rice (Table 12.1: No. 1). In one of his *Satires*, Horace has a doctor suggest to his rich and mean patient to give up poor or shoddy food and to eat rice for his stomach ache (Table 12.1: No. 2). In his *De medicina*, Celsus said rice is a good medicine for people who suffer from a disease named *peripnevmonikón* (περιπνευμονικόν) (Table 12.1: No. 3). Also, Pliny the Elder (PLIN., *NH*, XXVIII, 108–110) acknowledged rice's curative peculiarities: using the faeces of starlings nourished only with rice, it was possible to cure some skin imperfections (Table 12.1: No. 4).

Again, Marcellus Empiricus, a doctor at the court of Theodosius I who lived between the fourth and fifth centuries, proposed in *De medicamentis liber* to use rice for different illness such as those of the nose (Table 12.1: No. 5), throat (Table 12.1: No. 6), as an ingredient for dermatological liniments (Table 12.1: No. 7) and for all stomach illness (Table 12.1: No. 8).

The use of rice as a medicine is also apparently testified, albeit extremely rarely, from archaeological data, since rice grains have been discovered in buildings that could be identified as hospitals. From Novaesium, an ancient

[1] In this chapter, I do not list all ancient authors that mention rice in their works, nor do I include every occurrence of the crop in the writings of the authors cited.

legionary fortress in the province of Germania Inferior built around the beginning of the first century AD and located near modern Neuss in Germany (Rowan 2015: 5), archaeologists found 198 rice grains from a building dated to the first quarter of the first century AD and identified as a hospital (Knörzer 1966: 433; Morales and van der Veen 2015: 60). In Mursa, a city located in Pannonia Inferior near the border with ancient Dacia in modern Osijek, Croatia, five charred rice grain fragments were identified from sample 391 (SJ391), recovered from pit 3, which was dated to the first phase of the colony (from its foundation in AD 120–130 to the second half of the second century). This pit was located inside a large building in *insula* A. Based on associated remains (animal bones and other wild and cultivated plants), the excavators propose to identify this wooden building (the largest in this *insula*) as a large house, guesthouse or tavern (Leleković and Reed 2019: 271–273; Reed et al. 2022: 2–3). However, based on its size and the discovery of rice, it could be also interpreted as a hospital.

Nevertheless, in this scenario there are two cases that we could consider as exceptions, i.e., Mogontiacum and Egypt, because rice grains were discovered there in contexts that cannot be linked to medicine or healing. In Mogontiacum, modern Mainz in Germany, archaeologists found a single rice grain, dated to the second half of the second century AD, during the Magna Mater temple excavation (Zach 2002: 104–105). In Egypt, rice grains were found during excavations in Berenike and Myos Hormos, modern Quseir al-Qadim, two ports on the Red Sea. For the Roman period[2], in Myos Hormos 22 rice specimens (hulled grains, grains and hull fragments) were found (Cox et al. 2011: 46) and Berenike produced between 11 and 50 rice specimens dated between the first and second centuries A.D. and about the same quantity from the fourth to fifth centuries (Cappers 2006: 161). Berenike and Myos Hormos were the most important ports on the Red Sea, as we can deduce also from ancient writers such as the *Periplus of the Erythraean Sea*'s anonymous author and Strabo who mention only these two ports on the Red Sea (Table 12.1: Nos. 9 & 10). These ports were the 'gates' for merchants who arrived from India (Nappo 2018: 27, 43, 50, 53). This role may explain why in these two sites rice grains were discovered in a different context in comparison with the other sites (except for Mogontiacum). The small quantity of the specimens shows once again how rice was a marginal commodity during the Roman period—even at the two ports where merchants and goods arrived from India and where Roman merchants bought rice (see below). Moreover, it is possible that these rice specimens were linked more to Indian merchants as food than as a trading commodity inside the Roman Empire as suggested by Muthukumaram (2014: 3), because at both sites rice grains were recovered with Indian pottery (Myos Hormos) and Indian pulses (Berenike).

[2] Berenike was occupied from the first to sixth centuries AD, Myos Hormos from the first to third centuries AD and again from the eleventh to fifteenth centuries (Morales and van der Veen 2015: 56).

However, if ancient authors agree with the function of rice as a medicine, they do not specify from where they learnt this information, although all writers, both Roman and Greek, link rice to India. If Strabo in the *Geografia* (Table 12.1: No. 11), Diodorus Siculus in his *Bibliotheca Historica* (Table 12.1: No. 12), or Pliny the Elder in his *Naturalis Historia* (Table 12.1: Nos. 13 & 14), give notices about Indian customs, the anonymous author of the *Periplus of the Erythraean Sea* explicitly states that Roman traders bought rice from India (Table 12.1: No. 15).

In India, rice has a long history and scholars date its introduction there to between the seventh or sixth and second millennia BC (Bates et al. 2017: 194–195, 200), the exact chronology being much debated (e.g., Fuller et al. 2006: 29; Fuller and Murphy 2016: 345, 350–351; Spengler et al. 2021: 3). As in Southeast Asia, rice became a staple food in India. Today in India, rice feeds some 800 million people and the country produces 112,000,000 tons of rice (Bisen et al. 2020: 8) cultivated on 43.78 million hectares of land (Indian Ministry of Agriculture and Farmers Welfare: https://agricoop.nic.in/en/annual-report). But food is not the sole use for rice in India; it also has an important role in medicine. Ancient Indian medical knowledge is present in the Ayurveda texts, *Charaka Samhita* and *Sushruta Samhita.* These are the end result of a long writing process which integrated the oral Vedic tradition (Bocianowski *et al.* 2018: 2). The four Veda books (*Rigveda*, *Sāmaveda*, *Yajurveda* and *Atharvaveda*) – Ayurveda texts are considered a supplementary subject of the Atharva Veda – were composed between 1500 and the ninth century BC (Narayana 1995: 101). In these books, rice is the main or one ingredient for different health dishes – as tradition defines these foods – such as *idli* (fermented rice product), *dosa* (another fermented product), *dhokla* (dish from rice and Bengal gram fermentation) or *halu kashaya* (a healthy herbal drink) (Choudhary 2015: 98–100, 105). Moreover, in *Sushruta Samhita* (Table 12.1: No. 16) it is suggested to feed a pregnant woman with rice, especially during the third month of gestation. This practice originated from the idea that 'digestive product of the mother's food, is listed as one of the five factors responsible for procreation, together with mother, father, soul, and wholesomeness' (Weiss-Amer 1993: 7) and thus that correct nutrition is very important.

Even today, traditional medicine including rice has a role in Indian medicine. At the Gujarat Ayurved University, students and scholars study Ayurveda texts and conduct research on this topic (Kurup 2001: 14). Rice in India is used in extremely varied situations, for example as an ingredient to prepare medicaments which are used to cure inflamed surfaces, as *Pinda sweda* (a poultice with hot medicated balls of cooked rice) or *Annalepa* (smearing of cooked and medicated rice on the whole body) or to prevent skin infection (Chandra 2001: 150; Bhowmik 2012: 8–9). Rice is also used to prevent high blood pressure or (as in the ancient period in Mediterranean basin) for digestive disorders (Bhowmik 2012: 9). Different medicinal effects are ascribed to different rice varieties and it is believed they can cure different pathologies, for example *Aalcha* for dermatological uses, *Laicha* and *Maharaji* for pregnancy and safe delivery, *Baisoor* for epilepsy and also dermatological uses, *Karung kavuni* for cholesterol, elephantiasis and to treat dog bites, or *Gathuwanor* for rheumatism (Bhattacharjya and Roy 2012: 65; Mohanty et al. 2012: 547; Mythili et al. 2019: 2093–2094).

12.3. From Luxury Good to Staple Food

For a long time in the Mediterranean basin and in Europe, rice was a luxury and rare commodity, imported from Asia. Its two main uses in these lands were as a medicine (as described in the previous section) and as a 'special' ingredient to prepare expensive recipes.

The high cost of rice and its luxury status during the ancient period is testified by the *Edictum de Pretiis* (Table 12.1: No. 17), promulgated in 301 by the emperor Diocletian (244–313). If its price is compared with other goods in the same category, the price of rice is 200 denarii, while barley (at the same unit of measurement) is 60 denarii (*Edict. Diocl.*, 1, 2), wheat 100 denarii (*Edict. Diocl.*, 1, 1a), peas 60/100 denarii (*Edict. Diocl.*, 1, 13–14), lupins 4/60 denarii (*Edict. Diocl.*, 1, 19–20), and beans 100 denarii (*Edict. Diocl.*, 1, 21). In this list, rice is listed after pulses and not with other grains. Only worked spelt, sesame and cumin have the same price (Table 12.1: Nos. 18, 19 & 20). We are able to deduce that rice was very expensive during the Roman period also from Horace's passage mentioned previously, where the rich and mean Opimio is complaining about the price of the medicine he considers too expensive. Perhaps from this passage we could deduce another aspect: rice was a medicine in the ancient world, but probably only the rich could afford such luxury, as suggested by Tabacco (2019: 622) and Jagusiak et al. (2015: 128). Further support for rice as a luxury good comes from a *Historia Augusta* passage about Elagabalus (Table 12.1: No. 21). Here the text describes his expensive and eccentric dietary habits and how he fed his animals. The coupling of rice with pearls may suggest the former was considered expensive enough to be eaten in a course like this, a dish presented to an emperor known for his eccentric and profligate customs. Nevertheless, some scholars do not agree with this supposition, believing that in this passage rice is not characterised as a luxury good with only the pearls (and amber, gold, etc.) being regarded as luxurious and very expensive (as well as eccentric) dishes (Tabacco 2019: 628).

We find again rice used as a metaphor of opulence in Augustine of Hippo's treatise *De moribus ecclesiae catholicae et de moribus Manichaeorum* where the author criticises bitterly Mani and the religion he founded, Manichaeism. In one passage (Table 12.1: No. 22), Augustine compares (disapproving and criticising it) the typical vegetarian and expensive diet of the Manichaean Elects with a frugal and more simple diet.[3] Among the foods the Elects were fed there is rice, in this passage symbolic of a wasteful diet and, more generally, for a life practice that was '*perversius* ... *delirum*', that is to say 'perverted' and 'deluded' (AUG., *Manich.*, XIII, 30).

The use of rice in cooking is witnessed by some authors, for instance Apicius in his *De re coquinaria* (Table 12.1: No. 23), although always marginally, as in Apicius where rice is used to prepare sauce or as subsidiary ingredient.

The trends described above did not change in Italy during the Middle Ages. Perhaps Sicily could be an exception, if we follow Muslim authors and note that in 875 the Arab governor fixed rules for trading rice, which was also taxed (Ballarini 2014: 19–20; Vassallo 2015: 21). Nevertheless, archaeologists have not found archaeobotanical evidence of rice dated to this period and among scholars there is not unanimity as to whether rice was cultivated on Sicily from the ninth (Ferrero and Vidotto 2010: 349) or only from the eleventh century (Mariani 2013: 90; Fois and Poli 2019: 35). If we accept the hypothesis that rice was already cultivated by around the ninth century, it could be a possibility that rice cultivation failed and was re-introduced at a later time, perhaps around the sixteenth century, when its production increased. In my opinion there is no plausible reason to refute the conclusion that rice was grown on Sicily from the ninth century. The Arabs introduced rice cultivation in all their territory, including Spain where it was cultivated from the tenth to eleventh centuries (Watson 1997: 114; Gavinelli and Gilardi 2017: 423). It continued to be grown in Valencia after the Christian reconquest of the thirteenth century, especially in marshlands (Igual Luis 2016: 61–62, 67). In Sicily, furthermore, there are marshlands and other areas where rice could be cultivated without competition from other cereals, for instance in the southeast of the island (Polto 2012: 117). Thus, if the Arabs introduced rice to Spain, it is highly likely that they also cultivated it in Sicily, where the climate is the same and suitable areas existed for rice cultivation. However, rice cultivation always had a marginal role in Sicily; even during the nineteenth century when rice cultivation was more widespread, its production did not exceed about 7700 ha and today it has practically disappeared (Ferrero and Vidotto 2010: 349; Cormegna et al. 2019: 12).

Physical evidence of rice is extremely rare in Italy, with the first (controversial) archaeobotanical data from the Italian peninsula dated to the eighth to tenth centuries when archaeologists found a rice husk and a rice grain fragment from Classe in Emilia Romagna in stratigraphic layer 4052, the fill of layer 4078 (Augenti et al. 2006: 155, 157). However, if some scholars accept this chronology (Castiglioni and Rottoli 2013: 133), other do not, considering this sample as contamination (Augenti et al. 2006: 157). We have to wait until the twelfth century for

[3] On the Manichees' diet see, for example, Prosperi (2008). On ecclesiology, see Leurini (2004).

the first uncontroversial evidence of rice: ten rice chaffs or caryopses from Torre Civica, Pavia (Bernacchia 2018: 48). Nevertheless, in both cases scholars agree about their introduction to Italy from another land (Castiglioni and Rottoli 2013: 140; Corvino 2013/2014: 19).

We can apparently date the first cultivation of rice in Italy to the thirteenth century. We do not have archaeobotanical data, only bibliographical sources: a document where Cistercians from the Abbazia di Santa Maria di Lucedio, founded in 1123 in modern Vercelli province (Piemonte), declared it cultivated 1732 ha of rice (Comba 2009: 525; Ballarini 2014: 20). However, rice was still a luxury and marginal crop in Italy, used in connection with its role as a medicinal herb or to make cakes (Fois and Poli 2019: 35–36; Arcieri and Ghinassi 2020: 520). In Savoia's *Libro delle Spese* ('The Book of Expenses') there is a note to buy some hundred grams of rice from Vercelli's hospital in 1250 (Fabbri 2010: 54; Arcieri and Ghinassi 2020: 520)

In the Kingdom of Naples, rice was probably introduced by Spanish soldiers after its conquest by Alfonso V of Aragon (Ballarini 2014: 22). In the territory of the Republic of Florence, we have a report that Leonardo Colto de Colti established a paddy field near Pisa and asked permission to use water to grow rice in 1468 (Ballarini 2014: 22; Arcieri and Ghinassi 2020: 520–521). However, it was only between the fifteenth and sixteenth centuries that rice was fully adopted as a staple food, thanks to the Duke of Milan, Galeazzo Maria Sforza. He planned rice cultivation in the territory under his rule, encouraging the spread of the crop (Gavinelli and Gilardi 2017: 423). Moreover, he partially encouraged the spread of rice into other Italian states such as the Duchy of Ferrara when he sent three letters to Nicolò de Roberti, Ercole I d'Este's ambassador, with twelve bags of rice seeds, advising him to start rice cultivation in the marshlands (Bertazzini 2011–2014: 6).

Rice cultivation in Italy was regulated by the authorities and, for example, in the Duchy of Milan it was under state control from the beginning, with laws prohibiting the exportation of rice seeds out of the State. In the Republic of Venice, after a chaotic initial spread, the Senate later also regulated its cultivation (Ballarini 2014: 20–21). However, private initiative also had an important role in the spread of rice (Braudel 2010: 56).

12.4. Conclusions

As regards the question of where ancient Romans and Greeks learnt of the medical use of rice, although ancient authors do not give any information about this, it is likely that the knowledge came from India. Indeed, all writers agree with an Indian origin. If rice had a medical use in India for centuries before Roman merchants arrived with their ships, it is probable that knowledge of this use was transmitted from India to the ancient Mediterranean.

As regards the second question considered in this chapter, why the Romans didn't introduce rice cultivation in their empire and why rice only became widely grown and used as a staple food between the fifteenth and sixteenth centuries, the reasons are less clear. As mentioned, the ancient writers do not deal with this topic but we can nevertheless perhaps obtain some information from them. Once again, it is a physician who could enlighten us. Galen, in his *De alimentorum facultatibus* (Table 12.1: No. 24), suggests to use rice as a medicine (for dysentery), but he also assigns it a 'negative connotation', declaring that rice is harder to digest, less nutritious, and generally of lower quality in comparison with wheat. Earlier, Dioscuride had also declared that rice has a lower nutritional value (see section 2). Can, therefore, we see a cultural hindrance to the spread of rice in the Mediterranean basin during ancient times? It is possible that the ancient people would not have felt the need to grow a new cereal they considered to have a lower nutritional value. We could summarise this aspect in a duality even now effective: rice in the East and wheat in the West (Jasny 1950: 231). We have to remember that wheat was the base of the Roman diet (Dazzi *et al.* 2014: 155, 160; Gandolfo 2015: 35; Battistini et al. 2020: 2) and generally in the ancient Mediterranean basin (Bonacci 2013/2014: 4–6), with the provinces of Sicily (Table 12.1: No. 25), Africa and Egypt considered at different historical moments as Rome's granaries (DE ROMANIS 2003: pp. 691–692). Again, it was wheat that fed the *plebs* of Rome through the so-called *frumentationes* (as the *lex Sempronia frumentaria* passed from Caio Gracco as tribune) (Bonacci 2013/2014: 93–94). Perhaps the spread of rice in the Roman empire was hindered because cereal production was sufficient and thus another crop was not required. Another factor could have been the difficulty of obtaining rice seeds from India.[4]

By contrast, between the fifteenth and sixteenth centuries we can single out a particular and 'favourable' socio-economic conjunction to perhaps explain why the greatest obstacle to the spread of rice—its 'uselessness' or, better, the idea that rice cultivation was 'superfluous'—broke down. From the second half of the fifteenth century, there was a new population increase in Italy (although with different results and areas from north to south, especially when compared with the thirteenth century) and by the mid sixteenth century the population of Italy was comparable to that of the central centuries of the Middle Ages (Ginatempo and Sandri 1990: 97–99; Tognetti 2012: 761; Pinto 2014: 52–53). Between the fifteenth and sixteenth centuries, demographic pressure was again high. To cite only a few cases, if in Venice in the middle of the sixteenth century there lived around 150,000 inhabitants—showing that Venice had recovered and exceeded the population it had before the Black Death (around 100,000) (Ginatempo and Sandri 1990: 83; Pinto 2014: 39, 53)—in Milan in the same period there were about 100,000 people, evidence that it hadn't recovered the same population

[4] Rice was grown also in Mesopotamia during the Roman period (Nesbitt et al. 2010: 322, 325–326), but Mesopotamia was ruled by the Parthians and then the Sasanians, both enemies of Rome and Romans could not introduce rice from there.

it had at the beginning of the fourteenth century when Milan had more than 100,000 and perhaps around 150,000 inhabitants (Pinto 2014: 39; Loi 2021: 218). Before the Black Death, there were around 50,000 people in Padua. The city's population dropped to 32,000 after the Black Death, then to 18,000 following pestilence in 1405 and 1411 (Ginatempo and Sandri 1990: 81; Pinto 2014: 41). In the thirteenth and fourteenth centuries, between 20,000 and 40,000 people had lived in Naples, which became one of the biggest cities in Europe and the biggest in Italy in the mid-sixteenth century with about 200,000 inhabitants (Tognetti 2012: 762; Pinto 2014: 41, 53). This population growth was, therefore, considerable, especially in the cities (Braudel 2010: 56; Alfani 2016: 25, 37). This generated difficulties to feed the growing population and there were frequent famines in this period in Italy (Tognetti 1995: 269, 284; Alfani 2016: 29–30; Alfani et al. 2017: 27).[5] To feed the population and consequently avoid famines, different governments conducted tillage to improve the soil, cut down forests, and carried out reclamation works (e.g., Castagnetti 1977: 71–72; Andreolli 2002: 138–139; Bevilacqua 2012: 10; Avanzini and Salvador 2015: 56).

In particular, we have to focus on the marshlands located in previous centuries in Italy from north to south, for example the marshlands along the Adige[6], in the Val di Chiana (e.g., Marrocchi 2017), the Paludi Pontine or in southeastern Sicily (Polto 2012: 117). The marshlands and swamps were not barren lands (anything but that) and they were used for centuries by the people who lived there for their subsistence, to fish, hunt and to collect plant resources (e.g., Cherubini 2000: 45; Bevilacqua 2012: 111; Galetti 2012/2013: 100; Marrocchi 2017: 76; Gardoni 2021: 123), even though the populations of those areas may not have been very high (see e.g., Niccoli 2013–2014: 4). However, the widespread population growth of the late medieval era led to the search for new lands for farming. As a result, different states from the north to south of Italy drained (or attempted to drain) marshlands and swamps (Bevilacqua 2012: 18). Nevertheless, draining these areas was not easy: political, economic and technological factors led to failure in many cases. As Braudel (2010: 54) points out, the technological and economic difficulties involved with these works required large amounts of capital and resulted in failure in many cases. Citing a source, Braudel (2010: 55) writes that in 1534 speakers from Brescia in front of the Senate of Venice complained that many citizens went bankrupt when they attempted to drain marshlands and swamps. Another, later example was the reclamation work planned by Duke Bonifacio Caetani in the zone between Sermoneta e Sezze (in the Paludi Pontine) around 1565 that failed because it was too expensive (Gallinelli 2019/2020: 117–118; Bevilacqua 2012: 122). Generally, only centralised authorities (Braudel 2010: 55) or larger groups of entrepreneurs (Cazzola 1991: 18; Bevilacqua 2012: 19) possessed the huge amounts of capital required to drain these wetlands. In sum, we could assume a connection between the spread of rice and this socio-economic context: rice could have been a 'new method' to feed a rapidly growing population to use all the lands available, including marshlands and swamps where only rice, among the various cereals, can grow. Therefore, the expensive reclamation works and works to preserve the soil from flooding would become superfluous.

In this context, how could we interpret the introduction of rice by the Cistercians to the Abbazia di Santa Maria di Lucedio's lands? Indeed, the historical context is not so different when compared with the situation in the fifteenth and sixteenth centuries. In the mid thirteenth century (i.e., when the rice introduction to the abbey is dated – see section 3), Italy (and Europe in general) came from two centuries of economic and population development. But during the thirteenth century, this economic development (and the connected general social optimism) began to enter into crisis. Food began to be scarce and population growth brought about famines (Palermo 2013: 57–58; Magni 2015: 103). Famine struck Florence around 1280 (Magni 2015: 103), Italy between 1328–1330 (Alfani 2017: 383), England in 1293–1295, and Europe as a whole from 1315–1318 and again from 1321–1322 (Mueller 2008: 560). Famines were caused not only by population growth, but also by climatic change and the difficulty to supply the people (Palermo 2013: 63; see also Mueller 2008: 560 and bibliography). But why was there this crisis after a period of high economic development that had lasted two centuries? We can find one answer in agricultural technology. In western Europe in the early Middle Ages, after the Western Roman Empire collapsed the landscape changed, with forests and marshlands reoccupying lands that had been cultivated during the Roman period. Fields, forests and marshlands became waste lands: '*ovunque si intersecavano integrandosi*' (Galetti 2012/2013: 97; Ciancioso 2008: 155; Wickham 2008: 211). When the 'boom' began around the tenth century (Petralia 2008: 291), people first needed only to reoccupy waste lands, but after two centuries these had almost disappeared in some areas, as in the Mantova plain where, in 1223, ecclesiastics declared (with amazement) that the waste lands they owned in Casaletto near the Mincio river were cultivated (Gardoni 2021: 106). In the 'second stage', people started to reclaim wetlands and to deforest in a process we could call a 'war on the uncultivated'; especially from the twelfth century, this process spread everywhere in Italy (Canzian 2014: 231–232; Cianciosi 2008: 103, 107; Rao 2013: 211). For instance, in the territory under the ruler of Verona and Mantova, along the Allegrezza channel the forests were cut off from ecclesiastics (Verona) or private citizens (Mantova) (Gardoni 2021: 118). However, we shouldn't think that wastelands—here meaning wetlands, marshlands and forests—disappeared everywhere. These

[5] These famines increased in severity, especially during the second half of the sixteenth century. In Naples alone in this period, there were six famines (1560, 1565, 1570, 1584, 1585 and 1591) (Braudel 2010: 630) and the last one was the most serious everywhere in Italy (Trubbiani 2007: 312; Alfani 2009: 63–64).

[6] These marshlands were reclaimed between the thirteenth and fifteenth centuries by town authorities, bishops, monasteries or lords (Franceschini 2016: 261–262, 267, 269).

lands could not be adapted to the cereals grown, or they were cultivated only for a short period and then once more become waste lands, as in the Po Valley (Gardoni 2021: 106–107) or the Paludi Pontine that were reclaimed only during the Fascist era, or they were protected as forests (Bertoni 2013: 92–93; Rao 2013: 217–218, 220).

Nevertheless, during the thirteenth century, the recovery of new areas of cultivation became a hoary problem for the different communes and governments and they started to explore and use unproductive lands, such as those in mountainous terrain (Montanari 2002: 75–76). Cuneo, in modern Piemonte, reached its peak in population in the mid thirteenth century, and to feed its population the rulers had already organised the occupation of waste lands from the eleventh century. During the fourteenth century, however, because of the need for new fields, unproductive mountainous terrain was also deforested and used for cultivation (Bertoni 2013: 89, 91, 95–96).

During the Middle Ages there was not a real improvement in technological knowledge regarding agriculture, neither in terms of farming tools, nor in terms of improved cultivation. Fields were fertilised using manure, though not always, and crop rotation was applied, above all the two-field system (half fallow and half planted). Only from the end of the fifteenth century did agronomists suggest use of the three-field system, though for a long period this fell on deaf ears. For agricultural tools, compared to the Early Middle Ages, steel was more widespread and asymmetric mouldboard ploughing and wheeled ploughing were introduced but not widespread (Montanari 2002: 67–69). Thus, population growth was 'supported' during the Middle Ages (from around the tenth century) more by the wide availability of uncultivated fields than by improvements in technological knowledge. When in the thirteenth century all available lands were already cultivated, therefore, and there was no opportunity to further expand agricultural output, famines increased in frequency and severity. In this socio-economic context, the introduction of rice by the Cistercians could be interpreted as an experiment to feed the overcrowded cities and countryside, as happened between the fifteenth and sixteenth centuries. However, this first test might have failed because of the Black Death: demographic pressure decreased and this 'experiment' could have been abandoned, only to be revisited in the fifteenth and sixteenth centuries.

Table 12.1. Original quotations from ancient authors.

No. 1 Dsc., II, 95	*Ὄρυζα…Τρόφιμος δέ ἐστι μετρίως καὶ κοιλίας σταλτική.* – 'Rice… is moderately nutritious and it binds the bowel.'
No. 2 Hor., *Sat.*, II, 3, 142–157	*…Opimius…quondam lethargo grandi est oppressus…Hunc medicus multum celer…addit et illud:… 'Deficient inopem venae te, ni cibus atque ingens accedit stimacho fultura ruenti. Tu cessas? Agendum, sume hoc tisanarium oryzae'. 'Quanti emptae?' 'Parvo'. 'Quanti ergo' 'Octossibus'. 'Eheu quid refert, morbo an furtis pereamque rapinis?'.* – '…Opimio…un giorno fu oppresso da una catalessi pesante…Un medico molto sollecito…aggiunge questo:…'Sventato tu resterai, poveraccio, se il cibo buono e un sicuro ristoro non accederanno allo stomaco che ormai crolla. Indugi? E dai, prendi questo decotto di riso'. 'Quanto costa?' 'Un nulla'. 'Ebbene, quanto?' 'Otto assi'. 'Oddio! Che importa se per malattia o per furti e rapine io muoio?'.
No. 3 Cels., IV, 14	*…et in primis ad pulmonem veniendum, ex quo vehemens et acutus morbus oritur, quem περιπνευμονικόν Graeci vocant…Quod ad cibus vero pertinet, huic nec salsis opus est, necque acribus, neque amaris, neque alvum adstringentibus, sed paulo lenioribus. Ergo primis diebus danda est sorbitio ptisanae, vel alicae, vel oryzae…* – '…ed in primo luogo ai polmoni nei quali può svilupparsi una violenta ed acuta malattia che i Greci chiamano peripneumonia…In quanto all'alimentazione, per questi malati non sono adatti i cibi salati o acri o amari o astringenti, ma al contrario lo sono quelli piuttosto delicati. Quindi nei primi giorni bisogna dare decotti d'orzo o di alica o di riso…'.
No. 4 Plin., *NH*, XXVIII, 108–110	*…crocodileam…lentigines tollit ac varos maculasque omnes. Et contra comitiales morbos bibtur ex aceto mulso binis obolis. Adoposita menses ciet…Adulterant amylo aut Cimolia, sed maxime sturnorum fimo qui captos oryza tantum pascunt.* – '…sterco di coccodrillo…fa sparire le lentiggini, i brufoli e tutte le chiazze. Inoltre contro l'epilessia si beve l'ossimele, alla dose di due oboli. Applicata in pessari provaca le mestruazioni…Viene sofisticata con l'amido, con la creta di Cimolo, ma soprattutto con lo sterco degli storni alimentati in cattività unicamente col riso.'.
No. 5 Marcell. med., X, 17–18	*Grauidines narium sedabuntur, quae ex sole aut perfrictione generantur…Sorbitiunculae quoque prosunt, quae fiunt ex amulo uel tisana uel oryza.* – 'Stockschnupfen, der infolge von Sonneneinstrahlung oder Erkältung entsteht, wird dadurch zum Abklingen gebracht, daß man… Auch Suppen nützen, die aus Stärke, Gerstenschleim oder Reis zubereitet werden.'.
No. 6 Marcell. med., XIV, 42	*Sorbitionibus uti oportebit eos, qui de uua et faucibus laborant, factis ex amulo, tisana, oryza, alica, farina triticea, faba alba, urtica tenera cum lacte decocta…* – 'Diejenigen, die am Zäpfchen und im Rachen Schmerzen haben, müssen Suppen zu sich nehmen, die aus Stärke, Gerstengrütze, Reis, Speltgraupen, Weizenmehl, weißen Bohnen und zarter Brennessel, mit Milch gekocht, zubereitet worden sind…'.

No. 7 Marcell. med., XIX, 8	*Ad inpetigines detergendas lapathi radices cum aceto coques ad tertias et in eo iure iterum oryzam discoques et sic ad perducendum corpus uteris.* – 'Um Hautausschläge zu beseitigen, läßt man Sauerampferwurzel mit Essig auf ein Drittel einkochen, kocht in dieser Brühe wiederum Reis und verwendet (es) so, um den Körper zu bestreichen.'.
No. 8 Marcell. med., XX, 25–26	*Prosunt etiam quaedam in aqua macerata stomacho...bene facit etiam sucus malorum cydoneorum per se cocliario sumptus uel cum alfita aut alica aut trago aut oryza...Solutum autem stomachum et omnia reicientem initio quidem curari oportet...Cibos uero dari oportet huiusmodi:...item lenticulam cum beta uel oryza incoctis illic malis Cydoneis...* – 'Dem Magen nützen auch manche Ingredienzien, wenn sie in Wasser eingeweicht worden sind...gut wirkt auch Quittensaft, für sich mit einem Löffel eingenommen oder zusammen mit Schrotmehl, Speltgraupen, Weizengrütze oder Reis...Der magen aber, in dem die Zersetzung zu stark ist und der alle Speisen wieder von sich gibt, muß zu Beginn... Es müssen aber Speisen folgender Art gereicht werden:...und einem anderen derselben Art, ebenso Linsen mit Mangold oder Reis, worin Quitten...'.
No. 9 Strab., XVII, 1, 45	*Ἐντεῦθέν ἐστιν ἰσθμὸς εἰς τὴν Ἐρυθρὰν κατὰ πόλιν Βερενίκην, ἀλίμενον μέν, τῇδ'εὐκαιρίᾳ τοῦ ἰσθμοῦ καταγωγὰς ἐπιτηδείους ἔχουσαν.* – 'Da lì un istmo mette in collegamento col Mar Rosso all'altezza della città di Berenice, che non ha porto, ma tuttavia dispone di punti di approdo che, considerando il vantaggio di poter poi utilizzare l'istmo, ben si prestano a tal funzione'. *Οὐκ ἄπωθεν δὲ τῆς Βερενίκηςἐστὶ Μυὸς ὅρμος, πόλις ἔχουσα τὸ ναύσταθμον τῶν πλοιζομένων...* – 'Non lontano da Berenice si trova Myos Hormos, una città in cui sbarcano quanti risalgono quel tratto di mare...'.
No. 10 Pme., 1	*Τῶν ἀποδεδειγμένων ὅρμων τῆς Ἐρυθρᾶς θαλάσσης καὶ τῶν περὶ αὐτὴν ἐμπορίωνπρῶτός ἐστιν λιμὴν τῆς Αἰγύπτου Μυὸς ὅρμος, μετὰ δὲ αὐτὸν εἰσπλεόντων ἀπὸχιλίων ὀκτακοσίων σταδίων ἐν δεξιᾷ ἡ Βερνίκη· ἀμφοτέρων <δὲ> οἱ λιμένες ἐν τῷἐσχάτῳ τῆς Αἰγύπτου κόλποι [δὲ] τῆς Ἐρυθρᾶς θαλάσσης κεῖνται.* 'Of the designated harbors of the Erythraen Sea and the ports of trade on it, first comes Egypt's port of Myos Hormos and, beyond it, after a sail of 1800 stades to the right, Berenicê. The ports of both are bays of the Red Sea on the edge Egypt.'.
No. 11 Strab., XV, 1, 13	*Ἰνδική...ἐν μέν οὖν τούτοις τοῖς ὄμβροις λίνον σπείρεται καὶ κέγχρος, πρὸς τούτοις σήσαμον, ὄρυζα, βόσμορον...* – 'L'India...Durante le piogge, comunque, si seminano il lino e il miglio, e oltre a questi il sesamo, il riso e il bosmoro...'.
No. 12 D.S., II, 36, 3	*...φύεται κατὰ τὴν Ἰνδικὴν πολλὴ...ὄρυζα...* – '...cresce in India molto...riso...'.
No. 13 Plin., *NH*, XV, 29	*Indi e castaneis ac sesima atque oryza facere dicuntur...* – 'Si dice che gli Indiani ricavino olio dalle castagne, dal sesamo e dal riso...'.
No. 14 Plin., *NH*, XVIII, 71	*Primum ex omnibus frumentis seritur hordeum...Hordeum Indis sativum et silvestre, ex quo panis apud eos praecipuus et alica. Maxume quidem oryza gaudent, ex qua tisanam conficiunt, quam reliqui mortales ex hordeo. Oryzae folia carnosa, porro similia, sed latiora, altitudo cubitalis, flos purpureus, radix geminae rotunditatis.* – 'Di tutti i cereali l'orzo è quello che si semina per primo... In India c'è un orzo coltivato ed uno selvatico, da cui si ottengono un pane ed una farinata ottimi. Ma il loro cereale preferito è l'oriza, con cui fanno quella tisana che tutti gli altri uomini fanno con l'orzo. L'oriza ha foglie carnose del tutto simili a quelle del porro, ma più larghe, è alta un cubito, ha il fiore color porpora e la radice con una doppia rotondità.'
No. 15 Pme., 31	*Συνεχρήσαντο δὲ αὐτῇ καὶ ἀπὸ Μούζα τινὲς...ἐπιβάλλοντες ὄρυζά τε καὶ σῖτον καὶ ὀθόνιον Ἰνδικὸν...* – 'Trade with it used to be carried on by some of the shippers from Muza...would exchange rice, grain, cotton cloth...'.
No. 16 *Sushruta Samhita*, II, *Śárira Stháma*, X	'During the first three months of pregnancy...Several medical authorities recommend a food made of Shashtika rice with milk, to be given to her specially in the jhird month of gestation, with curd in the fourth, with milk in the fifth and with clarified butter in the sixth month of pregnancy.'.
No. 17 *Edict. Diocl.*, 1, 23	*[O]ri<z>ae mundae k(astrensem) mo(dium) (unum) ducentis* – 'Riso brillato 1 moggio militare den. 200'.
No. 18 *Edict. Diocl.*, 1, 25	*[A]lice mundae k(astrensem) mo(dium) (unum) ducentis* – 'Spelta brillata 1 moggio militare den. 200'.
No. 19 *Edict. Diocl.*, 1, 26	*[S]esami k(astrensesem) mo(dium) unum ducentis* – 'Sesamo 1 moggio militare den. 200'.
No. 20 *Edict. Diocl.*, 1, 32	*Cymini mundi k(astrensem) mo(dium) unum ducentis* – 'Cumini puliti 1 moggio militare den. 200'.
No. 21 *Elag.*, XXI, 1–3	*Canes iecinoribus anserum pavit...Misit et uvas Apamenas in praesepia equis suis et psittacis atque phasianis leones pavit...Exhibuit et sumina apruna per dies decem tricena cottidie cum suis vulvis, pisum cum aureis, lentem cum cerauniis, fabam cum electris, orizam cum albis exhibens.* – 'He fed his dogs on goose-livers...He sent grapes from Apamea to his stables for his horses, and he fed parrots and pheasants to his lions...For ten successive days, moreover, he served wild sows' udders with the matrices, at the rate of thirty a day, serving, besides, peas with gold-pieces, lentils with onyx, beans with amber, and rice with pearls.'

(*Continued*)

No. 22 Aug., *Manich.*, II, XIII, 30	*Quid porro insanius dici aut cogitari potest, hominem boletos, orizam, tubera, placentas, caroenum, piper, laser, distento ventre cum gratulatione ructantem, et quotidie talia requirentem, non inveniri quemadmodum a tribus signaculis, id est a regula sanctitatis excidisse videatur; alium vero fruges vilissimas fumoso obsonio condientem, tantumque hinc assumentem, quantum refectioni corporis sufficit, et tres cyathos vini sustentandae valetudinis gratia sorbentem, et ab illo victu ad istum transeuntem, certo supplicio praeparari?* – 'What could be more insane than to think that a man who is belching with contentment, his stomach loaded with choice mushrooms, rice, truffles, cake, sweet beverages, pepper and rare herbs, and who feasts in this manner daily, cannot be accused in some way of violating the three symbols, that is, the rule of sanctity, yet another man who seasons his dish of common vegetables with a bit of smoked meat, taking only enough of this poor fare to sustain his body, and who drinks three small portions of wine in order to preserve his health, would, for giving up the first diet in favor of this one, be inviting inescapable punishment?'.
No. 23 Apic., II, 2	*Piper teres pridie infusum, cui subinde liquamen...cui defritum admisces quod fir de cotoniis... Quod si non fuerit, vel caricarum defritum mittes quod Romani colorem vocant, ac deineps amulum infusum adicies vel oryzae sucum...* – 'Trita del pepe messo ad ammollare il giorno prima; ad esso aggiungerai la salsa...a questo unirai dello sciroppo di cotogne...Se non l'hai, adopera sciroppo di fichi che i Romani chiamano 'colore', e prendi poi infuso d'amido e succo di riso...'. *Ossucla de pullis exbrommas. Deinde mittis in caccaburn porros, anetum, salem. Cum cocta fuerint, addes piper, apii semen, deinde oridiam infusam, teres...* – 'Spremi il succo da ossicini di pollo e metti in un tegame dei porri, dell'aneto e del sale. Quando saranno cotti, aggiungi pepe, seme di sedano e riso ammollato e ben tritato...'.
No. 24 Galen., I, XVII	*Semine hoc omnes in fiftenda aluo vtuntur, ipsum eodem modo, quo alicam, coquentes; difficilius tamen quam alica conficitur, et minus alit, ceu vtique et in edendi suauitate ab ea quam plurimum relinquitur.*[7] – 'Todos dizem usar-se este grão para adstringir os intestinos, fazendo uma fervura como se faz com a sêmola. Porém, este é menos digestivo e menos nutritivo que a sêmola; e, da mesma maneira, fica muito atrás da sêmola no que toca ao (sabor) aprazível.'.
No. 25 Cic., *De imperio Cn. Pompei*, 34	*Qui nondum tempestivo ad navigandum mari Siciliam adiit, Africam exploravit, in Sardiniam cum classe venit atque haec tria frumentaria subsidia rei pubhcae firmissimis praesidiis classibusque munivit.* – 'He landed in Sicily on a sea not yet seasonable for sailing, reconnoitred Africa, came to Sardinia with his fleet, and safeguarded those three suppliers of the commonwealth's corn with the toughest garrisons and fleets.'

References

ANCIENT AUTHORS

Apicius, *De re coquinaria*, II, 2. Translated by F. Introna (2018) Ariccia: RL S.p.A.

Augustine, *De moribus ecclesiae catholicae et de moribus Manichaeorum*, II, XIII, 30. Latin text edited by J. B. Bauer (1992) Vienna: Academiae Scientiarum Austriacae; translated by D. A. Gallagher and I. J. Gallagher (1966) Washington: The Catholic University of America Press.

Celsus, *De Medicina*, IV, 14. Translated by R. M. Ciulla (1990) Palermo: Sellerio Editore.

Cicero, *Oratio de imperio Cn. Pompei (pro lege Manilia)*, 34. Latin text edited by J. C. Nicol (1899) Cambridge University Press; translated by I. Gildenhard and L. Hodgson (2014). Cambridge: Open Book Publishers.

Dioscorides, *De Materia Medica*, II, 95. Greek text edited by M. Wellman (1958) Berlin: Berloni apud Weidmanmos; translated by L. Y. Beck (2005) Hildesheim: Olms-Weidmann.

Diodorus Siculus, *Bibliotheca Historica*, II, 36, 3. Translated by G. Cordiano and M. Zorat (2004) Milano: Biblioteca Universale Rizzoli.

Edictum Diocletiani et Collegarum de Pretiis Rerum Venalium in Integrum Fere Restitutum e Latinis Graecisque Fragmentis. Translated by M. Giacchera (1974) Genova: Istituto di Storia Antica e Scienze Ausiliarie.

Galen, *De Alimentorum Facultatibus*, I, XVII. Latin text edited by Martino Gregorio interprete (1554) Lyon: Philibertus Rolletius & Bartholomaeus Fraenus; translated by N. S. H. Ferreira (2020) Coimbra: Coimbra University Press.

Historia Augusta. Translated by D. Magie (1980) Cambridge MA: Harvard University Press.

Horace, *Satire*, II, 3, 142–157. Translated by L. De Vecchi (1981) Roma: Carocci Editore.

Marcellus Empiricus, *De medicamentis liber*, X, 17–18. Translated by M. Niedermann, E. Liechtenhan, J. Kollesch and N. Diethar (1968) Berlin: Akademie-Verlag.

Marcellus Empiricus, *De medicamentis liber*, XIV, 42. Translated by M. Niedermann, E. Liechtenhan, J. Kollesch and N. Diethar (1968) Berlin: Akademie-Verlag.

[7] But in a different Latin text version we have: *Semine hoc omnes in fiftenda alvo utuntur, instituentes similem ipsius ut alicae concoctionem. Difficilius quam alica, alica in ventriculo conficitur, et minus alit: ut et in edendi suavitate ab ea quam plurimum* relinquitur. (Galen, I, XVII, Latin text Apud Asingam de Fries, edited by Martini Gregoii, Lucduni Batavorum, Leiden , 1633).

Marcellus Empiricus, *De medicamentis liber*, XIX, 8. Translated by M. Niedermann, E. Liechtenhan, J. Kollesch and N. Diethar (1968) Berlin: Akademie-Verlag.

Marcellus Empiricus, *De medicamentis liber*, XX, 25–26. Translated by M. Niedermann, E. Liechtenhan, J. Kollesch and N. Diethar (1968) Berlin: Akademie-Verlag.

Periplus Maris Erythraei. Translated by L. Casson (1989) Princeton University Press.

Pliny, *Naturalis Historia*, XV, 29. Translated by A. Aragosti (1984) Torino: Giulio Einaudi Editore.

Pliny, *Naturalis Historia*, XVIII, 71. Translated by F. E. Consolino, (1984) Torino: Giulio Einaudi Editore.

Pliny, *Naturalis Historia*, XXVIII, 108–110. Translated by U. Capitani and I. Garofalo (1986) Torino: Giulio Einaudi Editore.

Strabo, *Geografia*, XV, 1, 13. Translated by N. Biffi (2005) Bari: Edizioni del Sud.

Strabo, *Geografia*, XVII, 1, 45. Translated by N. Biffi (2005) Bari: Edizioni del Sud.

Sushruta Samhita, II, Śárira *Sthána*, X. Translated by K. K. L. Bhishagratna (1911) Calcutta: Bharat Mihir Press.

MODERN AUTHORS

Alfani, Guido (2009) Crisi demografiche, politiche di popolazione e mortalità differenziale (ca. 1400–1630). *Popolazione e Storia. La Popolazione all'Alba dell'Era Moderna* 10(1): 57–76.

Alfani, Guido (2016) La popolazione dell'Italia settentrionale nel XV e XVI secolo: scenari regionali e macro-regionali. In Guido Alfani, Angela Carbone, Beatrice Del Bo and Riccardo Rao (eds.), *La Popolazione Italiana del Quattrocento e Cinquecento*, pp. 19–40. Udine: Forum.

Alfani, Guido (2017) Crisi demografiche e crisi economiche nell'Italia preindustriale (ca. 1300–1800). *Storia Economica* XX(2): 377–394.

Alfani, Guido, Luca Mocarelli and Donatella Strangio (2017) Italy. In Guido Alfani and Cormac Ó Gráda (eds.), *Famine in European History*, pp. 25–47. Cambridge University Press.

Andreolli, Bruno (2002) L'uso dei boschi e degli incolti. In Giuliano Pinto, Carlo Poni and Ugo Tucci (eds.), *Storia dell'Agricoltura Italiana. Il Medioevo e l'Età Moderna*, pp. 123–144. Firenze: Edizioni Polistampa.

Arcieri, Marco and Graziano Ghinassi (2020) Rice cultivation in Italy under the threat of climatic change: trends, technologies and research gaps. *Irrigation and Drainage* 69(4): 517–530.

Augenti, Andrea, Mila Bondi, Marialetizia Carra, Enrico Cirelli, Cecilia Malaguti and Maddalena Rizzi (2006) Indagini archeologiche a Classe (Scavi 2004): primi risultati sulle fasi di età altomedievale e dati archeobotanici. In Riccardo Francovich and Marco Valenti (eds), *IV Congresso di Archeologia Medievale: Scriptorium dell'Abbazia, Abbazia di San Galgano (Chiusdino-Siena) 26–30 settembre 2006*, pp. 153–161. Borgo San Lorenzo (FI): All'Insegna del Giglio.

Avanzini, Marco and Isabella Salvador (2015) I boschi delle Valli del Leno (Trentino meridionale): evoluzione storica del rapporto tra ambiente naturale ed attività antropica. *Dendronatura* 1: 55–72.

Ballarini, Giovanni (2014) Introduzione. In Giovanni Ballarini and Silvia De Lorenzo (eds.), *Itinerari di Cultura Gastronomica. La Cucina del Riso*, pp. 10–54. Milano: JDT.

Bates, Jennifer, Cameron A. Petrie and Ravindra N. Singh (2017) Approaching rice domestication in South Asia: new evidence from Indus settlements in Northern India. *Journal of Archaeological Science* 78: 193–201.

Battistini, Andrea, L. Bracanzi, P. Catalano, F. De Angelis, S. Di Giannantonio, G. Facchin, V. Gazzaniga, C. Lubritto, C. Martínez-Labarga, P. Ricci, O. Rickards, R. Santangeli-Valenzani and S. Varano (2020) Food at the heart of the empire: dietary reconstruction for imperial Rome inhabitants. *Archaeological and Anthropological Sciences* 12(10): 1–21.

Bernacchia, Roberto (2018) Il riso in occidente in età antica e medievale. In Centro UPM di Rovasenda (ed.), *Atti del XV Convegno Il Riso: Alimento Fondamentale per la Salute Umana*, pp. 47–51.

Bertoni, Laura (2013) Le attività economiche a Cuneo: metà XIII-metà XIV secolo. *Bollettino della Società per gli Studi Storici, Archeologici ed Artistici della Provincia di Cuneo* 148: 87–100.

Bevilacqua, Irene (2012) *I papi e le acque. Politiche del territorio nelle paludi pontine (XVI–XVIII)*. Università di Pisa: PhD Thesis.

Bhattacharjya, Sudeshna and Roy Shyamashree (2012) Medicinal rice-an ignored episode. *Advance Research Journal of Crop Improvement* 3(1): 64–66.

Bhowmik, Debjit, R. Pushpa, K. P. Sampathkumar and Manivannan Umadevi (2012) Rice-traditional medicinal plant in India. *Journal of Pharmacognosy and Phytochemistry* 1(1): 6–12.

Bisen, Jaiprakash, Nitiprasad N. Jambhulkar, Bipin B. Panda, Himanshu Pathak and Rahul Tripathi (2020) Eco-regional-based rice farming for enhancing productivity, profitability and sustainability. *NRRI (National Rice Research Institute) Research Bulletin* 22.

Blum, Winfried. E. H., Carmelo Dazzi and Bruce R. James (2014) Bread and soil in ancient Rome. A vision of abundance and an ideal of order based on wheat, grapes, and olives. In G.J. Churchman and Edward R. Landa (eds.), *The Soil Underfoot. Infinite Possibilities for a Finite Resource*, pp. 153–173. Boca Raton: Taylor & Francis.

Bocianowski, Jan, Mariet Jose, S L Kurmari, Remya Madhu, B. C. Patra, R. Dinesh Raj, J. C. Rana, O. N. Singh, George Thomas, George Varghese, M. R. Vinitha and Rashmi Yadav (2018) The prehistoric Indian Ayurvedic rice Shashtika is an extant early domesticate with a distinct selection history. *Frontiers in Plant Science* 9: 1–10.

Bonacci, Lucia (2013–2014) *La politica granaria di Roma imperiale. La Gallia come modello metodologico.* Università della Calabria: PhD Thesis.

Braudel, Fernand (2010) *Civiltà e Imperi del Mediterraneo nell'Età di Filippo II*, vol. I. Translated by Carlo Pischedda. Torino: Mondadori Printing S.p.a.

Canzian, Dario (2014) Acque, insediamenti e attività umane nella marca veronese-trevigiana (secoli X–XIV). In Enrico Lusso (eds.), *Attività Economiche e Sviluppi Insediativi nell'Italia dei Secoli XI–XV. Omaggio a Giuliano Pinto*, pp. 215–244. Cherasco: CISIM - Centro Internazionale di Studi sugli Insediamenti Medievali.

Cappers, René T. J. (2006) *Roman Foodprints at Berenike: Archaeobotanical Evidence of Subsistence and Trade in the Eastern Desert of Egypt*. Los Angeles: Cotsen Institute of Archaeology.

Castagnetti, Andrea (1977) La pianura veronese nel Medioevo. In Giorgio Borelli (eds.), *Una Città e il suo Fiume. Verona e l'Adige*, vol. 2, pp. 33–138. Verona: Banca Popolare di Verona.

Castiglioni, Elisabetta and Mauro Rottoli (2013) Broomcorn millet, foxtail millet and sorghum in North Italian early medieval sites. *European Journal of Post-Classical Archaeologies* 3: 131–144.

Cazzola, Franco (1991) L'espansione dell'area coltivata nell'Italia centrosettentrionale tra XV e XVII secolo. Qualche linea interpretativa. *Proposte e Ricerche* 27: 13–25.

Chandra, Shailaja (2001) Role of traditional systems of medicine in national health care systems. In Ranjit Roy Chaudhury and Uton Muchtar Rafei (eds), *Traditional Medicine in Asia*, pp. 135–157. New Delhi.

Cherubini, Giovanni (2000) L'approvvigionamento alimentare delle città toscane tra il XII e il XV secolo. *Rivista di Storia dell'Agricoltura* XL(1): 33–52.

Choudhary, Ruplal, Chanda Dhumal, DH Lohith Kumar, Shubham Subrot Panigrahi, Preetam Sarkar (2015) Traditional and ayurvedic foods of Indian origin. *Journal of Ethnic Foods* 2: 97–109.

Cianciosi Alessandra (2008) *L'insediamento medievale tra storia e archeologia: dal saltopiano al vicariato di Gallieria (IX–XIV Secolo)*. Università di Bologna: PhD Thesis.

Cohen, David J. (2011) The beginnings of agriculture in China. A multiregional view. *Current Anthropology* 52 (Supplement 4): 273–293.

Cormegna, M., F Sciorati and C. Simonelli (2019) Storia del riso: un cereale millenario tra i diversi continenti. *La Rivista di Scienza dell'Alimentazione* 3: 7–26.

Corvino, Cinzia (2013–2014) *Ricerche Archeobotaniche per la Ricostruzione dei Paesaggi Vegetali e dell'Uso delle Risorse Naturali e Agricole della Capitanata Medievale: Analisi di Resti Antracologici e Carpologici da Scavi nella Puglia Settentrionale.* Università di Foggia: PhD Thesis.

Cox, Alison, Jacob Morales and Marijke Van Der Veen (2011) Spices. Culinary and medicinal commodities. *Journal of African Archaeology Monograph: Consumption, Trade and Innovation. Exploring the Botanical Remains from the Roman and Islamic Ports at Quseir al-Qadim, Egypt* 6: 39–74.

De Romanis, Federico (2003) Per una storia del tributo granario africano all'annona della Roma imperiale. In Catherine Virlouvet, Brigitte Marin (eds), *Nourrir les Cités de Méditerranée. Antiquité-Temps Modernes*, pp. 691–738, Parigi: Maisonneuve & Larose.

Fabbri, Anna (2010) *Alimentazione in Area Estense tra '400 e '500*, Finale Emilia.

Ferrero, Aldo, Francesco Vidotto (2010) History of rice in Europe. In S. D. Sharma (eds), *Rice: Origin, Antiquity and History*, pp. 345–375. Enfield: Taylor & Francis.

Fois, Luca and Eleonora Poli (2019) Il riso. In Folco Vaglienti (eds), *Abbiate Cura di Ciò Che Vi è Stato Dato. Storia e Storie del Patrimonio della Ca' Granda*, pp. 35–39, Milano: Gamedit.

Franceschini, Italo (2016) Le paludi dell'Adige. Diritti di sfruttamento e tentativi di bonifica tra XIII e XV secolo. In Vito Rovigo (eds), *Il Fiume, le Terre, l'Immaginario. L'Adige come Fenomeno Storiografico Complesso, Atti del Convegno Rovereto, 21–22 febbraio 2013*, pp. 251–272. Rovereto: Edizioni Osiride.

Fuller, Dorian Q., E. L. Harvey, Basanta Mohanta and Rabindra K. Mohanty (2006) Early agriculture in Orissa: some archaeobotanical results and field observations on the Neolithic. *Man and Environment* XXXI (2): 21–32.

Fuller, Dorian Q. and Charlene A. Murphy (2016) The transition to agricultural production in India: South Asian entanglements of domestication. In Gwen R. Schug and Subhash R. Walimbe (eds.), *A Companion to South Asia in the Past*, pp. 344–357. Chichester: John Wiley.

Galetti, Paola (2012–2013) I paesaggi rurali italiani dell'alto Medioevo: sistemi insediativi e assetti di produzione. *Norba. Revista de Historia* 25–26: 89–103.

Gallinelli, Diego (2019–2020) *Trasformazione dell'Uso e della Copertura del Suolo, Dinamiche Territoriali e Ricostruzioni GIS nei Possedimenti Pontini della Famiglia Caetani (XIX–XXI Secolo)*. Università degli Studi Roma Tre: PhD Thesis.

Gandolfo, Lucina (2015) L'alimentazione in età romana. In Francesca Spatafora (eds), *Nutrire la Città. A Tavola nella Palermo Antica*, pp. 35–39. Palermo.

Gardoni, Giuseppe (2021) Metamorfosi del paesaggio nella pianura fluviale mantovana (secoli VIII–XIII). In *Historia Medieval. Anales de la Universidad de Alicante, Medio Ambiente, Recursos Naturales y Paisaje Agrario en los Espacios Medievales (Siglos VIII–XV)* 22: 105–131.

Gavinelli, Dino and Gilardi Thomas (2017) Dalla pianura lombardo-piemontese alle falde dell'Himalaya. Colture e culture risicole a confronto. In Irini Bajini, Maria Vittoria Calvi, Giuliana Garzone and Giuseppe Sergio (eds.), *Parole per Mangiare. Discorsi e Culture del Cibo*, pp. 419–434. Milano: Edizioni Universitarie di Lettere Economia Diritto.

Ghiardo, Luca (2020) Curarsi col riso: spunti di ricerca tra scienza e saperi popolari. *Palaver* 9(2): 189–207.

Ginatempo, Maria and Lucia Sandri (1990) *L'Italia delle Città. Il Popolamento Urbano tra Medioevo e Rinascimento (Secoli XIII–XVI)*. Firenze: Le Lettere.

Igual Luis, David (2016) Vom feld aufs schiff. Handel und export von reis aus Valencia nach Flandern im spätmittelalter. *Hansische Geschichtsblätter* 134: 61–95.

Jagusiak, Krzysztof, Maciej Kokoszko and Zofia Rzeźnicka (2015) Cakes and breads in Oribasius' Collectiones Medicae. *Symbolae Philologorum Posnaniensium Graecae et Latinae* XXV(1): 127–140.

Jasny, Naum (1950) The daily bread of the ancient Greeks and Romans. *Osiris. Commentationes de Scientiarum et Eruditiones Historia Rationeque* 9: 227–253.

Jiang, Leping, Gyoung-Ah Lee, Li Liu and Juzhong Zhang (2007) Evidence for the early beginning (c. 9000 cal. BP) of rice domestication in China: a response. *The Holocene* 17(8): 1059–1068.

Kim, Jangsuk and Jinho Park (2020) Millet vs rice: an evaluation of the farming/language dispersal hypothesis in the korean context. *Evolutionary Human Sciences* 2: e12.

Knörzer, Karl-Heinz (1966) Über funde römischer importfrüchte in Novaesium (Neuß/Rh.). *Bonner Jahrbücher* 166: 433–443.

Kurup, Paneenazhikath N. V. (2001) Ayurveda. In Ranjit Roy Chaudhury and Uton Muchtar Rafei (eds.), *Traditional Medicine in Asia*, pp. 3–16. New Delhi.

Leleković, Tino and Kelly Reed (2019) First evidence of rice (Oryza cf. sativa L.) and black pepper (Piper nigrum) in Roman Mursa, Croatia. *Archaeological and Anthropological Sciences* 11(1): 271–278.

Leurini, C. (2004) Per una Storia degli Studi sull'Ecclesiologia Manichea. *Varia Iranica. Serie Orientale Roma* XCVII: 137–181.

Loi, Maria Cristina (2021) Palazzi e case dell'ultimo gotico a Milano tra XIV e XV Secolo. Il 'gran Palagio' della famiglia Pusterla nella Contrada della Balla. *Lexicon. Storie e Architettura in Sicilia e nel Mediterraneo* speciale 2: 217–224.

Magni, Stefano G. (2015) Politica degli approvvigionamenti e controllo del commercio dei cereali nell'Italia dei comuni nel XIII e XIV Secolo: alcune questioni preliminari. *Mélanges de l'École Française de Rome: Moyen* Âge 127(1): 97–114.

Mariani, Luigi (2012) Il riso. In Luigi Mariani and Massimo Pirovano (eds), *Il cibo e gli uomini. L'alimentazione nelle Collezioni Etnografiche Lombarde*, pp. 88–97. Cremona: Fantigrafica.

Marrocchi, Mario (2017) Lo sfruttamento di un'area umida: comunità locali e città nella Val di Chiana Centrale. *Riparia* 3: 58–94.

Mohanty, Rabindra K., Taranisenpanda Panda and B. K. Tripathy (2012) Medicinal rice varieties of India need urgent attention. *Current Science* 102(4): 547.

Montanari, Massimo (2002) Colture, lavori, tecniche, rendimenti. In Pinto, Giuliano, Carlo Poni and Ugo Tucci (eds), *Storia dell'Agricoltura Italiana. Il Medioevo e l'Età Moderna*, pp. 59–81. Firenze: Edizioni Polistampa.

Morales, Jacob and Marijke Van Der Veen (2015) The Roman and Islamic spice trade: new archaeological evidence. *Journal of Ethnopharmacology* 167: 54–63.

Mueller, Reinhold C. (2008) Epidemie, crisi, rivolte. In Francesco Benigno, Carmine Donzelli, Carlo Fumian, Salvatore Lupo and E. Igor Mineo (eds), *Storia Medievale*, pp. 557–584. Roma.

Mythili, S. R., T. C. Nikitha and Ramamoorthy Pushpam (2019) Medicinal rice and its medicinal values. *International Journal of Current Microbiology and Applied Sciences* 8(10): 2090–2095.

Nappo, Dario (2018) *I Porti Romani nel Mar Rosso da Augusto al Tardoantico*. Napoli: FedOAPress.

Narayana, A. (1995) Medical science in ancient indian culture with special reference to Atharva Veda. *Bulletin of Indian Institute of History of Medicine* XXV(1–2): 100–110.

Nesbitt, Mark, St John Simpson and Ingvar Svanberg (2010) History of rice in western and central Asia. In S. D. Sharma (eds), *Rice: Origin, Antiquity and History*, pp. 308–340. Enfield: Taylor & Francis.

Niccoli, Francesco (2013–2014) *Forme del Popolamento e Dinamiche Insediative nei Territori di Calci e Vicopisano tra Età imperiale e Medioevo*. Università di Pisa: Graduation Thesis.

Palermo, Luciano (2013) Di fronte alla crisi: l'economia e il linguaggio della carestia nelle fonti medievali. In Pere Benito I Monclus (eds), *Crisis Alimentarias en la Edad Media. Modelos, Explicaciones y Representaciones*, pp. 47–67. Lleida: Milenio.

Petralia, Giuseppe (2008) Crescita ed espansione. In Francesco Benigno, Carmine Donzelli, Carlo Fumian, Salvatore Lupo and E. Igor Mineo (eds.), *Storia Medievale*, pp. 291–318. Roma: Società Tipografica Romana.

Pinto, Giuliano (2014) Tra demografia, economia e politica: la rete urbana italiana (XIII-inizi XVI secolo). *Manifestaciones del impacto urbano en la Baja Edad Media, Edad Media. Revista de historia* 15: 37–57.

Polto, Corradina (2012) Le Zone umide della Sicilia sud orientale: dalla bonifica alla tutela. *Humanities* 1(1): 116–126.

Prosperi, I. (2008) Alimentarsi di false credenze. Agostino e la critica dei precetti alimentari Manichei. *Doctor Virtualis: Rivista Online di Storia della Filosofia Medievale, Tracce di Debolezza nel Pensiero Medievale* 8: 7–55. (https://riviste.unimi.it/index.php/DoctorVirtualis/issue/view/18, 2008).

Rao, Riccardo (2013) Una civiltà del castagno: uomini e boschi nell'Appennino ligure-piemontese durante l'apogeo del Medioevo (secoli XII-metà XIV). *Archivio Storico Italiano* 171(2): 207–227.

Reed, Kelly, Tino Leleković, Lisa Lodwick, Rhona Fenwick, Ruth Pelling and Helmut Kroll (2022) Food, farming and trade on the Danube frontier: plant remains from roman Aelia Mursa (Osijek, Croatia). *Vegetation History and Archaeobotany* 31: 363–376.

Rowan, Erica (2019) Same taste, different place: looking at the consciousness of food origins in the Roman world. *Theoretical Roman Archaeology Journal* 2(1): 1–18.

Spengler, Robert N., Sören Stark, Xinying Zhou, Daniel Fuks, Li Tang, Basira Mir-Makhamad, Rasmus Bjørn, Hongen Jiang, Luca M. Olivieri, Alisher Begmatov and Nicole Boivin (2021) A journey to the West: the ancient dispersal of rice out of East Asia. *Rice* 14: e83.

Tabacco, Raffaella (2019) Il riso nelle fonti latine: cereale pregiato di importazione e di uso medico. In Silvia Condorelli and Marco Onorato (eds.), *Verborum Violis Multicoloribus. Studi in Onore di Giovanni Cupaiuolo, Collana di Studi Latini*, pp. 617–644. Napoli: Paolo Loffredo Editore Srl.

Tognetti, Sergio (1995) Prezzi e salari nella firenze tardomedievale: un profilo. *Archivio Storico Italiano* 153(2): 263–333.

Tognetti, Sergio (2012) L'economia del regno di Napoli tra Quattro e Cinquecento. Riflessioni su una recente rilettura. *Archivio Storico Italiano* 170(4): 757–768.

Trubbiani, Andrea (2007) La carestia 'grandissima et universale' del 1590–1592 a Montecassiano. *Proposte e Ricerche. Economia e Società nella Storia dell'Italia Centrale* 59: 311–324.

Vassallo, Guglielmo (2015) Il riso in Sicilia. In Silvia De Lorenzo (ed.), *La Cucina del Riso. Relazioni e Ricette*, pp. 20–24. Roma: Digitalialab Srl.

Watson, Andrew M. (1997) La conquista islámica y los nuevos cultivos de Al-Andalus. In Patrice Cressier, Joaquín Gómez-Pantoja Fernández-Salguero and José Morilla Critz (eds), *Impactos Exteriores sobre el Mundo Rural Mediterráneo: del Imperio Romano a Nuestros Días*, pp. 111–124. Madrid: Artes Gráficas Gala, S.L.

Weiss-Amer Melitta (1993) Medieval women's guides to food during pregnancy: origins, texts, and traditions. *Canadian Bulletin of Medical History* 10(1): 5–23.

Wickham, Chris (2008) Economia altomedievale. In Francesco Benigno, Carmine Donzelli, Carlo Fumian, Salvatore Lupo and E. Igor Mineo (eds), *Storia Medievale*, pp. 203–226. Roma: Società Tipografica Romana.

Zach, Barbara (2002) Vegetable offerings on the Roman sacrificial site in Mainz, Germany. Short report on the first results. *Vegetation History and Archaeobotany* 11: 101–106.

Index

www.ingramcontent.com/pod-product-compliance
Lightning Source LLC
LaVergne TN
LVHW071214210926
848482LV00015B/811